VALUE
Its Measurement, Design, and Management

VALUE

Its Measurement, Design, and Management

M. Larry Shillito

David J. De Marle

A WILEY-INTERSCIENCE PUBLICATION

John Wiley & Sons, Inc.

NEW YORK / CHICHESTER / BRISBANE / TORONTO / SINGAPORE

Copyright © 1992 by John Wiley & Sons, Inc.

Library of Congress Cataloging in Publication Data:

Shillito, M. Larry, 1939-
 Value: its measurement, design, and management / M. Larry
Shillito, David J. De Marle.
   p.   cm.
 "A Wiley-Interscience publication."
 Includes bibliographical references and index.
 ISBN 0-471-52738-6
 1. Value analysis (Cost control)   2. Quality control.   3. Value.
I. De Marle, David J.   II. Title.
HD47.3.S53   1992   91-26987
658.15′52—dc20   CIP

Printed and bound in the United States of America by Braun-Brumfield, Inc.

10 9 8 7 6 5 4 3 2 1

CONTENTS

PREFACE

Value is a goal of everyday life and is something we all seek. A book on value, properly conceived and written, should be of interest to many people. This book was written for people in a variety of different professions. Engineers, designers, consultants, entrepreneurs, architects, economists, managers, and politicians should find this book interesting and valuable. We believe that anyone who is interested in understanding value or who is interested in the measurement, design, or management of value will find this book useful.

Writing this book has not been an easy task. Value is a large, complex, and abstract subject, and simply defining it required a great deal of thought. Fortunately, once its nature is defined, it is easier to treat. To simplify the subject, we have arranged the book in four sections. The sections follow a logical order, each section building on material from previous sections. We hope this will make study easier. For serious students, we include a large number of references. The sections are

1. The Nature of Value,
2. The Measurement of Value,
3. The Design of Value, and
4. The Management of Value.

The first section is philosophical and describes the nature of value. A new concept of value is proposed in which value is described as a behavioral force. Value is seen as a form of energy analogous and convertible with other forms of energy. People perceive value subjectively and are motivated to obtain items that satisfy their needs. The value of items is a function of their relative importance and cost. A *value force* is described, and equations are developed with which its magnitude can be ascertained subjectively. Theories of human motivation and personal growth based on value are discussed. The value force is seen as the basis for evolution, including the evolution of products, services, and societies to fill human needs.

The second section moves from philosophy to practicality. This section contains a comprehensive description of value measurement. The basis of value measurement is the ability to gauge the value of elements in a system using subjective measurements of relative importance and cost. Value measurement techniques are described that are used to quantify the elements of importance and cost. Assigning numerical weights to the value of components in a product or service establishes their value and uncovers areas where improvements are needed. Value indices, graphs, and targets are described that enhance communications and act as vehicles for interdisciplinary teamwork and design improvement. The construction of dynamic models to portray changes in value with time is described and illustrated. To overcome signal-to-noise problems, subjective judgments undergo structured dialogue and testing.

The third section deals with design and is devoted to a description of function analysis and function matrices. Patterned on the engineering adage that form follows function, this section describes different ways to map functions and create system designs of improved value. A total value concept based on satisfying customer, retailer, and producer needs is described. Function analysis diagrams and indentured function matrices are used to portray how, why, and at what cost these needs are met in every element of a design. Detailed "how-to" descriptions of three new leading edge function design matrix systems, quality function development, technology road maps, and customer-oriented product concepting, are provided. Value measurement techniques are used to quantify the value of new system designs.

The final section describes several methods for managing value. Valuism, an evolving worldwide movement for value improvement is described. Value analysis, value engineering, and value management methodologies, which are central to valuism, are described in detail. Value analysis, value engineering, and value management rely on interdisciplinary teams to analyze functions and improve the value of products and services. The organizational and behavioral aspects of this teamwork are treated in detail. Value planning, a powerful new methodology based on functional technology planning, is described and illustrated. We conclude the book with a forecast of the future of valuism and value management.

In publishing this book, we hope to infuse value and value thinking into industry, academia, and government. We hope that every reader will be able to apply value measurement and value management in their professions. We are convinced that value measurement should form the basis of prioritization and decision making and can lead to innovative solutions to complex problems worldwide. It is our belief that value thinking can effectively link people in organizations such as marketing, manufacturing, research, and engineering together. Value improvement can weld a strong link between productivity and quality.

Rochester, New York   M. LARRY SHILLITO
March, 1992   DAVID J. DE MARLE

INTRODUCTION

The chapters in this book are grouped in these sections and expand upon the book's title; *Value: Its Measurement, Design, and Management.*

Section 1
The Nature of Value

In Chapter 1, De Marle offers a new theoretical concept of the nature of value. A value force is postulated where value is treated as a motivational force. The magnitude of this force is a function of the importance of an item and its relative cost. The value force is compared to other physical forces and simple mathematical equations are developed to describe value. The dependance of value on need and its transience with satisfaction and time is described.

In Chapter 2, De Marle shows how the value force governs the evolution of products and services. Theories of personal and society growth are related to the value force. Value is proposed as the measure for Darwin's evolutionary "fitness" selection where only the strong survive. Numerous examples are used to illustrate the entropy of this evolution.

Section 2
The Measurement of Value

In Chapter 3, De Marle describes the process of value measurement. Using the equations developed in Chapter 1, the relative value of each component in a product is numerically measured. Value indices, graphs, and targets are calculated that form the basis for product improvement. The chapter illustrates how value measurement can be used to quantify value and describes the latest contributions from Japan and China in this area.

In Chapter 4, Shillito describes a large number of specific value measurement and screening techniques. The advantages and disadvantages of each technique are discussed, and detailed instructions are given on their use. Methods based on nominal, ordinal, interval, and ratio scales are included in this compilation of the most common value measurement methods.

In Chapter 5, De Marle describes how system dynamic modeling can be used to simulate changes that frequently occur in the value of items over time. Examples are used to illustrate hourly as well as seasonal value changes. The construction of dynamic models using *Stella* and *Ithink* computer software[1] is described and illustrated in detail.

In Chapter 6, Shillito and De Marle offer a theoretical framework for value measurement by introducing a new theory for value and decision making. Signal detection theory is used as a model to explain a person's processing of value stimuli. Numerical language, value standards, as well as new ways to evaluate ideas are discussed. The filtering effects of paradigms and conflict on value decisions is discussed using a case history.

Section 3
The Design of Value

In Chapter 7, Shillito and De Marle discuss the origin and evolution of function analysis. FAST diagrams are used to analyze how and why products and services provide value. Function matrices and new techniques such as dysfunction analysis and adjacency diagrams are described.

In Chapter 8, Shillito discusses a total product concept in which products are structured to meet customer needs. The quality function deployment process, which uses a set of indented function matrices, is described and illustrated. This process was developed in Japan and is used to control the quality and value of products and services.

In Chapter 9, Shillito discusses the construction and use of the technology road map, a function–technology matrix used for developing product concepts. It is especially useful for designing capital equipment, services, systems, and processes.

In Chapter 10, Shillito describes the customer-oriented product concepting process, a function matrix approach for developing products that meet customer needs. Customer, market, manufacturing, and design needs are used to optimize new product design concepts. The process is appropriate for consumer-type products, services, and systems.

Section 4
The Management of Value

In Chapter 11, De Marle describes valuism, a worldwide industrial management movement that creates products and services of improved value. Valuism uses many different methods to improve the effective performance of products and services while improving the efficiency and lowering the cost of their production and distribution.

In Chapter 12, De Marle describes the value analysis and value engineering (VA/VE) process. He traces the growth and evolution of VA/VE from its invention

---

[1]Stella and Ithink are copyrighted software programs of High Performance Systems Inc., 48 Lyme Road, Suite 300, Hanover, NH 03755.

at General Electric in 1947 by Larry Miles, to its worldwide use and the subsequent development of value planning and value management.

In Chapter 13, Shillito deals with the importance of the behavioral and organizational aspects of VA/VE. The success of VA/VE is dependent upon how people, politics, and organizations are considered and integrated into applications. There are dos and don'ts to address early in an application.

In Chapter 14, De Marle describes a value planning process that integrates methodology from technological forecasting with value engineering techniques used in Japan. A value planning job plan that combines forecasting and planning with traditional VA/VE is detailed.

In Chapter 15, Shillito and De Marle present their perception of the future of valuism and value management. We detail the importance of technology integration, intuition, and leadership and offer several templates for integrating VA/VE with other disciplines.

VALUE
Its Measurement, Design, and Management

I

THE NATURE OF VALUE

This section describes the nature of value. A new concept of value is proposed in which value is described as a behavioral force. Value is seen as a form of energy analogous and convertible to other forms of energy. People perceive value subjectively and are motivated to obtain items that satisfy their needs. The value of items is a function of their relative importance and cost. A "value force" is described, and equations are developed with which its magnitude can be ascertained subjectively. Theories of human motivation and personal growth based on value are discussed. The value force is seen as the basis for evolution, including the evolution of products, services, and societies to meet human needs.

1

THE VALUE FORCE

David J. De Marle

WHAT IS VALUE?

Value is mistakenly seen as a property of goods or services. This view had its origin in the philosophy of ancient Greece. The Greeks believed that certain primary or essential principles existed in our environment. These indwelling principles gave value to the items they inhabited. Thus ethics contained "the good"; religion, "the holy"; and aesthetics, "the beautiful." When the indwelling principle was present, the object had value, when it was absent, the object was worthless. Religion was perverted when it lost its holiness. Art was degraded when it lacked beauty.

The concept of value as an indwelling property of goods or services was typical of the anthropomorphic philosophy of that age. Things were given human characteristics. It was common for people to believe that spirits dwelled in the flora and fauna, the rocks and objects around them. Man's essence was in his soul, and when things went wrong you drove a devil from it.

Today the devil has lost his essence, and incantations are seldom uttered for his departure, yet the Greek idea of value dwelling within a product still permeates our thinking. Engineers and economists alike see value in terms of the features that a product or service has. Modern advertising even describes these features in animist terms, terms that liken cars to colts and stuffs tigers in gas tanks. Anthropomorphism is enshrined in Madison Avenue, and the belief that value is a property of an object continues unabated and unchallenged. Yet value is more than a property of matter, it is a force that governs our behavior. We need to discard the anthropomorphic concept of value and examine this force.

Value is the primary force that motivates human actions. It is dichotomous, centered in people and the objects they desire. Value is a potential energy field

3

between us and objects we need. It draws us to items in our environment that we find appealing. When this attraction is large, we expend our energy to acquire, possess, use, and exchange objects that are rewarding. In this way, the potential energy of value is converted to the kinetic energy of human action.

When we acquire possessions, we increase our wealth and influence. The things we acquire then transfer their utility and power to us. For some people, the acquisition of material goods is an end unto itself. For others, needs extend beyond material possessions to include information and knowledge. Such people work to increase their knowledge and understanding of things. As information is acquired, knowledge increases and people grow wise.

This process is fundamentally hedonistic; we feel good when we acquire things and bad when we loose them. Uncontrolled acquisition of wealth and power, however, is often detrimental to the freedom of others, absolute power corrupts absolutely. Community values moderate our hedonistic behavior. They conform our individual values to those of our friends and neighbors and prevent us from acting as predators among our associates. Community values influence our perception of right and wrong and focus our search for value upon objects that are acceptable in our community. Community values also define the goods and services that constitute a market. Competition for resources in this market and the right to obtain and retain property and goods obtained from it motivate people to work and reward those who work efficiently.

THE FORCE OF VALUE

Figure 1.1 is a diagram of the "force of value." A person with a need is represented at the center of the drawing. Surrounding this person are several circles labeled A through H. These circles represent external items that are capable of satisfying specific needs or wants the person has. The size of each circle represents the perceived ability of each item to satisfy specific needs. Large circles represent items that will satisfy more of the person's needs than items shown as small circles. The items vary in cost, which is represented by the distance from the person to each item. Items close to the person cost less than items further away. The dotted circle in the diagram passes through two items that have equal costs, but that differ significantly in their ability to satisfy needs. Item C will satisfy more needs than item F. Value exists as a force between the objects and the person. Like gravity or magnetic force, value is a force that can be quantified. The magnitude of this force depends on the interplay between needs, usefulness, and cost. When the force is active, it motivates the person to obtain and use those items that best satisfy personal needs.

Figure 1.1 can be used to illustrate many market situations. For a hungry person, the circles can represent different foods on a menu. For a student, the circles can represent different schools or different subjects in a curriculum. For an individual who needs transportation, the circles can represent different cars in a specific car market. In all of these markets, a force of value exists between the person and the goods available in the market. This force varies from person to person, because

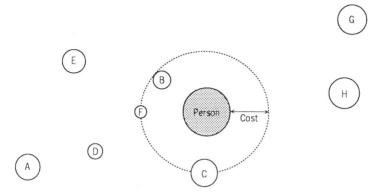

FIGURE 1.1 The force of value.

individuals have different needs and wants. It also varies from moment to moment, because needs vary with time. The force of value is much more complex and variable than a physical force like gravity. Yet value, like gravity, is not simply a property of matter, it is a form of energy that affects matter.

IS VALUE A FORM OF ENERGY?

Force is created when energy acts upon matter. If value is a force, what energy drives it? Is value an unrecognized form of energy? Or is value a force that focuses our own energy?

We do not think of value in terms of energy, because we classify energy in physical terms, and value in financial terms. Accordingly, we associate energy with physics and value with economics. Heat, light, electricity, and gravity are forces of nature that science has discovered and measured. Isaac Newton described gravity; Albert Einstein, atomic energy; and so on. Yet in every case, when scientists looked for a tangible "physical" something, they found nothing beyond the action of an invisible force on a physical substance. What is gravity? What is electricity? What is magnetism? What is energy? Is value energy?

PROPERTIES OF ENERGY

Physicists tell us that energy has four primary attributes—it is itself invisible, it affects matter, its effects have magnitude, and its energy is conserved when one form of energy is converted into another.

Invisibility

Energy seems to be something made of nothing that moves everything; a ghostly entity not unlike the anthropomorphic essence in which the Greeks believed. Light, heat, and temperature are names that we give to its effects. I have defined value as a potential energy field that separates people from objects they desire,[1] an invisible force, hidden in the space between our needs and the usefulness of objects around us. Although the objects are visible, the force that attracts us to them is not.

Causality

A primary property of energy is its ability to affect matter. Energy moves matter. Just as heat drives the pistons in an engine, value drives the actions of life. Value activates us and moves us to action. Value fashions the motions of life and society. Is not motion central to life? An amoeba needs food and moves to capture a bacterium, a tiger prowls the veldt looking for water to quench its thirst. A bulldozer levels a hill and prepares a roadway. Value powers each of these actions.

Attempts to understand the life force in purely chemical or physical terms have been largely unfruitful. While we know that our bodies obtain energy when we metabolize food, calories seem an inappropriate descriptor of the life force. This is because our taxonomy is wrong and lacks a requisite generalism. The wind is a function of air masses, not molecules, and people's behavior is a function of value, not calories.

Magnitude

Another fundamental property of energy is intensiveness. Energy, like matter, has magnitude. The temperature 1000 °C is "hotter" than 100 °C, and the potential caloric energy of butter is greater than the potential energy of eggs. The fact that energy is intensive has enabled people to construct measurement scales. Thus, we speak of footcandles, degrees Kelvin, and megatons. If value is a force, it must be intensive and measurable.[2] Indeed many methods exist to measure value[3] that will be described in detail later in this book. Using value measurement techniques, people can answer the question, "How much value does an object exhibit?"

Energy Conservation and Conversion

Through the past 20 centuries, the concept of energy conservation has evolved into a fundamental tenant of science. Table 1.1 lists events in the development of our concept of energy. Speculative thinkers in Ionia, a province of ancient Greece, pondered the nature of things that changed in their world. Why did things grow? What caused the heavenly bodies to move across the evening sky? Aristotle developed the concept of a fundamental force that caused all things to move. His Aristotelian dynamics supposed that a permanent invisible force was responsible for motion. This force affected matter and caused the stars to move in the heavens and clouds to move in the sky.

For centuries, energy was seen as the force that created motion. In the 17th century, René Descartes proposed that this force could be conserved by the conver-

TABLE 1.1 Evolution of the concept of energy

Date	Hypothesis
634–546 BC	Ionian thinkers speculate about the causes of change.
384–322 BC	Aristotle states that a hidden force causes and determines the motion of matter.
1126–1198	Averroes states that Aristotle's force is composed of a primary and secondary force.
1225–74	Thomas Aquinas postulates that a constant and undiminishing force (*vis infatigablis*) causes the heavenly bodies to move throughout the heavens.
1280–1322	Petrus Aureoli postulates that the velocity of a body comes from a capacitive force.
1644	Descartes notes that force causing motion is a product of the mass and velocity of the body. This is the first expression of the conservation of force (energy).
1686	Leibniz describes force in terms of energy (*vis a viva*), "a living force," and states that this force (later called *kinetic energy*) is equal to the velocity squared times the mass of the body.
1735	J. Bernoulli describes the conservation of energy in mechanics.
1738	D. Bernoulli describes the equivalence of mechanical and chemical energy by comparing the potential energy of coal to the production of human labor.
1796	Rumford shows the equivalance of mechanical and thermal energy in experiments on metal turning and cannon manufacture.
1796	Davy proves the conservation of mechanical and thermal energy in experiments on ice friction.
1803	Carnot defines potential and kinetic energy as a latent and an energetic force. His latent force $LE = W/d$ and the kinetic force $KE = m \times v^2$.
1829	Coriolis refines Carnot's formulation of kinetic energy to $KE = (m \times V^2)/2$.
1829	Poncelet defines Carnot's latent energy as potential energy and calculated work in terms of killogram meters.
1842	von Mayer relates heat to mechanical energy.
1885	Maxwell proposes that a field of magnetic energy can be converted to electrical energy. He shows the equivalency of various electromagnetic forms of energy.
1890	Ostwald proposes that energy in the various forms found in nature is the fundamental basis of the universe. He postulates that the conservation of energy as shown by the conversion of one form of energy to another shows energy rather than matter to be constant.
1905	Einstein postulates that matter and energy are interconvertable and demonstrates the equivalency of energy and inertia in his formula $E = m \times c^2$.

sion of the force to the velocity of a mass of moving matter. Gottfried Wilhelm Liebniz then proposed that this force was converted to a motive energy that he called the *living force, (Vis a Viva)*. In the 18th century, scientists postulated that the energy of motion might be conserved in other forms of energy. The Bernoullies postulated the conversion of energy from potential to kinetic forms and hypothesized an equivalency between human labor and the potential energy of coal. Scientists in the 19th century set about measuring the conversion of energy from one form to another. Constants were found that showed that energy was conserved as it was converted from one form to another. New forms of electromagnetic energy were discovered, and the dogma of energy conservation through conversion was well established. In the 20th century, as measurement techniques improved and began showing slight discrepancies in some conversions, Albert Einstein postulated the conversion of matter to energy.

Although the mechanical energy of a horse (its horsepower) was adopted as a standard by which people can measure the mechanical work of an engine, because of the mortality of living things, life itself was considered outside of the laws of physics. Death would apparently violate the law of energy conservation. Nonetheless, people's work energy was described in calories, and we can easily calculate the conversion of chemical energy or mechanical energy into hours of human labor. As long ago as 1738, Daniel Bernoulli stated that the force "latent in a cubic foot of coal if totally extracted would do the work of 8 or 10 men."

Yet the energy that people receive or expend is converted into value energy every day. People expend their energy in the construction of things of value. A painter estimates how long it will take to paint a house. Is this not a conversion of work to value? A construction firm uses the program evaluation review technique (PERT) to estimate the hours it will take the firm to complete the construction of a bridge. Is this not a conversion of human energy to the energy of value? If we ignore this conversion, life violates the law of energy conservation. If we include it, people become a dynamic point of energy conservation and conversion. Man's knowledge, inventions, and civilization are the product of his role in the creation of items of value. Motivated by the value force, humans convert their energy into artifacts and inventions that satisfy their need to live and grow.

Thus value possesses four requisite properties of energy—it is invisible, it affects matter, it is intensive, and its energy is convertable into other forms. Because value is intimately connected with our wants and needs, value moves us to action. Values centered in our mind, control our body, and govern our life.

VALUE AND GRAVITY

An analogy between value and gravity will help to describe the force of value. Isaac Newton described gravity as the physical force that attracts masses to each other. In his famous *law of universal gravitation,* he stated that the force of gravity is directly proportional to the product of the masses of the bodies and inversely proportional to the square of the distance between them. He found that gravity was independent of the physical or chemical state of the bodies and of the presence of intervening bodies. He expressed his law according to equation 1.1:

$$g = \frac{m_1 \times m_2}{d^2} \qquad (1.1)$$

where:

g = gravity
m_1 = mass of an object 1
m_2 = mass of a second object 2
d = distance between the objects

The force of value is analogous to the force of gravity and is defined in equation 1.2 as the product of the need for an object times its ability to satisfy this need, divided by the cost of the object.

$$v = \frac{n \times a}{c} \qquad (1.2)$$

where:

v = value of some object or service
n = the need for an object or service
a = the ability of an object or service to satisfy this need
c = the cost of the object

Equation 1.2 shows that value is directly proportional to the product of the need for an object or service and the ability of the object or service to satisfy this need and inversely proportional to the cost of the object.

Individuals rate the importance of items in terms of their ability to satisfy their needs. Importance is defined as the product of the intensity of a need and the ability of the item to satisfy this need. Thus;

$$I = n \times a \qquad (1.3)$$

where

I = importance
a = ability to satisfy need[5]
n = need

It is often convenient to measure a customer's perception of the importance of elements in a product. Substituting equation 1.3 into equation 1.2, yields the simple equation for value shown in equation 1.4:

$$V = \frac{I}{C} \tag{1.4}$$

where

V = value of some object $\overset{o\mathcal{R}}{\text{of}}$ service
I = the importance of an object or service
C = the cost of the object[6]

A wide variety of measurement techniques that compare the importance and cost of items can be used to measure value. These techniques allow people to establish numerical estimates of the importance and cost of items in a market. They also can use these techniques to evaluate the value of functions and/or components in a design. These estimates are normalized by calculating the relative importance or cost of each item as a percentage of the total rating given to all items. With this treatment, equation 1.4 is converted to equation 1.5.

$$V = \frac{\% \, I}{\% \, C} \tag{1.5}$$

where

V = the value of a thing
$\% \, I$ = the relative importance of this thing[7]
$\% \, C$ = the relative cost of this thing

Using this simple equation, people can measure and graph the value of many different items. This gives them a way to communicate their perceptions of value and allows them to target areas in a product or service that have poor value. Equation 1.5 is often used to screen the value of items that can then be measured in more detail.

The analogy between value and gravity is depicted in Figure 1.2.

Similarities Between the Forces of Gravity and Value

The force of value is directly related to the product of the need an individual has and the ability of the product to satisfy this need (importance). Similarly, gravity is directly related to the product of the mass of two bodies. Value is inversely related to the cost of an item, whereas gravity is inversely related to the square of the distance between the two masses. Cost acts as a deterrent to acquisition of a product just as distance acts as a deterrent to aggregation of the two masses. Cost represents the work that has been put into the creation of a product, and this cost must be recovered by an owner before an individual can obtain the item. The distance squared between the planet and its moon is a deterrent to aggregation, because the kinetic energy of the moon is a product of its velocity and its mass. This energy exactly balances the force of gravity and holds the moon in orbit around the planet.

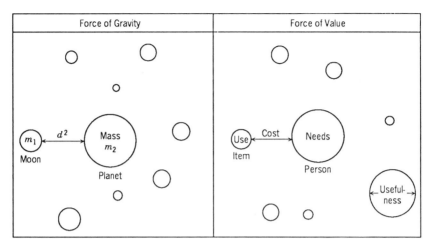

FIGURE 1.2 Comparison of value and gravity.

Differences Between the Forces of Gravity and Value

Although the gravitational force is a constant throughout the universe, its magnitude on any body can be calculated with great accuracy and precision. Table 1.2 shows how the mass of various bodies in our solar system affects the weight of objects on the surface of each body. The table compares the gravitational attraction on earth with that of each other body. For comparison, the weight of a cubic foot of ice on the surface of each body is shown. Although weight varies from body to body, it is a constant function of a gravitational force that is constant throughout the universe. The weight on each body is time invarient, for all practical considerations, because mass and distance do not vary.

VARIABILITY OF VALUE

Unlike a gravitational force, which is based on the constant interaction of mass and distance, the force of value is dependent upon the interaction of highly variable needs, utilities, and costs. This interaction often causes value to grow or diminish rapidly and creates great variability in its magnitude. Because value often varies from moment to moment, it is much less predictable than gravity or other physical forces. In Figure 1.3, the value of lighting is shown as a function of time. In the drawing, A depicts an individual's need for artificial light; the area and placement of circles B and C denote the usefulness (lumens) and electrical cost of a 75 W incandescent light (B) and a 25 W fluorescent lamp (C). At 2 P.M. there is no need for either the incandescent or fluorescent lamp, because daylight adequately illuminates the area where the person is. However at 11 P.M. when daylight is gone, the individual's need is large, as denoted by the size of circle A. At this time, the 25 W fluorescent lamp C furnishes more light than the 75 W incandescent lamp B and has a value more than three times

TABLE 1.2 The force of gravity on some bodies in our solar system

Planets and Satellites	Gravity on Earth / Gravity on Body	Weight of a Cubic Foot of Ice (lb)
Earth	1.0	57.1
Moon	6.04	9.5
Mars	2.64	21.6
Phobos	1851.6	0.03
Deimos	2644.3	.02
Jupiter	0.40	142.6
Io	5.43	10.5
Europa	7.36	7.8
Ganamede	6.86	8.3
Callisto	8.02	7.1
Saturn	0.93	61.4
Titan	7.08	8.1
Uranus	1.16	49.2
Miranda	110.6	0.52
Oberon	3.79	15.1
Neptune	0.87	65.4
Triton	12.5	4.6
Pluto	31.2	1.8

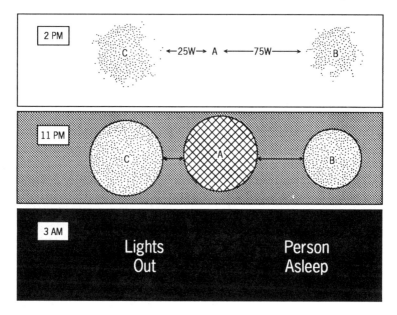

FIGURE 1.3 Variability of value with time.

that of B. At 3 A.M., the individual is sleeping and has no need for either lamp, which have both been switched off.

This artist's concept shows the variability of the value force. Later in this book we will describe how value can be measured and modeled. At this point, I merely want to indicate that value is much more variable than the forms of energy with which science has dealt. This is one of the reasons why people have treated value as a property of matter. This eliminates the need to treat the great variability of human needs. In simply describing fluorescent lamps as more energy efficient than incandescent lamps, however, we miss the point that our needs directly affect the force of value.

The lighting example illustrates the variability of needs as a function of time. We don't need lights when the sun is shining. When we sleep, most of our needs are dormant. We need lights in the evening after the sun sets and before we sleep. Compared to gravity, which is constant, the value of electric lighting is highly variable.

Often our needs change rapidly. Our appetite for food is a function of when and what we ate and governs our need and desire for when and what we will eat again. Some needs vary during the week or season. From Monday to Friday, many of our needs center on our employment. On weekends or holidays, our needs for leisure and recreation are active. The seasons bring different needs—heat to warm our homes in the winter, air-conditioning to cool them in the summer. As these needs change so do the value of goods to satisfy them. As our needs change, store managers alter prices and inventories. Some needs arise randomly. Illness and accidents bring special needs. Changes in needs alter the value of products and services to us and occur with a frequency that makes the force of value far more variable than physical forces like gravity.

Changes in Cost and Usefulness

The utility and cost of products is not constant, but varies with time. Products depreciate in value at different rates. Automobiles depreciate yearly; foods, weekly. Unless refrigerated, many foods can spoil in a day. Ownership alters value: Used merchandise looses value. Yet some products become more valuable with time. Wine and numerous collectibles appreciate in value over time. To accommodate these changes, retailers alter the price of goods. Day-old food is sold at reduced prices. The price of selected coins increases. Supply and demand alter the price of goods. Price increases when goods are scarce and decreases when goods are abundant. These changes alter the value of goods and services and make it difficult to quantify value.

Add to this complexity the fact that tastes and preferences vary from person to person. To accommodate this variability, restaurants provide their patrons with menus that list many different foods. The menus offer foods at different prices so a customer can select a meal to fit his particular needs. The breakfast, lunch, and dinner menus are different to accommodate tastes that vary from hour to hour. Food is only one product in a market. Think of the multitude of items that are manufactured and sold in other businesses such as clothing, transportation, housing, and entertainment.

In each of these fields, our preferences and tastes vary from individual to individual. Yet value governs behavior in each of these fields.

People have been able to measure and control constant and predictable forces like gravity, kinetic energy, and electricity, but they have failed to measure and control value. The complexity of value is heightened by the fact that individuals often view value in totally opposite ways. Our perception of value is a function of self-interest. Buyers and sellers see value in opposite and incongruous ways. A buyer, for example, finds value in the performance of a product that a seller has little use for. A seller finds value in the sale of goods for which she has little need. Thus, value takes on different meaning from person to person and group to group.

Confusing the Meaning of the Word Value

The word *value* and the word *worth* are used interchangeably in the English language.[8] Originally these words had slightly different meanings. The words came into English from different sources. The word *value* was derived from the Latin word *valare*, which had a market connotation. It expressed value in terms of trading and marketing. The word *worth* is a derivative of the Anglo-Saxon word *weorth*, which entered English as a consequence of the Norman conquest of England. *Weorth* had more of a personal connotation and was used to express the sentiment or esteem that people have for items that they possess. The Norman society was made up of small individual feudal estates, not large trading cities like Rome. *Worth* connoted individual value. In this book, we will attempt to use value and worth as they were originally used. We use *worth* to describe the worth of a product or service to an individual. We use the word *value* to describe an average of the worth that a group of people ascribe to a product or service. Thus;

$$\text{value} = \frac{\Sigma \, (\text{worth } 1 + \text{worth } 2 + + \text{worth } n)}{n} \tag{1.6}$$

Using these definitions of value and worth helps us to differentiate between group and individual perceptions.

People often confuse the meaning of value and quality. Quality is a characteristic of a product or service that can increase or decrease the value of an item. Quality, reliability, serviceability, maintainability, and so on are attributes of a product or service. They are not part of a person's psyche. Value can include all of these attributes and more. Confusing value and quality can lead to the purchase of over-priced items with unnecessary features.

People also confuse value with price, which can lead to the sale of inexpensive merchandise of poor quality and reliability. Manufacturers and retailers confuse value with profits or margins. When this occurs, they often continue to manufacture and inventory items that their customers find less attractive than they do. They will fail to recognize the threat of competitive products that satisfy their customer's needs in a new way. Thus, people watch the television news and drop subscriptions to evening newspapers, while newspaper publishers compete among themselves for a share of a market that has lost value.

VALUE MEASUREMENT

Although the force of value is more variable than other forces, it can be measured. Figure 1.4 shows a number of simple ice-making devices available in today's market. These devices produce ice when filled with water and placed in a freezer. The ice is then used to cool drinks. Each device has different features that allow a customer to produce ice in different sizes, shapes, and quantities. The devices also vary in price and work in different ways. Figure 1.1 was drawn to represent my assessment of the worth of each item shown in Figure 1.4.

An aluminum tray (A) contains a metal ice separator that forms and releases ice cubes. A lever arm attached to the separator is used to break ice from the tray. This type of device was a standard ice cube tray in the 1960–1970s in the United States. It is still available today and can be purchased for $5.95. Its relative value to the author is shown by circle A in Figure 1.1. Today extruded plastic ice cube trays have largely displaced metal trays like A in the market. A small blue plastic ice cube tray (B) is priced at $1.00 while a larger white brand name plastic tray (C) costs $1.45. A small yellow plastic tray (E), which will produce a number of very small ice cubes, can be purchased for $1.79. Ice formed in these trays is removed by simply twisting the tray. Finally, a novel set of 12 multicolored (three yellow, three red, three blue, and three white), reusable water-containing spheres (D) is available for $3.25. Water is sealed inside each sphere so they can be frozen and reused after simple cleaning and refreezing.

All of these devices furnish ice, but have different features and use different design concepts to form and release ice. Each device has a different measurable value. In the

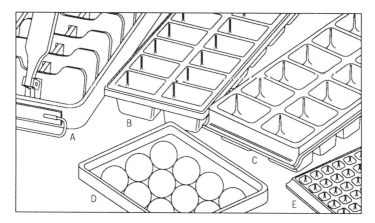

FIGURE 1.4 Ice making products.

market, customers ascertain the worth of each product and buy goods based on their perception of the value of items. The plastic ice cube trays have better value than the metal trays and have almost completely replaced them in the marketplace. The market value of the metal tray is low, because it costs more than the other items. It contains 17 separate parts and is expensive to manufacture and assemble. The manufacturer has to pass these costs on. Although the metal tray yields more ice than the plastic tray, this capacity is insufficient to overcome the cost of the unit. Indeed the plastic tray, even though it costs much less than the aluminum tray, is more reliable.[9] As in nature, products compete, and only the fittest (best value) survive.

EQUATIONS TO MEASURE THE VALUE OF A PRODUCT OR SERVICE

The value of these and other products or services can be represented by a set of positive and negative features present in the product. The positive factors are advantages that reward an owner, and the negative factors are disadvantages that accompany ownership.[10] The value of a product increases in direct proportion to its advantages over competitive products and decreases in proportion to its disadvantages. Product value is a function of the ratio between the advantages and disadvantages of similar products. Items that have high value will displace items of low value in the marketplace.

Equation 1.7 shows the performance price ratio that is often used to describe the value of a product. This equation assumes that a need for the product exists. Implied in the equation is the fact that performance is directed toward satisfying this need. If this is true, then the value of any product is a function of the product's performance and price relative to other products in the market. Performance is measured in terms of the functions the product performs, and price is measured by the monetary amount at which the product is sold.

$$\text{Customer value} = \frac{\text{performance}}{\text{price}} \tag{1.7}$$

Good value exists when a needed product costs little and performs well. Performance and price are interrelated in any product and depend upon the way a product is designed and manufactured. Each feature and function carries a specific cost that adds to the product's price. The cost of providing these functions varies from one producer to another and from one product design to another. The need for features and functions varies according to the needs of individuals in the market. On average, the value of a product or service is a function of a number of different performance features; each with an ability (a) to satisfy a need (n) and each with a cost (c). In equation 1.8, the numbers 1 and 2 through a final number (designated by a question mark) represent different functions and function costs. Each of these functions should satisfy a need that the consumer has.

$$\text{Value} = \Sigma \left[(n \times a/c)\,1 + (n \times a/c)2 + \cdots + (n \times a/c)? \right] \tag{1.8}$$

Consider the value of the ice cube trays pictured in Figure 1.4. A consumer needs to remove ice from the tray and place it in a drink. The function of separating ice is furnished in different ways in each device. Depending on the design, the cost of the separate ice function will vary. The way this function is provided adds substantially to the cost of design A. The ice separator is expensive to manufacture and assemble. The ease with which ice can be removed also depends on the design. It is hard to remove from A and easy to remove from the other designs. The market value of A is low, because it is difficult to use and expensive.

Definition of Value in Value Analysis/Value Engineering

Value analysis (VA) and value engineering (VE) are formal methodologies devoted to the study and improvement of value. They define value as the ratio of function to cost. In VA/VE, "maximum value is achieved when essential function is obtained for minimum cost."[11] This can be written mathematically as

$$V_{max} = \frac{F}{C_{min}} \tag{1.9}$$

This definition is similar to the performance/price definition and is based upon the assumption that the functions in a product are necessary and will satisfy the needs of a customer. Under this assumption, the numerator in equation 1.3, $(n \times a)$, reduces to the function F in equation 1.9. For many products this is a valid assumption. Equation 1.4 differs mainly in that it includes an expression for a consumer's need for functions that satisfy these needs. However, a designer's view of need may differ dramatically from the actual needs of consumer's.

Consumer Value

Equation 1.10 is an example of a value equation that can be used to measure the consumer value of these products. Notice its similarity to equation 1.8. In this equation, the value of an ice-making device is seen as a function of how well each product will fill the consumer's need to cool drinks with ice. Performance features such as the cooling capacity of a product, the ease with which it can be used, and the quality and reliability of the product are evaluated when choosing among similar products. Each of these features may have a different impact on the overall ability of a product to meet consumer needs. If this is true, the relative importance of each feature should be weighted. Value measurement techniques described in Chapter 4 are used to assign importance ratings to criteria. Assume that the importance of these criteria was judged to be

cooling capacity = 30%
ease of use = 50%
quality/reliability = 20%

In equation 1.10, these weightings (0.3, 0.5, and 0.2) are multiplied by the relative ability of each device to satisfy each need. The individual ratings are then summed and divided by the price of each product to obtain a figure of merit that represents the relative value of each device.

$$v = \frac{(0.3 \times tc) + (0.5 \times eu) + (0.2 \times qr)}{p} \qquad (1.10)$$

where:

$$
\begin{aligned}
v &= \text{value} \\
tc &= \text{tray cooling capacity} \\
eu &= \text{ease of use} \\
qr &= \text{quality and reliability} \\
p &= \text{price}
\end{aligned}
$$

Value measurement techniques described in this book are used to determine the value of consumer items. In the ice tray example, we find that

$$\text{consumer value A} = \frac{(0.3 \times 5) + (0.5 \times 1.5) + (0.2 \times 3)}{\$5.95} = 2.85/\$5.95 = 0.48$$

$$\text{consumer value B} = \frac{(0.3 \times 3) + (0.5 \times 3.5) + (0.2 \times 4)}{\$1.00} = 3.45/\$1.00 = 3.45$$

$$\text{consumer value C} = \frac{(0.3 \times 4.5) + (0.5 \times 4.5) + (0.2 \times 5)}{\$1.45} = 4.60/\$1.45 = 3.17$$

$$\text{consumer value D} = \frac{(0.3 \times 1.5) + (0.5 \times 5) + (0.2 \times 3.5)}{\$3.25} = 3.65/\$3.25 = 1.12$$

$$\text{consumer value E} = \frac{(0.3 \times 2) + (0.5 \times 3) + (0.2 \times 3)}{\$1.79} = 2.70/\$1.79 = 1.51$$

The greatest consumer value is supplied by product B. Product C is nearly as good, but costs too much. Product C performs better than B, but the consumer pays a premium for its higher quality and performance. The aluminum tray (A) costs far too much and is not as reliable as the other products. It has very poor value relative to the other products. Product D is clearly overpriced, but offers some interesting features that none of the other products do. If its price is lowered it may sell well, because it meets the customer's needs in a novel way.

Equations like these can be developed to model the customer value of any product or service. They serve as surrogates for the real value of each product, which is determined by actual market sales. Often these merit numbers are summed and used to calculate the relative normalized value of each product. Here we find that A = 5%, B = 35%, C = 32%, D = 12%, and E = 16%.

Retail Product Value

The value of a product to a retailer may differ significantly from the value of a product to a consumer. The retailer is primarily interested in a financial return from his investments. Retailers are interested in merchandise that they can sell quickly at a good profit. Retailers want maximum income from minimum investment. The actual cash return per sale is a very important to a retailer. If half of the money from a sale is returned to the retailer, the cash generated per sale for the products A to E is quite different. Product A would return nearly $3.00, whereas product B would return only $0.50. A store would have to sell six units of B to generate the same money it would get from one sale of A. At a fixed markup, higher-price items generate more cash per sale and require less sales and inventory.

A retail store will stock items that best fit its needs. Factors such as the shelf space required by a product, inventory investments, and the need to adjust inventories on a seasonal basis will influence a retailer's selection of items to sell. The attractiveness of a product, its aesthetic appeal, and the way it is displayed and packaged are important factors to a retailer, because they attract customers to a product. These factors often mean that retail value is different than customer value. Most stores evaluate the value of products in financial terms.

Equation 1.11 illustrates how a retailer could express value in terms of the income he or she could receive from items sold. Instead of calculating a cost–benefit ratio, retailers would estimate the sales revenues they would expect to receive from selling different products, deduct their expenses, and calculate the total return for each product.

$$\text{Retail Value} = \text{unit sales (unit price} - \text{unit cost)} \qquad (1.11)$$

The retailer's income is calculated by multiplying estimated sales by the dollar return per sale. Retail costs are calculated by multiplying the expected sales by the dollar cost per unit sale. Unit cost is made up of the total fixed and variable costs per unit sale. This cost varies from product to product and includes the retailer's inventory and sales costs. It is higher for seasonal products, because seasonal products require expensive inventory adjustments.

In this example, the need for products A, B, C, and E is seasonal with heavy picnic use in summer. Product D, plastic-coated ice spheres, is less seasonal than the other products, because it is used primarily to cool alcoholic beverages, which are consumed all year long. I have entered hypothetical data into equation 1.10 to show how we might measure the value of these products to a retailer. I assume that a store might sell a total of 1000 units in time t and that these sales will be apportioned by the customer's perception of the relative value of each item. If we multiply our estimate of the relative value of each item to a consumer by 1000, we have an estimate of the number of units of this product that will be sold in time t. The total money from the sale of these items is equal to the sale price times the units sold. With these assumptions, the income for each item is estimated in Table 1.3.

TABLE 1.3   Estimated income from products in Spring quarter

Product	Relative Value to Consumer (%)	Sales in Time t	Price per Unit	Income in Time t
A	4.9	49	$5.95	$291.55
B	35.5	355	$1.00	$355.00
C	32.6	326	$1.45	$472.70
D	11.5	115	$3.25	$373.75
E	15.5	155	$1.79	$277.45

This income would be for a store in a specific period of time, which we will assume is the spring quarter of the year. The expenses during this period would be equal to the fixed and variable costs for the quarter. For simplicity, let's assume they are equal to the retail purchase price of an item times an overhead burden rate. In this case, the retail price equals half the sale price, because the markup is 100%. Assuming a burden of 36% for seasonal products and 18% for nonseasonal products, the retail expense for each product is calculated by multiplying sales times retail purchase price times the burden rate. Profits are then determined by subtracting these expenses from total sales income per item. For the spring quarter the profits are estimated in Table 1.4.

The last column in Table 1.4 shows the relative retail value (calculated as a percentage of total profit) for each of these products. Note that these values are quite different than the consumer values we calculated earlier. The percentages were obtained by summing the theoretical profitability of the items and calculating a percentage value for each product. Converting these figures to a percentage allows us to compare the value of each product to a consumer and to a retailer.

In this example, an item's profit in the spring quarter was used to calculate its retail percentage value. We need to determine the effect of seasonal needs on the profitability of these products. If there are differences in the need for products, this will affect sales and profits. We assume that item D is nonseasonal and that its sales remain constant throughout the year. The other products are seasonal with few sales

TABLE 1.4   Estimated profit from products in Spring quarter

Product	Retail Price	Burden (%)	Total Expense	Profit	Relative Value to Retailer (%)
A	$2.975	36	$198.25	$ 93.30	15.5
B	$0.500	36	$241.40	$113.60	18.9
C	$0.725	36	$321.44	$151.26	25.2
D	$1.625	18	$220.51	$153.24	25.5
E	$0.895	36	$188.67	$ 88.78	14.9

in the winter and a heavy demand in the summer. A retailer should analyze the effect of this seasonality on the annual sales and profits for each product.

In Table 1.5 the quarterly and annual profits for each of these products are compared. The seasonality figures show the relative sale from quarter to quarter throughout the year. Sales in the spring are used as a base and multiplied by the seasonality to determine sales in each other quarter. This analysis assumes that a retailer will stock D throughout the year and will not stock the other items during the winter. Hence, the reusable ice cooler sells throughout the year and has the largest profit. If the other products were stocked throughout the year, more units would sell, but not enough to warrant their inventory cost and sales relative to other items.

If you compare the relative value of these products to consumers and retailers, you find many significant differences. Consumers strongly favor B and C and see much less value in A or D. Retailers find D most appealing, followed by C and B. The reusable plastic coolers should appeal to a retailer, because they offer the best overall profit. This plus the fact that the product does not take up much space and is packaged in an attractive manner make it attractive. However, the high customer price and the small cooling capacity will detract customers from buying it. Seasonality and profitability have a strong influence on retailers and little effect on consumers.[12] Under these conditions, a store owner would want to stock the reusable cooler annually with C or B on a seasonal basis.

Table 1.6 shows how the value of these products varies from consumer to retailer over time. Value measurement was used to estimate the force of value and describe some interesting differences in market value. Section II of this book elaborates on value measurement.

This example helps explain how and why products perform in the market. Today, product A is difficult to purchase. It is available mainly from retailers[13] who specialize in unique niche-type products. Twenty years ago, before the development of plastic ice cube trays, it dominated the market. It has been replaced by products like B and C, which are available in most large department stores, often stocked on a seasonal basis. Product D is new and is available in specialty shops and some large stores. Its price is falling. The product is offered in a wide variety of novel shapes and sizes. Item E is often hard to find. The distribution of these products in this market is easy

TABLE 1.5 Retail profits by quarter

Quarter	Seasonality		Profit				
	A,B,C,E,	D	A	B	C	D	E
Spring	1	1	93.30	113.60	151.26	153.24	88.78
Summer	1.6	1	149.28	181.76	242.02	153.24	142.05
Fall	0.4	1	37.32	45.44	60.50	153.24	35.51
Winter	0	1	0	0	0	153.24	0
Total	3	4	279.90	340.80	453.78	612.96	266.34

TABLE 1.6 The difference between consumer and retail value

Product	Percentage Customer Value	Percentage Retailer Value	
		Spring	Annual
A	4.9	15.5	14.3
B	35.5	18.9	17.4
C	32.6	25.2	23.2
D	11.5	25.5	31.4
E	15.5	14.8	13.6

to understand, based upon an analysis of the seasonal value of these products. Differences in retail and consumer value dictate how and when products are offered for sale. Nearly all markets are dynamic and characterized by rapid change. As prices and products change in a market, the value of products change. In Chapter 5, methods to model the dynamics of value that changes with time will be described.

Manufacturing Value

The manufacturer of a product for retail or direct consumer sale represents another level of the product hierarchy. Like a retailer, a manufacturer will likely focus on the dollar return he can expect from a product. However, a manufacturer's costs are likely to be quite different than a retailer's. Producers have high capital costs associated with their manufacturing equipment and plant. They have different overhead structures, which include research and development, engineering, administration, marketing, and distribution costs. Capital intensive costs force many manufacturers to keep people and machines producing products most of the time. Hence, the extruded plastic ice cube trays will be part of a larger line of extruded parts, which may have very different markets. One manufacturer may concentrate exclusively on plastic extrusions, another on metal fabrication. In this case, they may be unwilling to change their production process from metal forming to plastic extrusion. The danger here is that the manufacturer may not recognize significant changes in the consumer or retail markets until they seriously erode profits in a number of different product offerings. Hence, a manufacturer needs to constantly monitor developments in both of these markets. Equation 1.12 shows the general form of an equation that will allow a manufacturer to determine manufacturing value.

$$\text{Manufacturing value} = \frac{\Sigma \text{ [customer and retailer benefits } (1 \text{ to } n)] + \text{profit}}{\text{costs}} \quad (1.12)$$

Large companies are made up of departments that have different needs and wants. In these companies, value can vary within the company. Value is seen differently in the accounting, marketing, manufacturing, research, and engineering areas of a corporation. Depending on the department, products can offer different rewards and/or punishments. Research may see value in new technology that advances science yet may do little to satisfy marketing needs. This is one of the reasons that value studies in large companies use teams of people from different departments in the company.

Although I have used business corporations and consumer products to illustrate how value can differ within a large organization, similar situations exists in government and service organizations. Consider how value and values change between organizations like the Environmental Protection Agency, the Internal Revenue Service, or the Small Business Administration.

It should be obvious that value, like beauty, is in the eye of the beholder. Often we assume that others "see" things as we do and value things as we do. We often assume that "people get what they pay for" and that price is synonymous with value. Little wonder that the true nature of value has escaped us. This book uses value measurement and standards to quantify value. With this in mind let's return to the concept of value as energy.

CALCULATING THE ENERGY NEEDED TO CREATE VALUE

Energy is used to produce all of our products and services. Today energy consumption is a matter of growing concern. The United States imports over 50% of its oil, the cost of energy is rising, and we are becoming increasingly dependant on foreign supplies. The situation is similar in Europe and Japan, which are heavily dependant on outside supply. Adding to our concern is the greenhouse effect. The level of carbon dioxide, produced when fuel is burned to create energy, has been rising in the earth's atmosphere for over a century. Some scientists fear that further increases in carbon dioxide may trap heat in the earth's atmosphere and lead to global warming. This greenhouse effect could disrupt life on earth.

Energy conservation is increasingly important; it will reduce these dysfunctions. Value is created when we use energy to form goods and services to fill our wants and needs. People can determine the energy content of any product or service using a process of *energy accounting.* Instead of determining the cost of a product or service, energy accounting converts the value of a product into energy units, such as kilogram calories or kilowatt hours. The energy content of a product or service can be used to

1. Express product value in terms of energy.
2. Create energy standards for value.
3. Determine the efficiency of product designs.
4. Compare the efficiency of different designs.
5. Eliminate the ambiguity of monetary standards.

Energy accounting determines the total energy invested in a product. Chapter 6 shows how the energy content of products and services can be determined, and how

TABLE 1.7 Energy content of ice-making products

Product	Dry Weight Container (g)	Kilogram Calories to Make a Container	Capacity in Cubic Centimeters	Kilogram Calories to Make Ice
A	292.5	2220.5	600	47.8
B	230.0	103.5	700	55.8
C	236.0	106.2	900	71.7
D	20.2	9.1	400	31.9
E	167.5	75.4	280	22.3

the kilocalories of energy in each of the cooling products described above was calculated.

Table 1.7 shows the energy content of these products expressed in kilogram calories and represents the *minimum* energy content for each product. In this example, the containers differ significantly in the amount of energy required to make them. Most take more energy to make than the ice they hold will absorb in melting.

The energy content of a product or service can be used as a standard for value measurement. Because value is itself a form of energy, expressing value in terms of the energy content of goods and services allows people to determine the efficiency with which value is produced. Energy provides a better measure than monetary value, because it is constant while monetary value is transient, changing over time.

SUMMARY

Value is a form of energy, a force that satisfies our needs and motivates us to work. When people think of value as a property of an object or service they confuse cause with effect and overlook the energetic nature of value. Just as temperature is an effect of heat on matter, *valuableness* is an effect of value on matter. Value is directly related to the ability of a product or service to satisfy our needs and is inversely related to cost. Although value is transient and highly variable, it can be measured and modeled. By expressing value in energy terms analysts can bring the accuracy and rigor of science to economics and management. They can quantify and model the value of products and services and create new standards for value.

REFERENCES AND ENDNOTES

1. DeMarle, D. J., "A Metric for Value," *Proceedings of the Society of American Value Engineers,* pp. 135–139, Vol. 5 1970.
2. DeMarle, D. J., "The Nature and Measurement of Value," In *Proceedings, 23rd AIIE Conference,* May 1972 pp. 507–512.

3. DeMarle, D. J., "Value Measurement," In *Proceedings, Pacific Basin VE Conference, Honolulu, Hawaii,* April 24–27, 1988, pp. 57–68. The Lawrence D. Miles Value Foundation.

4. The word *need* is used in a broad sense. It connotes a set of needs or wants that are active and create a desire in the individual to satisfy these needs.

5. I will often use the words *utility* and/or *usefulness* to describe the ability of an item to satisfy needs. In this case, $I = n \times u$.

6. A purist might argue that I should substitute an expression for the utility of money in place of cost, since I have related importance to the utility of objects and services. If cost is used alone, it does not reflect the utility of money, because the purchasing power of money is not an arithmetic function on the quantity involved. Experiments performed by E. Galanter at the University of Pennsylvania indicate that the utility of money is approximately equal to the square root of the quantity of money involved: $u = \sqrt{m}$, where u = the utility of money and m = money expressed in dollars or similar monetary units. Using this concept, $V = I/\sqrt{c}$.

7. The % I symbol indicates that we have compared the relative importance of items to each other. The % C symbol indicates a similar comparison of costs.

8. Webster defines *value* as "a fair return or equivalent in goods, services, or money for something exchanged; the monetary worth of something." *Worth* is defined as "monetary value; the value of something measured by the quality or esteem in which it is held."

9. Improvements in plastics have largely eliminated the plastic breakage.

10. Later in this book we describe how advantages and disadvantages can be diagrammed as functions and dysfunctions.

11. Department of Defense Joint Course Book, *Principles and Applications of Value Engineering,* pp. 1–4. Superintendent of Documents, U.S. Gov. Printing Office, Washington, D.C.

12. These coolers are a one-time purchase that will last for a number of years. Individuals need them more at certain times of the year, but simply make use of them as needed.

13. Thomas Scientific, *Scientific Apparatus and Reagents,* Thomas Scientific, Swedesboro, N.J.

2

VALUE, GROWTH, AND EVOLUTION

David J. De Marle

INDIVIDUAL GROWTH

Dr. Abraham Maslow has developed a multiple need theory of human motivation.[1] He believes that motivation proceeds through a hierarchy of wants and that satisfaction of any particular want merely serves to activate a higher-order want. Inspection of this theory shows it to be related to the concept of a value force described in Chapter 1. Maslow's theory furnishes a hierarchy of needs that an individual uses as he searches for value and personal growth. It provides a general orientation for the acquisition of things to satisfy human needs. Figure 2.1 shows the relationship between Maslow's need hierarchy and a value acquisition concept. In this figure Maslow's needs are arranged in order of importance. The most important unmet needs are shown closest to a person. Individuals will satisfy their need for food, clothing, and shelter before they attempt to satisfy belonging or esteem needs. As people obtain items to satisfy these needs, they increase their own value and grow. Property and ownership increase wealth and a person with possessions can do more things than a person without them.

A home satisfies a basic need for shelter and, once owned, will increase a person's security, status, and wealth. The individual now seeks to acquire things to satisfy a need to belong. Perhaps he or she will join a social or fraternal group. Although the individual's need for food will constantly reoccur, the home owner can purchase an appliance like a refrigerator that allows food to be preserved. The decision to buy a specific refrigerator is made on the basis of the relative value of different refrigerators. Once again, value governs the individual's acquisition of things. Depending on the buyer's specific needs and the features of different refrigerators, a selection will be made of the refrigerator that best satisfies the customer's needs.

26

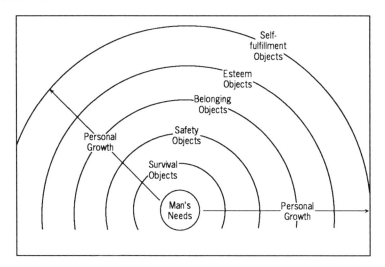

FIGURE 2.1 Maslow hierarchy of needs.

Employment provides wages and the possibility of a career. Once employed, value will govern an individual's behavior on the job. Dr. Frederic Herzberg has published a concept that relates theories of job satisfaction to a needs concept of human behavior.[2] Figure 2.2 is an adaptation of Herzberg's *job wheel*. Herzberg believes that each job contains *hygiene factors* that do not motivate an employee unless they are removed from his job. These hygiene factors are present in every job and represent factors owned by the employee. Thus they do not motivate an employee to expend further effort. However employees will work harder to enlarge their job and their position at work. Figure 2.2 is a value diagram with the job itself as a center of growth. A person working in a job is surrounded by items that he will work to obtain. In doing so, the employee grows and enhances his value. A person's wages are a case in point. Current wages are a hygiene factor, something that is owed to the individual and taken for granted.

Although Maslow's and Herzberg's theories provide a needs hierarchy and a focus for growth, both of their theories are based on a belief that people are motivated to satisfy their needs. Their theories offer insights into the way people grow by obtaining items of value.

PRODUCT EVOLUTION AND GROWTH

Because new product features act as surrogates for human needs, Maslow's theory of needs can be applied to new product development. Figure 2.3 diagrams a hierarchy of goals that a new device or service must meet. Maslow's basic goal categories are used as a framework for specific product objectives. A new product has a hierarchy of survival, safety, belonging, esteem, and fulfillment needs like those of a person.

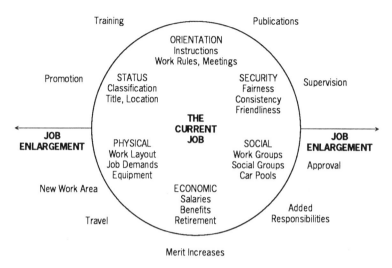

FIGURE 2.2 Herzberg job satisfaction and enlargement.

Profitability and usefulness are basic survival needs of a new product. Similarly, patentability is a safety need, and a product's fit with a firm's manufacturing and marketing facilities is a belonging objective. There is a degree of real anthropomorphism here.

Survival of the Fittest

Charles Darwin, in his book *Origin of Species,* postulated that a natural selection process is at work in nature in which only the *fittest* animals and plants survive. In this chapter, the author hypothesizes that products, like biological species, are evolv-

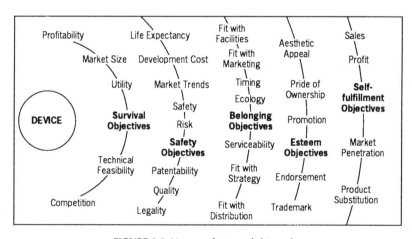

FIGURE 2.3 New product needs hierarchy.

ing. However, in products and services the measure of fittest is value. Products and services evolve in a competitive market in which those that best satisfy our needs survive. Value is the directive force underlying an innovation process that continuously creates improved products and services.

Value and the Evolution of Products

Products have been evolving since the dawn of human history. In this century the process is accelerating as people apply new technology to product design. I recall with some nostalgia the task of chipping pieces of ice from the large blocks of ice my family used to cool an icebox. Holding a sharply pointed ice pick with a wooden handle in one hand and the ice box top in the other, pieces of ice were chipped from the 50-pound ice cake by repeatedly thrusting the pick into the ice. A shower of ice usually scattered around the kitchen, then was painstakingly collected, mixed with salt, and used to chill ice cream made in our ice cream churn. The ice block and ice pick were the ancestors of the metal ice cube tray and lever arm described in the previous chapter. Of course our ice box has been replaced by a mechanical refrigerator/freezer that holds its supply of ice cream and ice cubes. Product substitutions such as refrigerators for iceboxes, dairy-made ice cream for homemade ice cream, and ice cubes for ice picks, occur rapidly.

Today, plastic ice cube trays have replaced metal trays, and reusable cooling devices are competing with plastic trays. As this evolution proceeds, the value of products increases. New products satisfy needs at lower cost and with less energy input. Functions and features are incorporated into a product, and separate components and procedures are merged into simple and more efficient designs. A one-piece plastic tray provides all of the functions and features of a metal tray that required the manufacture and assembly of 17 separate parts. Reusable ice coolers contain their own water, water that has to be supplied to other ice cube trays. Incorporating water in a reusable container eliminates nine operations performed by home owners when they fill and use ice cube trays.[3]

Calculating the energy involved in creating value allows us to measure the efficiency of innovations and set standards for the value of things. Measuring value in terms of energy is important today, because rising fuel prices and fears of a greenhouse effect make energy conservation desirable. If we plot the amount of energy required to manufacture these products versus time, we find that a dramatic decrease in energy use occurred as the products evolved. Note a logarithmic scale had to be used in Figure 2.4 to accommodate the rate of this reduction.

Improvements in the Value of Ice Making Devices. A similar picture appears when we examine the value of these products. Table 2.1 shows the value of these products arranged chronologically. Value is increasing with time. The plastic trays are substantially better than the metal trays, and the value of the reusable cooler is about equal to that of the plastic trays. If it drops in price and increases in capacity, it may develop into a replacement for many of the current plastic trays.[4]

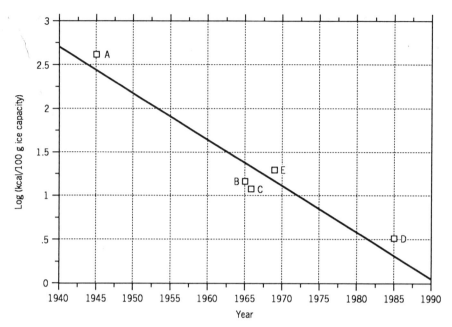

FIGURE 2.4 Reduction in the energy needed to make products.

Improvements in the Value of Artificial Lighting Devices The improvement of artificial lighting devices offers another example of how the value of products has increased. Improvements made in the efficiency of artificial lighting devices from 1850 to 1960 are plotted on a graph in Figure 2.5. The dramatic improvement that began with the invention of filament-based incandescent lamps accelerated as gas discharge and fluorescent lamps were invented. Note that this increase, like the increase in Figure 2.4, is exponential, and the data is plotted on a semi logarithmic scale.

TABLE 2.1 Improvement in value of ice cooling devices with time

Product	Customer Value (%)	Retailer Value (%)	Average Value (%)
Metal tray	5	14	9.5
Plastic trays			
B	35	17	26
C	32	23	27.5
E	16	14	15
Average	27.7	18	22.8
Reusable cooler	12	31	21

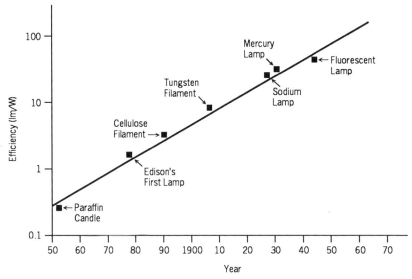

FIGURE 2.5 Improvement in the energy efficiency of lamps. Reprinted by permission of the publisher, from TECHNOLOGICAL FORECASTING METHODOLOGIES by H. W. Lanford, © 1972 AMACOM, a division of American Management Association. All rights reserved.

Figure 2.6 is an envelope curve[5] graphing the improvements made in the efficiency of incandescent and fluorescent lamps. This envelope curve plots the lighting efficiency of each design approach versus time. Each approach has a specific range over which it can operate. Note how efficiency improves slowly at first, then accelerates, and finally slows down as it approaches a limit. Plotted against time the efficiency line looks like a letter *S* that slants to the right. This S curve shows how the *provide light function* of each lamp increased to a limit. As new ways are found to provide a function, they create a series of S curves that stack upon each other. Similar improvements are occurring in nearly every field.[6]

The improvements documented in Figures 2.5 and 2.6 are part of a much longer history of improvement of artificial lighting devices. Throughout history our ancestors invented a number of devices that they used provide light. Table 2.2 chronicles the development of artificial lighting devices. Note how each new development represented an improvement in the technology of the time.

Improvements in the Value of Writing Devices Only useful and cost-effective products survive the competition for improved value. In time, a genealogy of products and services similar to a biological genetic tree is produced. A genetic product tree is shown in Figure 2.7. This genetic tree shows how writing devices evolved. Each branch of this tree shows how new forms of writing implements evolved from previous devices. Pencils evolved from pens. The modern felt-tip pen descended from the reed pen used by the Egyptians. Each new device evolved to overcome a problem or problems existing at the time. The reed pen was fashioned from hollow reeds that

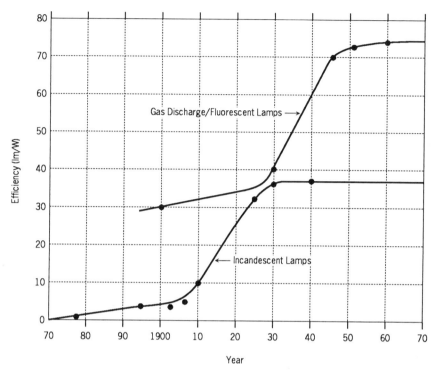

FIGURE 2.6 Efficiency of incandescent and fluorescent lights. Reprinted by permission of the publisher, from TECHNOLOGICAL FORECASTING METHODOLOGIES by H. W. Lanford, © 1972 AMACOM, a division of American Management Association. All rights reserved.

grew along the Nile River. Cut on a 45° bias to create a point, they were dipped in a vegetable dye and used to mark papyrus. The fragile reeds broke easily and made writing difficult. The reeds held little ink and had to be inked constantly. This difficulty was overcome with the invention of the pencil. The pencil furnished dry pigment in the form of a lead stick and represents a "mutation" in the genetic tree. It illustrates the principle of function consolidation mentioned previously. Function consolidation occurs when external functions are incorporated into a "parent" device. A pencil has an internal supply of pigment and does not require an outside supply of ink. This reduces labor involved in writing and makes writing more efficient. Later, fountain pens were devised which allowed ink to be stored inside the pens.

Each new writing device represented an improvement over its predecessors. Quills were pliable and less fragile than reeds. They held more ink and were readily available. With industrialization, steel pens were developed that allowed finer lines to be drawn on paper. With their improved legibility and durability, they soon were substituted for quills. Substitution was driven by the superior value of new products. Sometimes substitutions were caused by shortages. When graphite from the Barrodale graphite mine began to run out, it became more expensive. Shortages increase the cost and reduce the value of products. Soon synthetic graphite was developed.

TABLE 2.2 Improvement of lighting devices

Time	Device	Improvement
Prehistory	Wood fire	Lighted cave
	Torch	Portable light
3000 B.C.	Stone lamp (a porous stone placed in a pan of oil)	Used in Mesopotamian tents and sod huts
400 B.C.	Candle, made by dipping flax in tallow	Solid, portable lamp
15th century	Molded candle	Large, easy to make
16th century	Candelabra	Lighted large churches
18th century	Spermaceti candle[a]	Brighter, long lasting
1792	Gas lantern	Continuous fuel supply
1800–1855	Candle-molding machine	Mass production
1800–1860	Natural gas and coal gas lanterns	Street lighting[b]
1825	Braided candle wick	Nonsmoking candles, eliminated snuffing[c]
1850	Paraffin candle	Cheap, easy to manufacture
1850	Kerosene lantern	Portable, inexpensive light for home use
1876	Carbon arc electric lamp	High-intensity "white" light
1879	Gas discharge lamp invented	Colored light
1880	Edison and Swan develop first incandescent light bulbs	Efficient "white" light, low maintenance
1904	Tungsten-filament light bulb	Durable filament
1930	Sodium and mercury vapor lamps	Increased efficiency
1938	Tubular fluorescent lamp	Increased efficiency, "whiter" light

[a]Spermacete is a crystalline substance obtained from the head of the sperm whale. One candlepower is based upon the illumination provided by a spermaceti candle weighing a sixth of a pound that burns at a rate of 120 grams per hour.
[b]By 1820, streets in London, Paris, and Baltimore were lighted by natural gas.
[c]Candles were in such wide use that professional "snuffers" were required to constantly trim the wicks of candles in churches or public buildings. Until the invention of the braided wick, carbon deposited in the top of the wick caused the candle to flicker and smoke. A braided or plaited wick bends and is consumed in the flame.

Made from a mixture of clay and lampblack and baked in a kiln into a thin pencil "lead," it rapidly replaced natural graphite. The fabrication of synthetic graphite allowed the development of the mechanical pencil. Necessity is the mother of invention and leads to the creation of new devices of improved value. Without exception, only the fittest products and services survive. Today, ballpoint pens with erasable inks compete with a variety of wooden and mechanical pencils and pens, and extruded plastic casings are replacing wood casings in pencils.

Consider how value acted as a evolutionary force in the development of these devices. In Table 2.3, the relative value of nine of the writing devices in Figure 2.7 is shown. Raters used the criteria analysis technique described in detail in Chapter 4

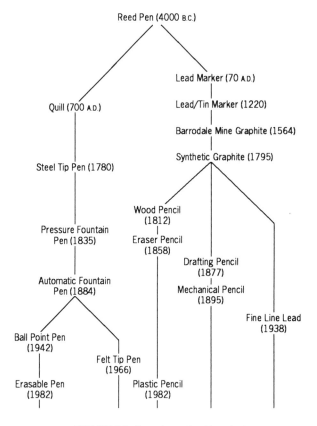

FIGURE 2.7 Genealogy of writing devices.

to measure the performance characteristics of each device.[7] In this analysis, the performance of each device was rated relative to the number of functions the device allowed (column A), the reliabiltiy of the device (column B), the quality of the writing (column C), and the ease of holding and operating the device (column D). These criteria were weighted, and the weightings were multiplied by how well each device met a specific criteria. The results of this subjective appraisal are shown in Table 2.3.[8]

This analysis indicates that the performance of writing devices has increased consistently in each of the four rating categories. It still continues today. Erasable inks have increased the sale of ballpoint pens, which are now equipped with erasers. These new inexpensive pens are displacing pencils in the market.

Improvements in the Value of Commercial Aircraft Most product changes that occur today happen at a very fast rate. Changes in aviation illustrate this well. Figure 2.8 shows some of the major changes that took place as product innovations altered U.S. commercial aircraft from 1935 to 1975. Early aircraft were made of wood, metal, and canvas. As commercial aircraft was developed, light weight, all-metal structures was substituted for these materials.

TABLE 2.3 Increasing value of writing devices

Product	Ratings[a]				
	A	B	C	D	Average
1. Reed pen	2.6	4.9	4.9	3.6	4.2
2. Natural graphite	2.0	11.0	4.9	2.2	5.7
3. Quill pen	4.4	4.1	6.0	6.0	6.1
4. Lead/tin marker	4.2	10.2	6.7	6.0	7.1
5. Wooden pencil	10.5	9.5	10.3	13.5	10.8
6. Fountain pen	10.5	9.6	14.6	12.5	12.0
7. Felt-tip pen	13.5	11.9	11.4	13.6	12.3
8. Ballpoint pen	13.2	12.5	14.1	14.3	13.5
9. Mechanical pencil	17.1	12.5	12.9	13.9	13.7

[a]The performance of each product was rated: A, the number of functions of the product; B, the reliability of the product; C, the quality of the writing; D, ease of holding and operating the product.

In 1936 there were a total of 146 commercial aircraft in the United States.[9] Twenty-nine of these aircraft were recently developed DC-3 aircraft, which were able to carry about twice as many passengers as the Ford or Boeing Trimotor aircraft that flew at that time. Compared to these aircraft, the DC-3 saved 33% in operating costs. The all-metal structure of the DC-3 provided greater strength and could be mass produced. In 4 years, the number of DC-3s grew to 232; the aircraft sold for about $110,000 and represented over 75% of the total fleet.[10, 11]

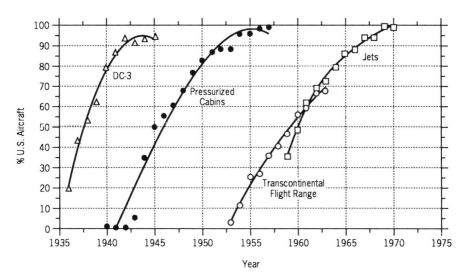

FIGURE 2.8 U.S. commercial aircraft innovations.

The DC-3 was a propellor-driven aircraft of improved value that rapidly replaced its competitors and dominated commercial aviation for several decades. It was eventually replaced by a number of long-range aircraft including the Continental DC-6, 7, and 8; the Lockheed 1049G and 1649; and the Boeing 707-120B and 720B. These aircraft had pressurized cabins and could fly at high altitudes. Pressurized cabins made it possible for aircraft to fly in the stratosphere and reduced the fuel required to transport passengers.[12] At high altitudes, airplanes encounter less air resistance and can fly faster. At the same power consumption, an aircraft can fly almost 40% faster at 30,000 feet than it can at sea level.

With high-altitude flight came long-range aircraft capable of nonstop transcontinental flight from New York to Los Angeles. In 1930 it took 36 hours to fly across the country. Twenty-two hours of air time were interrupted by 11 refueling stops and an overnight stay in Kansas City. The development of the DC-6 in 1950 made possible the first nonstop service between these two cities. The flight took 9 hours and 20 minutes. Successive aircraft used jet engines, had higher speeds, larger capacity, and cut travel time. The DC-7 took 7 hours and 33 minutes to cross the country in 1953.

Jet aircraft displaced propellor-driven aircraft rapidly in the 1960s. In 1952, Boeing began work on a prototype of the first 707 jet liner. By 1958, commercial jet aircraft were in widespread service in the United States. The jets were larger and faster than piston-driven aircraft, and reduced the operating costs of carriers.

In 1959, jets flew almost 10 of the total of 28 million seat miles per hour flown by all aircraft in the U.S. commercial airline fleet. Ten years later, jets flew 122.2 million seat miles compared to only 1.1 million seat miles for piston driven aircraft. The rapid takeover of airline transportation by jet aircraft was accompanied by several important advances in the efficiency of jet engine designs. Turbofan engines were developed that were more efficient than the original turbojets. By passing additional air outside the engine thruster, turbofans increased the compressor airflow while reducing the net velocity of the jet. This increased the fuel efficiency of the engine. A first generation of turbofan engines were substituted for conventional jet engines in the mid-1950s. A second generation of turbofan engines that shunt more air outside of the compressor were invented and entered service in 1968. They proved to be considerably more fuel efficient than the first generation and rapidly displaced sales of the original design. If we use 1 as an index of the fuel consumption of a turbojet engine operating at full power, Figure 2.9 shows the improvement in specific fuel consumption for turbofan engines that occurred from 1956 to 1969.

These examples of innovations in the airline industry illustrate how increases in the value of product cause substitutions in the type of products purchased and used. The amount of energy required to furnish a function decreased with time, just as it did with the ice-making devices (Figure 2.4).

Predicting the Rate of Diffusion of New Innovations

Numerous investigators have attempted to predict how rapidly a product innovation will substitute for current products in the market. In 1903 Gabriel Tarde noted the rate

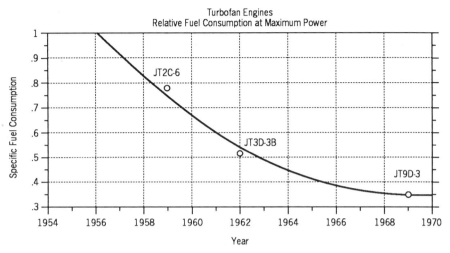

FIGURE 2.9 Improvement in the energy efficiency of aircraft engines.

at which ideas are adopted follows an S-shaped logistic curve.[13] In the 1920s, the U.S. Department of Agriculture studied the rate at which farmers adopted agricultural improvements proposed by farm extension agents. In the 1930s, studies were made of the spread of medical and educational innovations. Hundreds of studies of the diffusion of specific innovations in farm or other communities were conducted from 1930 to 1960, each attempting to define the effect of variables on the spread of the innovations. Until the 1960s, most of these studies were conducted independently in specific fields such as agriculture, anthropology, medicine, and sociology, with very little cross-communication.[14] Most of these social studies focused attention on a change agent who acted as an opinion leader and fostered adoption of an idea. Few talked of the economic or other benefits that the innovation possessed.

Since 1960, many studies of innovation have approached the study of the diffusion of innovations from an economic viewpoint. Mansfield conducted numerous studies that showed that the rate of adoption of an industrial innovation was positively correlated with the profitability of the innovation and negatively correlated with the capital investment involved.[15,16] He derived an equation for the rate of adoption that was a function of profitability, the size of the investment involved, and a term that was a function of the innovativeness of the industry involved. Blackman et al. used factor analysis to analyze Mansfield's innovation index.[17] They found that variables such as Research and Development funding, value added, and productivity growth in the industry formed the basis for an industry-specific innovation index.

Other researchers have found that noneconomic factors such as familiarity with a technology[18] and enforcement of antipollution laws[19] correlate positively with the rate of adoption of new innovations. In a study of the rate of diffusion of industrial innovations, Martino, Chen, and Lentz[20] examined 49 innovations in 14 different industries. They developed predictive models that attributed the rate of diffusion of a new innovation to the relative advantage of the innovation and the cost of introducing the new technology.[20] Their study includes more innovations than any previous study and characterizes the rate of diffusion of each innovation with a single rate constant obtained by fitting a Pearl substitution rate curve to market acceptance data. The study also rates the relative advantage of each innovation and normalizes the cost of each innovation.

Examination of the diffusion process by numerous researchers indicates that *the relative value of an innovation dictates its rate of acceptance.* Because value is a function of how well a design meets the needs of specific consumers, unique equations need to be written for each market innovation. Attempts to develop general statistical correlations to predict the rate of innovation for innovations in different fields have failed because of the unique needs and competitions in different situations. Nonetheless, these studies confirm the overall importance and direction of factors such as the cost of the new innovation (negatively correlated with rapid acceptance) and the relative advantage of the innovation (positively correlated with the rate of diffusion).

Value and the Evolution of Services

Services, like products, are growing and evolving to meet the needs of people. An explosion has occurred in the number and types of services that are available. Thousands of services offered today did not exist 10 years ago. This growth can confuse us, and we often need help in selecting those services most applicable to our needs.

The Evolution of Personal Banking Services In the last 20 years, banks have developed a large and frequently confusing number of new ways that their customers can invest their savings. Gone is the day when a customer simply deposited his money in a passbook savings account at 4 or 5% interest. Now a bank customer can invest in a wide variety of different financial instruments, each one of which has a different interest rate, maturity date, and risk factor. Some of the personal financial service products offered by a bank are listed in Table 2.4.

The alternative savings methods in (Table 2.4) evolved from the introduction of passbook savings in the United States in the 19th century. The first financial institution in the United States to allow personal banking was the Philadelphia Saving Fund Society, established in Philadelphia on December 2, 1816. Formed by 12 individuals, the purpose of the bank was to "receive and invest small sums of money as might be saved from the earnings of thrifty and industrious persons of modest means, and of affording such persons the advantage of security, and of interest."[21] The bank required a minimum balance of $5 before it would pay interest of 4.6%. Investors were required to make a minimum deposit of $1. This savings bank filled the needs of people who had previously been unable to make personal bank deposits, and its deposits grew rapidly. Four years later, its deposits totaled $64,263; 20 years later, they exceeded $1 million; 100 years later, they were over $130 million. Today the

TABLE 2.4 Saving investment alternatives

 1. Bankers' acceptance accounts
 2. Certificates of deposit
 3. Corporate bonds
 4. Deferred annuities
 5. Demand deposit accounts
 6. Government agency bonds
 7. Money market accounts
 8. Money market funds
 9. Negotiable order of withdrawal (NOW)
10. Passbook savings
11. Repro accounts
12. Savings bonds
13. Super NOW accounts
14. Treasury bills
15. Tax exempt bonds
16. Treasury notes

bank is one of the largest thrift banks in the United States with over $13 billion in deposits.

The Philadelphia Saving Fund Society started the passbook savings system, which was to dominate personal savings services for over a century. In a passbook account, a teller wrote the amount of each deposit made in a passbook owned by the client. Soon other banks began to offer the public passbook savings, and the banks used money from these accounts to offer their customers home mortgage loans. For 150 years, passbook savings and home mortgages dominated the savings and loan bank business. In the United States in 1966, deposits in passbook accounts represented over 88% of all the money held in personal savings accounts. However, from 1966 to 1979 deposits in passbook accounts dropped from 88% to 25.6% of all personal savings accounts! This precipitous change occurred as consumers transferred their money from passbook accounts to new savings accounts like certificates of deposit (CDs) and money market accounts, which payed higher interest. Savings and loan banks created these new high-interest accounts to attract deposits after the industry was deregulated. By paying higher interest rates, they attracted cash to the banks, which were now allowed to invest in higher-interest investments than mortgages. The new accounts were much less labor intensive than passbook accounts and helped the banks reduce their labor costs. With customer deposits insured by the government up to $100,000 per account, the depositors' and bankers' risks were minimal.[22] As competition between banks and other financial service agencies increased, a number of new services came into being. Examples are pay-by-phone services, direct deposits, and automatic teller machines.[23]

The rapid expansion of financial services in the United States is an example of the effect that government deregulation of an industry can have on the development of new services. For 150 years, regulation artificially restricted the evolution of new services and investments. With deregulation, an explosion occurred in this industry producing a cauldron of new high-interest services. Today the banks and banking services are feeling the effects of Darwin's law that only the fittest will survive.

Value of Personal Savings Services. Which of these services offer the best value? The answer to this question depends upon an individual's specific needs and the way each of these savings methods will fill these needs. A young married couple has different needs than a retired couple. In the spreadsheet in Figure 2.10, criteria analysis is used to measure the value of savings alternatives to these people. Five criteria were used to judge the merit of these investment options. These criteria are listed across the top of the spreadsheet—return on investment (ROI), tax savings, safety of an investment, access to the principle without undue penalty, and the amount of money required for an investment.

The importance of these needs vary for the young married couple and the retired couple. At the bottom of the spreadsheet, a set of weighting factors represent the relative importance of these needs. 100 points are alloted to these criteria based upon an assumption of the importance of each criteria to each party. Both parties view the importance of return on investment equally and 30 points show the importance of the return on investment criteria for both parties. The two parties view the safety of an investment quite differently: 40 points were allocated to the safety of an investment for the retired couple, whereas only 5 points were allotted to safety for the young married couple.[24]

The column on the left-hand side of the spreadsheet lists a number of different savings options. Numbers are used to show how the alternatives satisfy each criteria. The numbers in the matrix represent how an alternative satisfies a criteria—the higher

INVESTMENT	Return on Investment	Tax Savings	Safety	Access	Deposit Needed	Retired Person	Young Married
Bank CD's	3	0	4	3	1	311	185
Corporate Bonds	4	0	2	2	3	258	180
Deferred Annuities	4	2	3	2	2	321	215
Government Agency Bonds	4	2	2	2	3	278	210
Money Market Account	3	0	4	4	3	320	210
Money Market Funds	3	0	3	4	3	275	205
NOW Account	2	0	4	4	4	284	180
Passbook Savings	2	0	4	3	4	279	155
Tax Exempt Bonds	4	4	2	2	3	298	**240**
U.S. Savings Bonds	3	2	4	2	4	**332**	190
U.S. Treasury Bills (T Bills)	3	2	4	3	1	**331**	215
U.S. Treasury Bonds	4	2	2	3	3	283	**235**
U.S. Treasury Notes	3	2	3	3	3	290	210
Zero-Coupon Securities	4	0	2	2	4	260	180
Criteria Weighting Retired Person	30	15	40	10	5		
Criteria Weighting Young Married	30	5	5	25	25		

FIGURE 2.10 Criteria analysis of different saving methods.

the number, the better the alternative satisfies the criteria. The numbers correspond to the following scales:

Criteria	Scale
ROI	4 = highest return, 3 = medium, 2 = low, 1 = poor
Tax savings	4 = state and federal, 3 = federal only, 2 = state only, 0 = none
Safety	4 = little risk, 3 = low risk, 2 = moderate risk, 1 = high risk
Access	4 = immediate access, 3 = good, 2 = fair, 1 = poor
Deposit needed	4 = $0–500, 3 = $1000, 2 = $5000, 1 = $10,000+

The two columns on the right side of the spreadsheet show the score of each alternative for each couple. This rating is the sum of a set of numbers obtained by multiplying the numbers in each cell of the matrix by the appropriate criteria weightings. The scores printed in bold type show the two top-rated alternatives for each couple. Different needs alter the value of the alternatives for each couple.

Loan Services. In addition to savings services, a wide variety of loans are available today. Traditional forms of loans such as mortgages and personal loans have expanded into a diversity of new loan services. Today installment credit, revolving credit, credit cards, balloon mortgage loans, home equity loans, student loans, and car loans increase the type and variety of loan services. These loans fill different consumer and retailer needs. As a retailer, a bank profits from the money it loans. A bank calculates the value of a loan by determining the interest it earns from its investment. An individual who uses a loan to purchase a home that she would otherwise be unable to afford ascertains value in terms of the housing the loan makes possible. Values color our perception of the value of a loan. Some people feel that financial debt is bad and indicative of a lack of responsibility and thrift. Others feel that credit is good if used to acquire necessary goods in a timely fashion. Paradoxically, debt and credit are different words for the same thing, because to obtain credit is to incur debt.[25] Members of the United States House of Representatives learned this when the house bank closed in December 1991. When questioned about the similarity between check kiting an deficit spending, members might reply:

It was as we remember
something we did a lot.
Simply borrow money
from money we had not.

Although Congress may not understand, the value of loan services can be measured. The approach is identical to that shown for the savings investment alternatives.

The Evolution of Society

Just as value has governed the evolution of products and services, so too has it governed the evolution of society. An examination of human history shows that nations grow, prosper, and frequently die. Civilization is evolving societies in which all people can grow and share basic human freedoms. Value governs this evolution, because governments are aggregations of services that exist to serve the needs of people. History shows how people have evolved different types of communities to meet their needs.

Tribal Societies. Early peoples gathered together in small bands of hunter-gatherers who shared survival needs. Tribes could hunt prey and gather food more efficiently than individuals alone. Tribal chieftains and high priests were custodians of the laws that governed conduct. As the groups grew in size, they often depleted their supply of food, and it became harder to gather food. When a tribe grew too large to easily sustain itself, small bands of hunter-gatherers separated from the group and moved away from the original tribe's territory. As tribal knowledge of nature increased, the tribes were able to domesticate animals and farm the land. In time tribal life evolved into the village life of small farm communities.

Farms and Villages. Civilization was born about 10,000 years ago when people began to till the land. A number of farming communities were formed, and rules of property were developed. Possessions passed on to the eldest son in a family. Tribal chiefs and high priests evolved into village masters, feudal lords, rabbis, and bishops who became custodians of the laws that governed conduct. As the population of these communities grew, people developed energy sources and mechanized work. Horses, oxen, and other beasts of burden were domesticated to till fields first tilled by hand. Wood and animal oils provided fuels to warm tents made of hides. People used tents, sod huts, and log cabins for housing.

Cities and City-States As population grew, hamlets became cities. With agriculture and tillage, mechanization began. People invented aqueducts to carry water from distant dams to irrigate their crops. As urban centers developed, roads were built, and trade was established to support the growth of small city-states. Horse-drawn carriages carried goods over roads made of cobblestones, and large sailboats carried goods to distant ports. A market economy developed, and trade schools were created. Information was recorded in books, and libraries were built. As the population increased, trade routes were extended to cities in neighboring countries.

Nations. As time passed, a series of nation-states developed that governed large territorial areas. Distinct societies with unique languages and cultures evolved. Kings and popes became custodians of the laws that governed conduct. Figure 2.11 lists some of the national groups that grew as civilization spread throughout Europe. Growth was far from harmonious. Wars of conquest and religious crusades evolved from feudal and tribal wars. Nations became enemies, and men trained for war. Community wealth was hoarded in the hands of kings, lords, merchants, and aristocrats, who often became despots. Colonialism developed, and slavery flourished. Wealth and knowledge were the property of an aristocratic few.

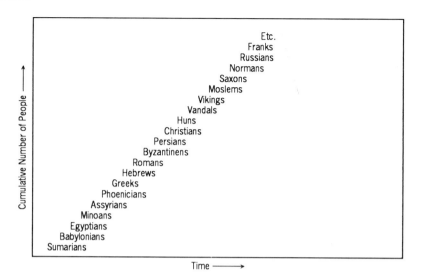

FIGURE 2.11 Growth of indo-european societies.

The value force caused the rise and fall of many nations. Figure 2.12 diagrams this process. Humans gather together in communities to satisfy their needs. The community works together, grows, and acquires possessions. As time passes, growth depletes resources, and a maldistribution of wealth and power concentrate in government and its leaders. The needs and wants of citizens grow unfulfilled. As resource dwindle, governments impose higher taxes to support their needs. Human rights are neglected as leaders are corrupted by power. As the original values of the community are degraded, dissent leads to revolution and/or separation and the formation of new communities. This process is repeated whenever a government fails to meet the needs of its people. Only the fittest communities survive.

Value acts to determine the fitness of nations. Dictatorships and despotism do not endure. Democracies that satisfy the rights of people replace them. New nations write into law the rights of free people, such as "We hold these Truths to be self-evident, that all Men are created equal and are endowed by their Creator with certain inalienable rights. That among these rights are life, liberty, and the pursuit of happiness."

VALUES

The cohesiveness of any society depends on concepts of right and wrong, which form the basis of our values. Community values focus the *force of value* on activities that benefit people and the community. *Values* represent a set of beliefs that individuals and groups use to control the behavior of people who reside together in a group.

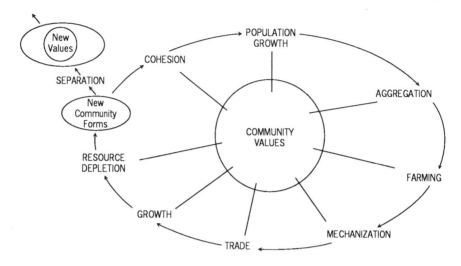

FIGURE 2.12 Historic aggregation and growth process.

Values are transmitted to children by parents and teachers and become part of a person's individual psyche. They help a person form a conscience of right and wrong. People who live together in families, communities, or nations usually share common values. These shared values are internalized and help people live together in harmony. Values are codified into laws to control the behavior of people who deviate from community norms. Community values often form a credo, or constitution, that defines the principles by which people are expected to live. The preamble to the U.S. Constitution defines the reason for community values. "We the people of the United States, in order to form a more perfect union, establish order, and insure domestic tranquility, . . . " Throughout history people's values have evolved and improved. Today people around the world share humanitarian values that endorse the rights of people.

SUMMARY

In this chapter I examined the process by which value influences growth. Darwin's law of the survival of the fittest applies to the survival of products, services, and societies. Value is the force that determines what will survive. A genealogy of products and services shows how human artifacts and societies have improved and evolved. All things are evolving toward better and fitter states. Value governs the creation, acceptance, and diffusion of innovations. People grow in influence and power when they acquire value. Community values moderate and control people and help them live together in harmony. Civilization is evolving toward a society in which all people share basic human freedoms that foster growth and value.

REFERENCES AND ENDNOTES

1. Maslow, A. H., *Motivation and Personality,* Harper & Row, New York, 1954.

2. Herzberg, F. *Work and the Nature of Man,* World Publishing Co., Cleveland, Ohio, 1966.

3. Function consolidation like this is a very common phenomena. Cameras that originally required separate light meters now control shutter and aperture settings with a light sensor built into the camera. Calculators that originally required batteries for power now operate on built-in solar cells. Lawns that originally required several separate applications of fertilizer now require a single application of a slow-release fertilizer. Television sets that originally had to be manually set now operate by remote control and can simultaneously show another program on a portion of the screen.

4. Refer to Figure 2.6, which plots some *S* curves for lamps. If the reusable ice cooler is at the beginning of a functional *S* curve, it may develop a substantial share of this market.

5. Martino, J. P. *Technological Forecasting for Decision Making,* American Elsevier, New York, 1972, p. 105.

6. Such as the replacement of propellor driven aircraft by jets or the replacement of vacuum tubes with transistors and integrated circuits.

7. De Marle, D. J., "Use of Value Analysis in Forecasting," presented at James R. Bright's Technology Forecasting Workshop, Castine, ME., Industrial Management Center Inc., Austin, Tex.

8. Cost was not used in this analysis, because it was not availabe. Thus, the assumption is that all of these devices were of equal cost at the time of substitution.

9. U.S. Civil Aeronautics Board, *Handbook of Airline Statistics,* pp. 484–487, Washington, D.C., 1973.

10. Miller, R., and Sawers, D., *The Technical Development of Modern Aviation,* pp. 47, 301, and Appendix II, Routledge and Kegan Paul, London, 1968.

11. Martino, J. P., Chen, K., and Lenz, R. C., *Predicting the Diffusion Rate of Industrial Innovations,* National Technical Information Service, Report UDRI-TR-78-42.

12. Miller, R., and Sawers, D., *The Technical Development of Modern Aviation,* Routledge and Kegan Paul, London, 1968.

13. Tarde, G., *The Laws of Imitation.* Holt, New York, 1903.

14. Rogers, E., and Shoemaker, F., *Communication of Innovations.* The Free Press, New York, 1971.

15. Mansfield, E., *The Economics of Technological Change.* Norton, New York, 1968.

16. Mansfield, E., *Industrial Research and Technological Innovation.* Norton, New York, 1968.

17. Blackman, A., et al., "An Innovation Index Based on Factor Analysis," Technological Forecasting and Social Change. **4,** 3, 1973.

18. Bundeguaard-Nielsen, M., and Fiehn, P., "The Diffusion of Technology in the U.S. Petroleum Refining Industry," *Technological Forecasting and Social Change* **6,** 1, 1974.

19. Lakhani, H., "Diffusion of Environment-Saving Technological Change," *Technological Forecasting and Social Change* **7,** 1, 1975.

20. Martino, J. P., Chen, K., and Lenz, R. C., *Predicting the Diffusion Rate of Industrial Innovations,* National Technical Information Service Report UDRI-TR-78-42.

21. Basch, M., "Philadelphia Saving Fund Society, First US Savings Bank, Inspired by British Examples," *American Banker* (150th Anniversary Issue) p. 72, 1986.

22. Unfortunately as we now know, many savings and loan banks invested their deposits poorly in high-risk endeavors that later failed. The collapse of many savings and loan banks resulted in the present savings and loan bailout.

23. Sherman, M. C., "Automatic Teller Machines Explore New Frontiers," *Computers in Banking.* November 1986.

24. It is important to stress that this is an example of how the value of these services can be measured. People should not act on the basis of the results of this specific analysis, as the value of investments can change rapidly with time, and their own needs may vary considerably from those depicted. This analysis represents conditions in 1988.

25. Nichols, D., and Erdevig, E., *Two Faces of Debt,* Chicago Reserve Bank of Chicago, August 1978.

THE MEASUREMENT OF VALUE

This section contains a comprehensive description of the process of value measurement. The basis of value measurement is the ability to gauge the value of elements in a system using subjective measurements of relative importance and cost. Value measurement techniques are described that are used to quantify the elements of importance and cost. Assigning numerical weights to the value of components in a product or service establishes their value and uncovers areas where improvements are needed. Value indices, graphs, and targets are described that enhance communications and act as vehicles for interdisciplinary teamwork and design improvement. The construction of dynamic models to portray changes in value with time is described and illustrated. To overcome signal-to-noise problems, subjective judgments undergo structured dialogue and testing.

3

VALUE MEASUREMENT

David J. De Marle

INTRODUCTION

Value measurement is a process in which numbers are used to quantify the value of elements in a system. It can be used to measure the value of many different items, such as products in a market, departments in an organization, components in a product, and tasks in a job. The process consists of several steps. First, a system is identified and its components are named. Second, needs are analyzed and the functions of the components are identified. Third, value measurement techniques and indicies described in Chapter 4 are used to quantify the valve of components in the system. Finally, these value indices are graphed, and various value, performance, or cost targets are derived and used as references in value improvement efforts. This chapter describes the value measurement process and illustrates its use.

Although value measurement is often used to quantify the value of components in complex systems, a simple example, with few components, was chosen to illustrate the value measurement process. This chapter first describes how value measurement was used to quantify the value of components in a decorative candle. It then describes how numerical value improvement targets were devised and used to improve the value of new candle designs. The process is straightforward and simple and illustrates the value measurement process. At the end of the chapter value graphs and indices of more complex systems are shown.

VALUE OF COMPONENTS IN A DECORATIVE CANDLE

Candles are an interesting example of a product whose use has changed with time. The invention of gas lanterns and Thomas Edison's invention of the incandescent

49

light bulb eliminated the need for candlelight and altered the form and function of candles. Although their main use has changed, significant numbers of candles are still manufactured. Over 40 million lb of paraffin is used for candle manufacturing in the United States annually. Today, candles are much larger and more attractive than the slender tapers of 50 years ago. They are decorative and fragrant, come in a wide variety of shapes and sizes, and make popular seasonal gifts. Modern candles can be purchased in gift shops around the world. Candles have esthetic appeal, and large numbers of them are still used in homes, restaurants, and religious services. However, increases in the cost of their ingredients have made them more expensive and lowered their value.

Quantifying Consumer Wants and Needs

The analysis began when focus group sessions were conducted in which consumers examined decorative candles and described the characteristics of candles they would buy. From these interviews, six important functions that a candle should provide were identified. People were then asked to rate the relative importance of these six functions. Each individual was asked to distribute a total of 100 points to the six functions in proportion to their perception of the relative importance of each function. If an individual felt that providing light was as important as all of the other functions combined, he would assign it 50 points and then distribute the remaining 50 points among the five remaining functions. If another individual felt that this function was not important at all, she might assign it no points, leaving 100 points to be distributed to the remaining five functions on the basis of her perception of their respective importance.

As was expected, wide variations in ratings were obtained. Some people felt that the emit light function was important, whereas others did not, and different individuals emphasized different criteria. Distributions were compiled and graphed, and the mean of each function's distribution was used to represent the importance of that function. To ensure that adequate testing was done, the rating process was repeated several times with different groups. Interestingly, although large variations continued to exist between individuals in a group, the average rating given to each function by each of the three groups was reasonably reproducible. The average rating by function of three different groups are shown in Table 3.1.

In each group, the emit light function had a bimodal distribution. About half of the raters gave high ratings to the importance of this function and half did not. Figure 3.1 shows the emit light rating distribution. Apparently, some candles are so decorative that people don't want to light and destroy them, whereas others see beauty in the candle flame itself. Figure 3.2 is a plot of the importance of fragrance, and Figure 3.3 is a plot of the importance of the attractiveness function. The distributions are typical of those encountered in subjective evaluations.

The Delphi method,[1] which is described in Chapter 14, can be used to improve the reliability of such ratings. This method is iterative, and each rater is furnished with a plot of the group ratings together with his own initial responses. If their ratings of an item are in the upper or lower quarter of the rating distribution, the raters are asked

TABLE 3.1 Importance of candle functions

Group:	1	2	3
Number of raters:	15	31	21
Functions		Average Percentage Importance of functions	
Emit light	20.7	16.3	16.2
Look good	26.5	31.7	31.8
Discharge fragrance	11.4	14.0	11.3
Maintain light	12.8	11.3	14.8
Prevent dripping	11.9	12.1	11.6
Prevent smoking	16.7	14.6	14.3

to explain why they rated the item as they did. This new information is compiled and furnished to each rater with a request to re-rate the importance of each criteria. Ratings are anonymous, and the raters are not identified. Delphi is an effective method that can be used to improve the reliability of subjective ratings. It is often used in forecasting and planning activities.

Quantifying the Importance of a Candle's Components

Once the importance of customer needs has been determined, the importance of each component in the candle can be ascertained. This is done by analyzing the function of each component and relating its importance to the role it plays in the overall

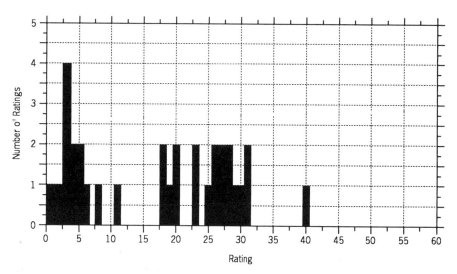

FIGURE 3.1 Ratings of the importance for a candle to give light.

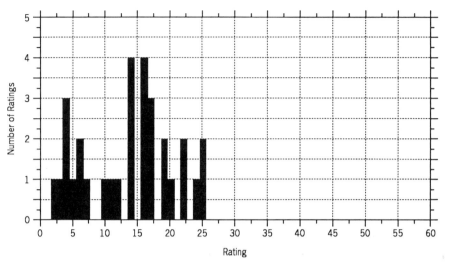

FIGURE 3.2 Ratings of the importance for a candle to be fragrant.

performance of the candle. Criteria analysis is a value measurement technique that can be used to determine the importance of components in a design. In criteria analysis a matrix is constructed that arrays design components against the user's needs. In each cell of the matrix, the ability of a component to satisfy a need is noted. Numbers are used to show how well a component satisfies a need. Category scales are often used to describe how well a component satisfies a specific need. A relative scale that describes how a component effects the attractiveness of a candle might consist of a number scale, where

5 = most important component
4 = very important component
3 = important component
2 = somewhat important component
1 = relatively unimportant component
0 = unimportant component

Using this scale, a rater can assign numbers to the effect of each candle component on the attractiveness of a candle. Thus, a dye added to a candle might be rated 4 or 5; a wick might rate a 4; and stearic acid, which can induce a slight luster, might be rated 2. Similar evaluations are performed throughout the matrix to show the contribution of each component to each user need. This analysis (see Figure 3.6) shows that visual attractiveness of a candle comes from several components—the wax, wick, and dye, whereas the fragrance of a candle comes primarily from a perfume that is added to the wax.

A thorough understanding of the behavior of the item under analysis is needed before rating the contribution of the components to the criteria. Although a candle appears simple, it really is fairly complex. Figure 3.4 is a cause–effect diagram that shows how different items in the design control the performance of a candle.[2] Components in the candle interact, and not only the components, but their interac-

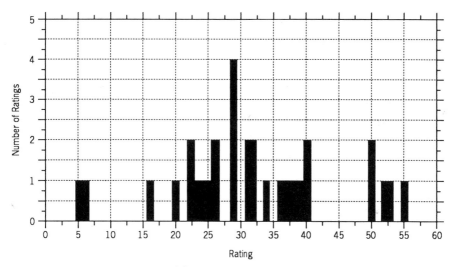

FIGURE 3.3 Ratings of the importance for a candle to be attractive.

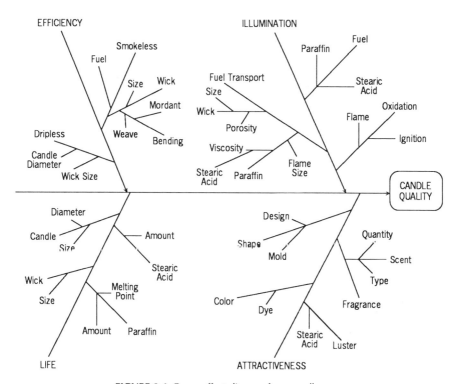

FIGURE 3.4 Cause effect diagram for a candle.

tions need to be understood. Thus, the wick's size and porosity control the amount
of wax that fuels the flame. The flame gives off heat, and the size of this flame
controls the amount of wax that fuels the flame. The amount of stearic acid added to
the wax controls the melting point and viscosity of the wax. The size of the wick
needs to be adjusted to match the diameter of the candle. Thus, the components in
a candle interact in a dynamic system and control a number of important feedback
loops that govern the performance of the candle. These factors make it important to
regulate factors such as the candle diameter, wick size, and fuel melting point. If
necessary, other more powerful methods can be used to describe these relationships.
Chapter 8 describes how quality function deployment can be used to describe and
control the relationships that affect quality.

The interactive nature of these relationships is best depicted by system dynamic
models that can be used to display the dynamic interactions in a system like a burning
candle. Figure 3.5 is a diagram of a system dynamic model for a burning candle. The
diagram uses arrows to show interactions that occur as a candle is ignited. A supply
of solid paraffin shown in the upper right-hand box is converted to liquid wax by heat
produced by the combustion of liquid wax. The amount of heat is a function of the
wax fed to the flame by the wick. Excess heat melts more wax than the wick can carry
and causes dripping. Heat loss to the air also regulates the amount of heat and affects
the rate at which paraffin melts. Candlelight is a function of combustion. A burning
candle is a good example of a dynamic system in which a set of components interact
to meet a customer's needs. The value of a candle changes with time as it is
consumed. Wax lost through dripping reduces the candle's efficiency and decreases
candle life. Chapter 5 describes the construction of system dynamic models and
shows how they can be used to model the value of products that change over time.

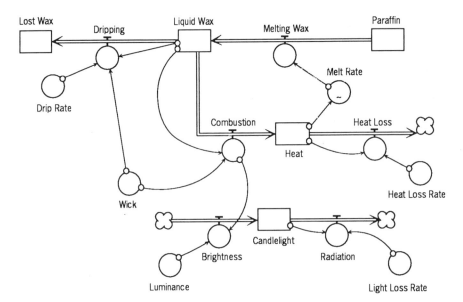

FIGURE 3.5 System dynamic diagram for a candle.

Once a thorough understanding of these interactive factors was obtained, the contribution of the candle components to the candle's performance was specified in the criteria analysis matrix shown in Figure 3.6. The criteria analysis method is described in detail in Chapter 4. Note that the cells of this matrix contain both the ratings (upper half of cell) and the product of the ratings (lower half of cell). The weighted criteria needs are shown in the weight percent column.

Calculating the Importance of Components in a Decorative Candle

Recall that in equation 1.3 in Chapter 1 we defined importance as the product of customer needs and wants and the ability of an item to meet these needs and wants.

$$I = n \times a \qquad (1.3)$$

where

I = importance
a = ability to satisfy need
n = need

The importance of a component in a criteria analysis matrix is calculated by multiplying each of its cell ratings by the importance of each of the consumer needs and

COMPONENTS IN CANDLE

CRITERIA	WT. %	WAX		WICK		STEARIC ACID		DYE		PERFUME	
Provide Light	.21	5	1.05	5	1.05	3	.63	0	0	0	0
Visual Attractiveness	.26	4	1.04	4	1.04	2	.52	5	1.30	3	.78
Fragrance	.11	1	.11	0	0	0	0	0	0	5	.55
Long Burning	.13	5	65	5	.65	3	.39	0	0	0	0
Dripless	.12	3	.36	5	.60	5	.60	0	0	0	0
Smokeless	.17	3	.51	5	.85	0	0	0	0	0	0
TOTAL IMPORTANCE	1.00	3.72 30% I		4.19 33% I		2.14 17% I		1.30 10% I		1.33 10% I	
TOTAL COST		28.6¢ 43% C		1.3¢ 2% C		10¢ 15% C		6¢ 9% C		20.7¢ 31% C	
VALUE INDEX =		.70		16.5		1.13		1.11		.32	

FIGURE 3.6 Criteria analysis matrix for a decorative candle.

summing these products. In the matrix in Figure 3.6, the importance of candle wax equals

importance of wax =
  [(contribution to illumination) × (importance of illumination)] +
  [(contribution to attractiveness) × (importance of attractiveness)] +
  [(contribution to fragrance) × (importance of fragrance)] +
  [(contribution to longevity) × (importance of longevity)] +
  [(contribution to driplessness) × (importance of driplessness)] +
  [(contribution to smokelessness) × (importance of smokelessness)]

importance of wax = (0.21 × 5) + (0.26 × 4) + (0.11 × 1) + (0.13 × 5) + (0.12 × 3) + (0.17 × 3)

importance of wax = (1.05) + (1.04) + (0.11) + (0.65) + (0.36) + (0.51)

importance of wax = 3.72

The value 3.72 represents a figure of merit that combines all of the contributions of wax to the set of consumer-weighted performance features that a candle should contain. Because the importance of an item is defined as the product of customer needs times the usefulness of an item in meeting these needs, the 3.72 rating represents the importance of wax in the candle's overall design. The importance of wax is compared to the importance of each of the other components in the candle in the total performance row at the base of the matrix. Note that the wick is the most important candle component with a performance rating of 4.19. The dye is the least important component with a total rating of 1.30 (all of which comes from its attractiveness rating). The relative importance of the components can be calculated by summing the importance ratings of all components and then calculating the percentage importance of each component. The percentage importance of each component is shown directly below each performance rating. Thus,

> % importance of wax   = 30%
> % importance of wick   = 33%
> % importance of stearic acid = 17%
> % importance of dye   = 10%
> % importance of perfume   = 10%

Quantifying the Cost of Components in a Candle

In a candle in which the total cost of the components was 67¢ the actual and percentage cost of each component was determined to be

> Wax   = 28.8¢ = 43%C
> Wick   = 1.3¢ = 2%C
> Stearic acid = 10.0¢ = 15%C
> Dye   = 6.0¢ = 9%C
> Scent   = 20.7¢ = 31%C

The actual cost of each component is a total cost that includes material and labor costs.

VALUE INDEX OF CANDLE COMPONENTS

After calculating the percentage importance and percentage cost of a component, a figure of merit—the ratio of importance to cost—is calculated for each component. The figure of merit is referred to as the *value index*.[3-5] The value index is a dimensionless number that expresses the relative value of items in a design. Recall that the value index was described in equation 1.5 in Chapter 1:

$$\text{value index} = \frac{\%\ \text{importance}}{\%\ \text{cost}}$$

The value index is often used to order a system of components by their perceived value. Generally, a value index greater than 1 represents good value. A value index less than 1 flags those components in a system that need attention and improvement. For the candle under study, the value index of the components is shown in Table 3.2. In this example, we used criteria analysis to measure the value of a candle's components. There are many other techniques for measuring the subjective parameters of the value index. In addition to criteria analysis a number of other useful subjective measurement techniques are described in Chapter 4. These techniques are used to quantify the importance, cost, and value of components in systems.

Correlation of Importance and Cost

In a design the cost of components should show a positive correlation with their importance. We want unimportant components to cost little, and we are willing to pay more for important components. By plotting the importance of parts against their cost, we can see how cost and importance correlate. You find no correlation between cost and importance when you use statistical regression analysis to study the correlation of a candle's components with their cost. A graph of this statistical analysis is shown

TABLE 3.2 Value of components in a candle

Component	% I	% C	Value Index
Wax	30	43	0.70
Wick	33	2	16.50
Stearic acid	17	15	1.13
Dye	10	9	1.11
Perfume	10	31	0.32

in Figure 3.7. Note that the best fit is a line across the middle of the graph that indicates that on average a component should have an importance of about 20%! The R^2 value of 0.000226 shows no correlation between the importance and cost of the candle's components.

THE VALUE GRAPH

A value graph is a plot of relative importance against relative cost.[6] Common practice plots percentage cost on the bottom of the graph and percentage importance on the side. If the x and y scales are uniform, then items with $\% I$ equal to $\% C$ will plot along a 45° line on the value graph. This line represents acceptable value, and items on the line have a value index of 1. The area above this line represents a region of good value, whereas the area below the line represents a region of low value. Items plotted to the bottom right of a value graph are prime candidates for improvement or elimination. Items plotted in the upper left corner of the graph offer exceptional value. Value graphs are widely used to portray the results of value measurement exercises. A value graph of the components in a candle is shown in Figure 3.8.

Five components are plotted on this graph. The value of each component is visually apparent, because any component plotted above the 45° line has relatively good value, whereas any component below this line has relatively poor value. Items far from the line represent extremes in value. Note that the wick offers exceptional value, whereas the perfume and paraffin wax offer relatively poor value.

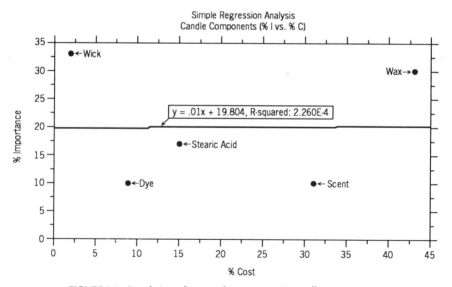

FIGURE 3.7 Correlation of cost and importance in candle components.

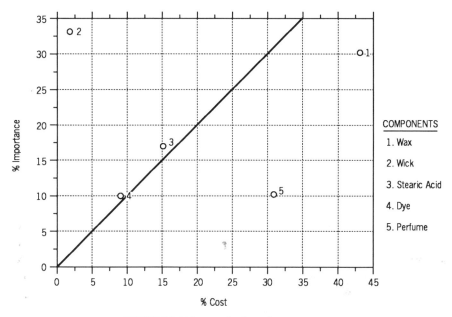

FIGURE 3.8 Value graph of candle components.

IMPROVEMENT TARGETS

The value index can be used to calculate three specific component targets; a cost target, an importance target, and a value target. The *cost target* is a cost at which a component's percentage cost will exactly equal its present percentage importance. An *importance target* is an importance rating at which a component's percentage importance will exactly equal its present percentage cost. A *value target* is a cost and importance goal obtained by averaging a component's present percentage cost and percentage importance. At the value target, the percentage cost and percentage importance of an item are equal. The value target is usually obtained by reducing the cost and increasing the importance of a component in a design.

Calculating Cost Targets

A *cost target* is obtained by multiplying the original cost of a component by its value index. If a component has a value index less than 1, multiplication by the value index will reduce its cost to a point where its new cost exactly equals its importance. Thus, if its value index is 0.5, its target cost will be exactly one-half of its original cost. Alternately, if its value index is greater than 1, multiplication will increase the cost of the component to a point where its new cost exactly matches its importance. In Table 3.3, the cost target for each component in the candle is shown. Note that the cost target for the wax and the scent are much lower than the current cost of these

TABLE 3.3 Decorative candle cost targets

Component	% I	% C	Value Index	Initial Cost (¢)	Cost Target (¢)	Δ Cost (¢)
Wax	30	43	0.70	28.8	20.2	−8.6
Wick	33	2	16.5	1.3	21.4	+20.1
Stearic acid	17	15	1.13	10.0	11.3	+1.3
Dye	10	9	1.11	6.0	6.7	+0.7
Perfume	10	31	0.32	20.7	6.6	−14.1

components, the target cost of the dye and stearic acid are slightly higher, and the target cost of the wick is *much* higher.

Cost targets are quite useful in cost reduction activities, because they tell us what a component should cost in a specific design. If the primary effort of design is cost reduction, we should attempt to find ways to reduce the cost of items with value indices lower than 1. We need not try to reduce the cost of any components with value indices above 1. If performance improvement is important, we could consider spending more on items with a value index greater than 1 to increase the functionality, reliability, or maintainability of these items. In the candle design, we could try to devise ways to lower the cost of wax and/or perfume. We could also consider if we should spend additional money on the wick. Cost targets allow designers to visualize how much money they should spend on components in a system. By comparing cost targets with current costs, they can determine exactly how much money they could subtract from or add to the cost of components in a design. Note that cost targets assume no change in the importance of the components.

Calculating Importance Targets

If cost reduction is not as important as performance improvement, the value index can be used to determine how the importance of components could be modified in a design. This is done by dividing the original importance rating of a component by its value index. If a component has a value index less than 1, division by the value index will increase its importance rating to a point where its percentage importance will exactly equal its present percentage cost. Thus, if its value index is 0.5, its importance rating will be exactly twice its score. Alternately, if its value index is greater than 1, division will decreae the importance rating of the component to a point where its new importance rating exactly matches its cost.

In Table 3.4, the importance target for each component in the candle is shown. Note that the importance target for the wax and the scent are much higher than the current ratings of these components, the importance targets of the dye and stearic acid are slightly lower, and the importance target of the wick is *much* lower. Target importance values help us to visualize how the importance of components can be altered to match their cost.

TABLE 3.4 Decorative candle importance targets

Component	% I	% C	Value Index	Importance Rating	Importance Target	Δ Rating
Wax	30	43	0.70	3.72	5.31	+1.59
Wick	33	2	16.5	4.10	0.25	−3.85
Stearic acid	17	15	1.13	2.14	1.89	−0.25
Dye	10	9	1.11	1.30	1.17	−0.13
Perfume	10	31	0.32	1.33	4.15	+2.82

Specific areas where the candle components' importance ratings might be modified are found on the criteria matrix of component importance ratings (Figure 3.6). Here we see that the importance rating of the wax would be improved if it could contribute more to the attractiveness, fragrance, driplessness, and smokelessness of the candle. In each of these areas it rated less than 5. However, even if it were rated a 5 in each of these areas, its overall rating would only be increased by 1.28, because the maximum rating is a 5.00. Thus, although it may be desirable to see if a different wax might improve the fragrance or combustion properties of the candle, it will still be necessary to reduce its cost in the design.

Importance and cost targets derived from the value index should not be used as absolutes: rather, because they are derived from subjective data, they should be used as pointers that allow designers to locate areas in a design where beneficial changes might be made. They allow us to look at a design in a new way and frequently lead us to think of altering a design in ways that we would not consider if we were unaware of the actual value of the design components.

Calculating Value Targets

In most situations improvements in importance and reductions in cost are both desirable. In this case the value index can be used to calculate a value target where cost and importance are improved simultaneously. In calculating the cost and importance targets described above, one variable was altered, whereas the second remained constant. In deriving a value target, cost and importance are both altered simultaneously. A new target is created in which the cost of a low value item is decreased, whereas the importance of the item is improved. Furthermore, the percentage change in cost and importance is equal, and in the resultant value target the percentage cost and percentage importance of the item are equal.

Cost, importance, and value targets are easy to portray on a value graph. The cost target of a component is equal to a cost obtained by drawing a line parallel with the cost axis from the component to the 45° value. In the value graph of the candle in Figure 3.9, this line is drawn from point 5 to the 45° line. It intersects this line at a point where the cost of the perfume would equal 10% of the total candle cost. In a similar manner, the importance of the perfume needs to be increased until it con-

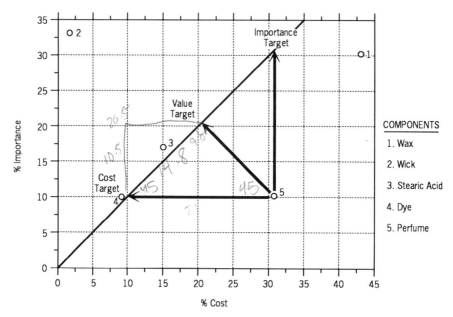

FIGURE 3.9 Importance, value and cost targets on a value graph.

tributes 31% of the total percentage importance of the candle. This point is the importance target for the perfume. It is depicted graphically by drawing a line parallel with the importance axis to the 45° value line. It may be possible to increase importance *while* reducing cost. If this is done, the design target will be any of a large number of points that lie on the 45° value line between the intersection of the cost and importance target points. A line drawn perpendicular to the 45° line from point 5 intersects this line at the value target point where the cost and importance both equal 20.5%. Here, the importance of the perfume has been increased and the cost has been reduced in an equivalent manner. Value targets for components in this candle are summarized in Table 3.5 which also shows the changes in importance ratings and actual cost required to meet the value targets.

TABLE 3.5 Decorative candle value targets

Component	% I	% C	Value Index	Value Target (% I and C)	Target Cost (¢)	Δ Cost (¢)	New Importance Rating	Δ Importance Rating
Wax	30	43	0.7	36.5	23.7	−5.1	4.52	+0.80
Wick	33	2	16.5	17.5	11.4	+10.1	1.42	−2.68
Stearic acid	17	15	1.1	16.0	10.4	+0.4	2.01	−0.13
Dye	10	9	1.1	9.5	6.2	+0.2	1.24	−0.06
Perfume	10	31	0.3	20.5	13.3	−7.4	2.73	+1.40

SELECTING ITEMS FOR VALUE IMPROVEMENT

Once the value of components in a system has been determined, components can be selected for value improvement. Which components in an item should be selected for further study? Should the lowest value items be selected? The highest cost items? The items with the worst performance?

It is often impractical to study each of the components in a design. In designs with many parts, it is a waste of time to try to reduce the cost and improve the performance of each component. In any design it is desirable to study the components that offer the best opportunities for product improvement first. Not every item in a design offers equivalent opportunities for cost reduction. The candle's wax and scent constitute 74% of the total cost of the candle. If we could reduce the cost of the wax and scent by 30%, we would reduce the cost of the candle by 22%. However, if we were to reduce the cost of the candle's wick and dye by 30%, we would only reduce the cost of the candle by 3.3%. Obviously, cost reduction of expensive components will produce larger cost reductions than similar cost reductions performed on low-cost items.

The same logic applies to performance improvement. Performance improvements are most needed in a few parts with low importance. If these components cannot be eliminated, perhaps we can improve their functionality. Perhaps several components can be combined. This would reduce the number of parts and increase the functional importance of the resultant parts. The candle's color and scent are provided by separate dyes and perfumes. Can a single organic chemical provide color and scent? What dye and perfume combination can be used to add to a candle's appeal? Would a green pine-scented candle molded in an evergreen shape attract attention and sales?

Perhaps functions can be added or extended in a component. The wick could be treated with fire retardants to ensure that it is not consumed quickly in the candle's flame. Mordants could be added to the wick to help ensure that the candle's flame is smokeless. Could a dye be used that would color the candle and allow it to glow in the dark? This might add to the candle's novelty and appeal. Adding functions to parts improves the importance of the parts and often simplifies the design.

Unlike most systems, a candle has so few parts that all of the low-value components could be examined to see if their cost could be reduced or their performance improved. High-value items like the wick might also be examined to see if more money should be spent to improve functionality and ensure reliability and maintainability. Yet even in a candle with only five components, it is best to work on some parts before we work on others. Fortunately, cost engineers in Japan and China have developed methods that allow designers to select those components that designers should examine first.

Optimal Value Zones

Dr. Masayasu Tanaka, professor of cost engineering at the Science University of Tokyo, draws lines on a value graph that help designers select items for study. Tanaka creates an area on a value graph that he calls the *optimum value zone*.[8] Two hyperbolic lines are drawn tangential to the 45° value line and inscribe and bound Tanaka's value zone. Tanaka places an optimal value zone on a value graph to focus attention on items in a device that most need to be studied. Items outside the value zone become targets for improvement, whereas items inside the optimal value zone are excluded from further initial analysis. Items inside the value zone have adequate value and if modified will provide little cost or performance improvement to the design.

The hyperbolic curves have the general formula $x^2 \pm y^2 = z^2$ and are drawn tangential to the value line.[7,8] Values of x and y represent the axial coordinates of points on these lines, whereas z represents the vertex of these lines on the x and y axis of the graph. The area of the optimal value zone is large near the origin of the graph and decreases as you ascend the 45° value line.

Figure 3.10 is a value graph for a candle that contains a Tanaka optimum value zone. Inspection of this graph shows that the wax and the perfume offer poor value, whereas the wick offers exceptional value. The optimum value zone directs our attention away from the stearic acid (3) and dye (4) to the perfume (5), paraffin wax (1), and wick (2).

The area within the optimal value zone varies depending on the choice of z. The value chosen for z is open to choice and often is a function of the items being analyzed. If the cost of items in a product vary widely, then a value of z can be relatively large. Note that z is set equal to a value of 10% on the candle graph. In this example, a value of 10 for z means that items that cost less than 10% of the total cost or contribute less than 10% to design importance will be excluded from initial analysis. This is proper, because the components range in cost from 2 to 43% of the

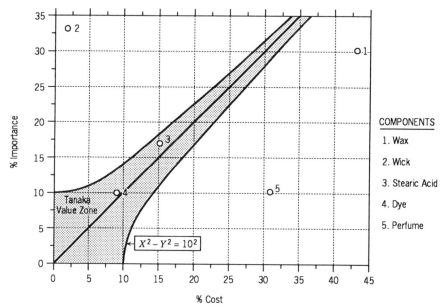

FIGURE 3.10 Optimum value zone on a value graph (Tanaka).

overall cost. If the components are more uniform in cost, then lower values of z are often used. Low z values are used by Professor Chi-Hsia Sun, who suggests that components in a design should first be sorted in a Pareto A, B, and C analysis.[10] Then the sorted groups are plotted on separate value graphs. She recommends A, B, and C sorts based on Pareto distributions of 72%, 50%, and 12.5%.[9]

Sun has simplified the construction of the optimal value zone proposed by Tanaka. Finding it time-consuming to calculate and draw Tanaka's hyperbolic lines on a graph, she constructs an optimum value zone bounded between two straight lines drawn from the x and y axis to the 45° average value line.[10] Sun draws a 55° line from z on the x axis to the 45° value line. A second line is then drawn from this intersection to the position of z on the y axis. These lines inscribe an area that Sun feels provides better value discrimination than Tanaka's hyperbolic lines. Professor Sun used analytic geometry to ensure that her value zones were nearly equivalent to those of Tanaka. Sun calls her method the *static symmetry method*. She also suggests that a *dynamic asymmetry method* be used to anticipate the changes that will result from subsequent cost reductions.

The dynamic asymmetry method adds a new value zone to the triangular area of the static symmetry method. It adds an area formed below a line from z on the y axis to a point twice the distance of the static symmetry intercept on the 45° value line. Because the wax used in the candle (point 1 on Figure 3.10) has low value and is an expensive component, brainstorming was used to produce a list of inexpensive noncombustible materials that could act as fillers in a candle. Figure 3.11 is a value graph of some brainstorming ideas that shows Sun's optimum value areas. Direct

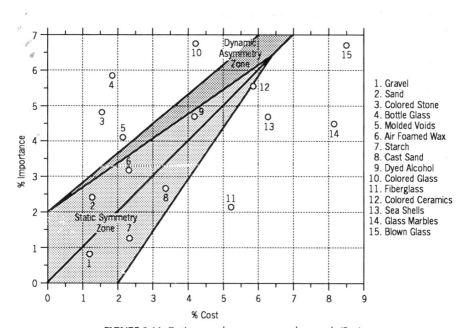

FIGURE 3.11 Optimum value zone on a value graph (Sun).

magnitude estimation, a value measurement method to be described in Chapter 4, was used to estimate the cost and relative importance of the suggestions. Note that on this graph, z was set equal to 2 to increase the selectivity of the optimum value zone in a design with more components. The graph indicates that air voids molded into the candle, as well as colored glass or stone, might be useful fillers in decorative candles.

Professor Ma Qing-Quo believes that De Marle's value index, Tanaka's optimal value zone, and Sun's dynamic asymmetry method may fail to show items in a design that should be changed. Ma Qing-Quo believes that items where the percentage cost equals the percentage importance may not represent acceptable value. Perhaps an item should have more importance than cost before it is acceptable. To make the process more selective, he preselects the components in the design that have the best value and uses them as a value criterion during redesign. Reasoning that the cost of high-value items need not be reduced, he uses them to calculate the *coefficient of an imaginary basic point*. This coefficient is obtained by dividing the percentage importance rating of several of the best components by their *actual* cost. Ma Qing-Quo then calculates a *basic value point index* for each component in the design by dividing the percentage importance of an item by its current cost and multiplying this ratio by the coefficient of the imaginary basic point. Target costs produced by multiplying the basic point value index by an item's current cost are more demanding than those produced from the value index and require substantial cost reductions. In effect, Ma Qing-Quo establishes a target cost for the total design and then proportions target costs according to the basic point value indices of each item to meet the cost reduction aim for the product.

Ma Qing-Quo's method is illustrated in Figure 3.12, a value graph of the candle filler ideas.[11] Here, the normal 45° equal value line has been replaced with a basic value point line that has a slope of 67°. Note that now the percentage importance of an item should be nearly twice its percentage cost before it has acceptable value. Percentage cost is used to make this graph analogous to those shown previously. Because items need to be nearly twice as important as they cost, only the colored stone (3) and bottle glass (4) plot above the new 67° value line and have good value. All of the other components now have low value.

Ma Qing-Quo's method does not include an optimum value zone, although one can be added by adapting Sun's or Tanaka's methodology to Ma Qing-Quo's new value line. This is easiest to do using Sun's linear intercept method. On a normal value graph, Sun creates an intercept line with a slope of 55° that is 10° larger than the 45° value line. In a similar fashion, an intercept with a slope of 77° is 10° larger than the 67° line. Originating at a Vertex of 2% on the x axis, this line bounds the lower portion of the optimum value zone on Ma Qing-Quo's value graph. An upper boundary is created by drawing a line from a vertex of 2% z on the y axis to the intercept of the 77° line on the 67° value line. A second line drawn to twice this intercept defines the upper boundary zone of this value graph. Of course, lines other than 67° can result from the basic point method. In these cases, Sun's optimum value method remains the same. In each case, 10° is added to the slope of the basic point line to determine the intercept.

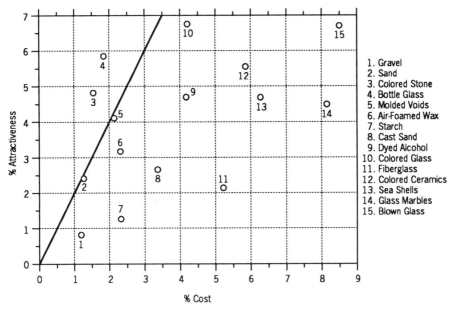

FIGURE 3.12 Non 45° value line (Ma Quing-Quo).

We believe that Ma Qing-Quo's method has special utility when used in VE rather than VA studies. It is also useful as an idea-screening method during the evaluation phase of VA/VE studies. In Figure 3.13 a number of ideas for improving the performance of a candle wick are listed. Although these ideas increase the cost of the wick, they add new functions to the wicks performance that enhance the value of a candle. They illustrate that value measurement can be used to cut cost *and improve performance.*

High-Value Design Concepts

Figure 3.14 shows four different candles that illustrate some new candle design concepts. Candle A has air voids molded into the candle in a random manner. It is made by pouring wax into a mold that contains ice. The ice creates voids in the candle wax. As the ice melts air voids are created, and the resulting candle is quite attractive. The ice is less expensive than the paraffin and quickly cools the wax increasing the rate of candle production. The candle can be provided with a special wick that will produce a colored flame.

Candle B is a round hurricane candle formed like candle A with ice. The center of this candle is hollow. A small tea candle is placed inside of this candle and light from this candle shines through the hurricane candle's outer shell. Dye placed in this candle's outer shell colors the light. The shell protects the inner candle from the wind and the candle is reusable.

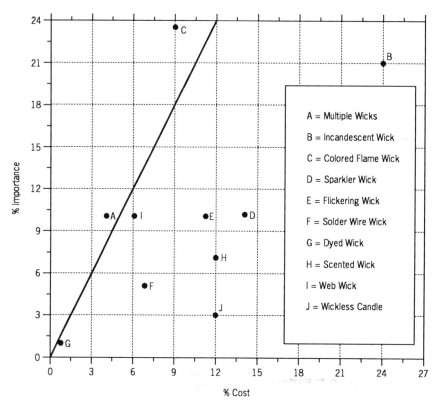

FIGURE 3.13 Value Graph—candle wick improvement ideas.

Candle C is a traditional birthday candle and should be compared to candle D, which is made by extending the birthday candle's wick and reducing the thickness of its wax coating. The resulting candle is a tall thin candle that is used as a substitute for a birthday candle. It is easier to insert and remove from a cake and uses less wax. It is called a *celebration candle* and is commercially available.[12]

Other Uses and Examples

Value graphs can be used to illustrate the value of components in any system. Figure 3.15 shows the relative value of a number of different types of candles. Sixty people were allowed to examine 10 different prepriced candles. They were then asked to select three candles from the set that they would be most likely to buy and three candles they would be least likely to buy, using the Pareto voting method described in Chapter 4. Their choices were recorded and used to scale the relative importance of each gift item. The average cost and importance of each item was then calculated and plotted on Figure 3.15, a value graph with Tanaka lines originating at a vertex of 7.5. In this graph the system is a set of candles rather than components in a candle.

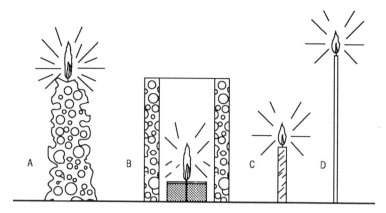

FIGURE 3.14 New candle design concepts.

This market research study showed that participants preferred traditional candles over speciality candle types. Information of this type is important as it shows the value of products in market niches.

Figure 3.16 shows the relative value of a number of common kitchen appliances. People used the Pareto voting method described in Chapter 4 to select gifts they would give to a young married couple. Their choices were recorded and used to scale the relative importance of each gift item. The average cost and importance of each

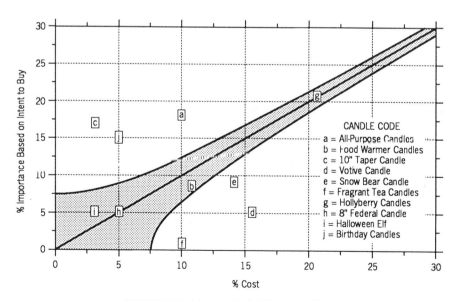

FIGURE 3.15 Value graph of different candles.

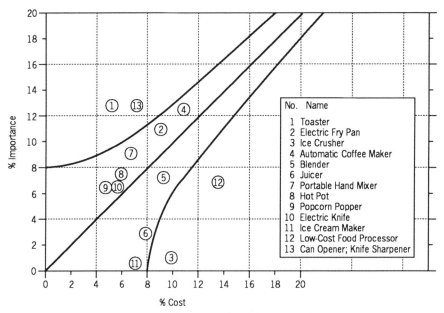

FIGURE 3.16 Value graph of small kitchen appliances.

item was calculated and plotted on the value graph with Tanaka lines of $z=8$, shown in Figure 3.16. In this graph the system is a market niche, and the components are different types of small kitchen appliances. A toaster and a combination can opener and knife sharpner had the best value, and a food processor and an ice crusher had the least value.

VALUE CONTROL

Value control is a process for controlling the value of components in a machine or process. It makes use of applied statistics to determine the standard deviation of the value index. Control charts are created that show how the value index of the components vary, and confidence intervals are plotted to show deviation of high- and low-valued items from an aim or mean.

The numerical value of value indices can vary widely because the index is obtained by dividing percentage importance by percentage cost. To accommodate this variability, the value index can be plotted on semilogarithmic paper, or graphs of the logarithm of the value index can be prepared. Figure 3.17 is a semilogarithmic plot of the value of the small kitchen items. Note the general similarity of this semilogarithmic plot to the arithmetic plot of the logarithms of the value indices shown in Figure 3.18. Although the shape of these plots are similar, note that average value is represented by the 1 line in Figure 3.17 and by the 0 line in Figure 3.18,

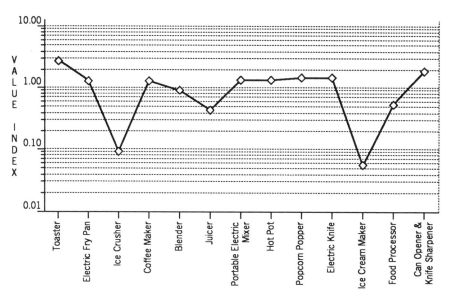

FIGURE 3.17 Value control graph for small kitchen appliances.

because the log of 1 equals 0. Figure 3.18 also shows the mean μ of the log of the value index estimates, as well as the 98% limits of these estimates. The 98% limits are value control limits that can be used to show variability beyond an acceptable deviation from a normal distribution of component value.

SUMMARY

Value measurement is used to quantify the value of items in a system. This chapter showed how people can measure value by assigning numbers to items in proportion to the importance and cost of the items. The value of an item equals the ratio of its importance and cost. A value index expresses the value of the item in numerical terms. Indices less than 1 indicate low value and above 1 reflect good value. Value graphs are plots of importance against cost that show the value of items visually.

Different design targets related to importance, cost, or value improvement are derived from the value index. Value graphs are often drawn with an optimum value zone to show items that are out of value control.

The value of items in a wide variety of different systems can be measured. Value measurement is often used to select products or processes for value improvement. It is then used to determine the value of items in these products or procedures and to determine the value of new product or process ideas. Value measurement can be used to determine the value of elements in any system, including markets, products, processes, procedures, organizations, and political systems. In Chapter 4 we will describe a wide variety of different numerical methods that are used to quantify value.

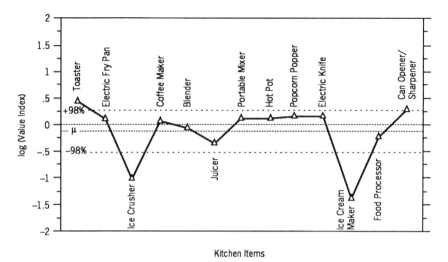

Kitchen Items

FIGURE 3.18 Value control chart with 98% confidence intervals.

In Chapter 5, we will apply value measurement to dynamic systems that change in value.

REFERENCES

1. Helmer, O., "Analysis of the Future: The Delphi Method," in *Technological Forecasting for Industry and Government* J. Bright, Ed., Prentice-Hall, Englewood Cliffs, N.J., 1968.

2. Inoue, M. S., and Riggs, J. L., "Describe Your System with Cause and Effect Diagrams," *Industrial Engineering,* pp. 26–31, April 1971.

3. De Marle, D. J., "A Metric For Value," *Proceedings, Society of American Value Engineers* **5,** 135–139, April 1970.

4. De Marle, D. J., "Criteria Analysis of Consumer Products," *Proceedings, Society of American Value Engineers* **6,** 267–272, May 1971.

5. De Marle, D. J., "The Nature And Measurement Of Value," in *Proceedings, 23rd Annual AIIE Conference,* May 1972, pp. 507–512.

6. De Marle, D. J., "Use of Value Analysis In Forecasting," presented at James R. Bright's Technology Forecasting Workshop, Castine, Me., Industrial Management Center, Inc., Austin, Tex.

7. Tanaka, M., "New Approach to the Function Evaluation System in Value Engineering" *International Journal of Production Research* **23**(4), 625–637, 1985.

8. Tanaka, M., "Evaluation of Function and Value Improvement by a Rating Approach," *Proceedings of the Society of American Value Engineers,* 1973 International Conference, **8,** 69–77, May 1973.

9. The Pareto A, B and C analysis method is described in *Principles and Applications of Value Engineering.* Department of Defense Course Book, Superintendent of Documents, U.S. Government Printing Office, Washington, D.C. 20402 pp. 3.35–3.36.

10. Sun, C., "The Principle and Application of the Dynamic Asymmetry Method," *Proceedings of the Society of American Value Engineers,* 1986 International Conference, **21,** 54–61, May 1986.

11. Quing-Quo, M., "Some New Methods in Value Engineering," *Proceedings of the Society of American Value Engineers,* 1983 International Conference, **18,** 115–119, April 1983.

12. Celebration candles are available from the American Greeting Corp., Cleveland, Ohio.

4

VALUE MEASUREMENT TECHNIQUES

M. Larry Shillito

INTRODUCTION

This chapter is not a treatise on measurement, scaling, or psychophysics. It is a description of value measurement techniques that have been successfully used in many and varied applications. It is a "how-to" document based on experience. We do not present the mathematical or statistical foundation for the techniques we describe. The mathematical and psychophysical basis for these techniques should be obtained from the bibliographies provided as appropriate. The chapter is a learning guide for practitioners involved in value workshops and/or projects. It provides quick access to needed measurement techniques and examples.

We will describe a variety of value measurement (VM) and value screening techniques that can be applied throughout the entire VA/VE process from origination to implementation. Obviously, some techniques lend themselves better to certain situations than others. The point to keep in mind is that if a technique appears useful for a particular application, regardless where one is in the value management (VM) process or even whether or not one is doing a VM project, use it!

The techniques have been divided into two sections, value measurement and value screening techniques. Value measurement techniques are used to evaluate items or attributes of items where a numerical measure is needed to indicate the relative strength of a respondent's perception of the item(s), for example, the importance of an item as perceived by an observer.

On the other hand, value screening techniques are more qualitatively oriented and are generally used to reduce a large list of items to a manageable size. They are not necessarily used to quantify the attributes of items where numbers and scales indicate some measure of importance. Value screening techniques are generally used to reduce a large list of items so that further attention and analysis can be focused on the vital few rather than the trivial many. The more discriminating value measurement techniques are generally applied to those items that survive the value screening process.

Numbers, as they are used in the value measurement and screening techniques discussed in this chapter, are used as a numerical language to aid communication. The numbers themselves will not provide a decision. They will, however, enhance communications that will allow groups and individuals to make more informed decisions. People can debate numbers and the reasoning behind their choice. The debates encourage multiple viewpoints among participants to challenge the reasoning needed for decisions.

Just as function analysis, with verb–noun descriptors and function analysis system technique (FAST) diagrams, allows participants to penetrate the confusion of semantics, so does value measurement focus debate on issues that are clouded by rhetoric and biased by paradigms. A value index and a value graph are the summary and graphic portrayal of these numerical debates. They more vividly represent the relative position of issues or subsets of issues.

VALUE MEASUREMENT TECHNIQUES

Simple Ranking

Description. This is one of the most common methods of determining the relative merit of a series of items. Once the items are ranked, numerical values can be assigned.

Method. Rank a list of items first, second, third, and so on in order of importance. Reverse numbering can also be used, where the highest number equals the best so that reverse ranks can also be used as scores.

Advantages

1. It is simple, easy, and intuitively appealing.
2. It allows one to put a number on things.
3. The process is fast and does draw out information from a group.
4. It also draws out differences of opinion. It makes differences of opinion visible; one can then re-rank (if desired) after group/team discussion of those differences
5. Participants don't need knowledge of math or detailed instructions on how to use it.
6. It can be used without a group leader.

Disadvantages

1. The process assumes linearity. It is easy to pick out the *extremes* (low and high ranks), but there is much ambiguity in the middle of the distribution of items.
2. Linearity also gives no real picture of *spread*, or separation, between judgments of different items. The interval or distance between ranks is constant. This uniformity or equidistance between ranks distorts reality. There is no way to determine how much more, or less, one item is from another along the continuum of attributes being ranked. No one really knows whether a scale has been established.
3. Participants become impatient and rush the process. The easiest way to get started is to identify the most important and least important items and then distribute the remaining items between these extremes.

Discussion

1. The technique is okay if one doesn't need ultimate accuracy or spread of the relative magnitude difference between items. It does have error.
2. The process should be used with less than 15 items, otherwise the process becomes exhausting and more noise develops in the middle of the distribution of ranks.

Alternate Ranking

Description. Alternate ranking is a method for developing a list of factors of any kind weighted to indicate relative importance. It can be used with groups of most any size, but preferably six to nine people seem to be a convenient number. Larger groups can be used by dividing them into groups of this size.

Method

1. The facilitator asks each person to write the items to be ranked on 3 × 5-in. cards. The cards are then placed face up in front of the participant.
2. The facilitator asks, If you had to throw away (*n*-1) of the items, which one would you keep? (*n* is the number of items being considered.) The participants mark this card with the number *n* and remove it from the rest and place it aside.
3. For the remaining items, the question asked is, If you could only keep (*n*-2) of these items, which one would you throw away? The participants are then asked to mark this card with a 1 and remove it from the rest. The questioning process is repeated, changing the numbers to reflect the number of cards that remain, until all cards are placed aside. As an example, assume there are five items to rank. The first question is, If you had to throw away four of the items, which one would you keep? Label this card 5 and set it aside. Next question: Of the four remaining items, if you had to throw one away, which one would it be? Label

this card 1 and set it aside. Next question: Of the remaining three items, if you had to throw two items away, which one would you keep? Label this card 4 and set it aside. Of the remaining two items, if you had to throw one away, which one would it be? Mark this card 2 and set it aside. The remaining card mark 3.

4. A tally is taken by asking each member to read off the number they have written for each item. The rank numbers are summed for each item, providing a total rank vote. The votes are discussed, and if necessary, the process is repeated until the vote reflects the true opinions of the majority of the group.

Advantages/Disadvantages

1. This method is quick and simple to use. It is an easy way of discriminating between a relatively large number of items.
2. The process needs a leader to lead the group through the process compared to regular ranking, where people can be left alone to do it.
3. The process is good for establishing anchor points, that is, tails of the distribution.
4. It is easier to work through all the noise in the center of the group of items. The zig-zag approach forces a decision.
5. It is easier to discriminate between a larger group of items than regular ranking.

Bibliography—Simple Ranking and Alternate Ranking. Delbecq, A. L., Van de Venn, A. H., and Gustafson, D. H., *Group Techniques for Program Planning, A Guide to Nominal Group and Deplhi Processes,* p. 59, Scott, Foresman, Glenview, Ill., 1975.

Successive Ratings

Description. The technique of successive rating elaborates upon the ranking process by assigning numbers to each ranked item. The items are first ranked from highest to lowest value. Numbers are then assigned to each item using a 0 to 100 scale. In a series of successive evaluations, the numbers are verified by comparison from both ends of the scale. The successive rating process converts a ranked list of items into a list of items with an interval scale.

Method

1. Rank items from highest to lowest.
2. Place the highest value item at the top of the list. This item is assigned a value of 100.
3. Assign numbers to the remaining ordered items by comparing each lower-ranked item to the highest-ranked item. Values are assigned so that each lower item is proportional to the highest-ranked item's value of 100. Continue this process until the lowest-valued item has a number assigned to it.

4. Starting with the lowest-valued item, repeat the procedure in reverse by working up the list. Compare the lowest-value item to each alternative above it in
the ordered list.

5. Generally the process is repeated for two full cycles.

Advantages

1. An interval scale of value is produced.
2. Unlike ranking, which maintains an equidistant separation between items, the
successive rating process separates items based upon their relative value.

Disadvantages

1. It should be used with less than 12 to 15 items.
2. It is quite time-consuming.

Bibliography. Churchman, C. W., Ackoff, R. L., and Arnoff, R. L., *Introduction to
Operations Research,* Wiley, New York, 1964.

Pair Comparison

Description. Pair comparison is a highly discriminatory rating-ranking technique
used to set a priority order and a relative magnitude to a number of related items. The
method presents items to be judged in all possible pairs and then asks for judgments
about each pair. A scale is created because each item is weighted against every other
item and a relative zero point results. The list of items will be ranked in order of merit.
Pair comparison provides a method of more accurately setting a priority order and
relative magnitude of a number of related factors than arbitrary ranking. The process
can be used to rate both positive criteria, such as importance, quality, and reliability,
and negative criteria, such as cost, maintenance, and downtime.

Method

1. Generate a list of the items to be ranked and establish a framework for the
comparisons to be made.
2. Prepare a matrix or graph to accommodate all of the entries being considered.
Items to be evaluated are arrayed against themselves in a triangular matrix. See
Figure 4.1.

Technique 1—Regular Pair Comparison

3. Compare items in pairs working across the rows until each item has been
compared with every other item. The rater must decide for each pair which item

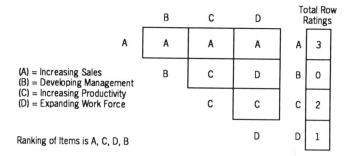

(A) = Increasing Sales
(B) = Developing Management
(C) = Increasing Productivity
(D) = Expanding Work Force

Ranking of Items is A, C, D, B

FIGURE 4.1 Regular pair comparison.

is more important (assuming importance is being rated). The code letter of the more important item is entered in the appropriate matrix cell representing the intersection of the two items.

4. Total the responses for each row. The item with the highest total frequency represents the concern with the overall greatest importance and so forth.

Technique 2—Scaled Pair Comparison

3. Design a set of preference weighting's to reflect different degrees of importance.

4. Compare items in pairs until each item is compared with all other items. In each comparison, the rater must decide which of the two items is more important. The appropriate letter signifying the more important item is recorded in the cell representing the intersection of the two items. In addition, a numerical weight chosen from the rating scale is also entered along with the letter. See Figure 4.2.

5. Total the numerical scores for each concern. The item with the highest total score represents the concern with the overall greatest importance. To express

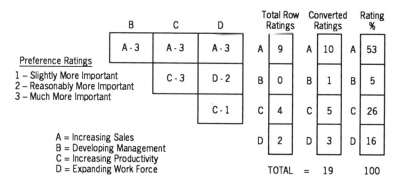

FIGURE 4.2 Scaled pair comparison.

each component as a percentage of importance of all of the items, sum the totals for each component row and divide this sum into each individual component score. An average value of component importance for a group of raters is obtained by averaging the value of all of the individual component ratings.

In Figure 4.2, as well as Figure 4.1 a zero occurs for item B. Many times people and computers have difficulty working with zero. This can easily be eliminated by adding 1 to each row total score for each item and summing the new totals. Likewise, a new percentage importance is calculated based on these new totals. This conversion could also have been used in Figure 4.1.

Usage

1. The process is useful in those applications where a high degree of subjectivity is present but where a need for one-to-one pair-wise comparison is essential.
2. It is useful for prioritizing items that are extremely close in importance and therefore difficult to separate and rank.

Advantages—Regular Pair Comparison

1. It is a highly discriminating prioritizing technique.
2. It forces the participant to make a choice—there can be no ties.
3. It is very thorough and methodical.
4. It forces pair-wise evaluation between every combination of pairs of items to be evaluated.
5. It is popular among those people who like a methodical approach.

Disadvantages—Regular Pair Comparison

1. It is very time-consuming, especially if there are 8 or more items to evaluate. For example, with 8 items, there are 28 comparisons; with 12 items, there are 66 comparisons; and with 36 items, there are 360 comparisons! The formula to calculate the number of comparisons is

$$C = \frac{n(n-1)}{2}$$

where n equals the number of items evaluated.
2. Due to the number of comparisons and time consumption, its use is best suited for less than 10 items.

Advantages/Disadvantages—Scaled Pair Comparison

1. It has the same advantages and disadvantages as regular pair comparison.
2. In addition, scaled pair comparison is more discriminating than other subjective rating methods including regular pair comparison. This is especially useful when it is used to separate seemingly equally important concerns.
3. It allows voters to express their spread in importance between pairs, that is, how much more important a choice is over another. Overall, there is a better indication of the magnitude difference between items. This can be seen by comparing the row totals of Figure 4.2 with Figure 4.1. Weighting factors or scales tend to amplify the extremes of the selections and, unfortunately again, leaves some noise in the middle of the distribution of scores. However, using the scale to express the magnitude of the difference between items is more discriminatory even though much of the information is contained in the initial choice.
4. Scaled pair comparison is more time-consuming than regular pair comparison.
5. Total scores, as seen in both examples, can be normalized. These percentage scores can then be used as weighting factors in other measurement techniques such as criteria analysis discussed later.

Discussion. In those cases where it is preferable to use pair comparison or scaled pair comparison for a large number of items (e.g., more than 15 or 20), a modified process called *partial pair comparison* can be used. The problem of excessive numbers of pairs is alleviated by using fractional statistical designs in which selected pairs are omitted from the entire analysis. A detailed description of these fractional array designs is beyond the scope of this book. The reader is referred to Chapter 7 in Bock and Jones's *The Measurement and Prediction of Choice.*

Bibliography

Bock, D. R., and Jones, L. V., *The Measurement and Prediction of Choice,* Holden-Day, San Francisco, 1968.

Mudge, A. E., "Numerical Evaluation of Functional Relationships," in *Proceedings, Society of American Value Engineers* 2, 111–123, 1967.

Mudge, A. E., *Value Engineering, A Systematic Approach,* McGraw-Hill, New York, 1971.

Pair Cost Comparisons

Description. The pair cost comparison method is a variation and extension of the scaled pair comparison method. The difference is that instead of using a simple rating scale to express the magnitude difference between pairs, real dollar amounts are used to express the difference between preferred and nonpreferred pairs.

Method

1. Listing the cost of every item, perform a pair comparison of all items.
2. With the listed costs in mind, select the preferred and nonpreferred items.
3. After deciding which item you prefer, state how much the cost of the least-preferred item should be reduced to change your preference. Record this new price as well as the unaltered price of your original preferred item.
4. For any pairs for which you have no preference, record their current (unaltered) price.
5. After completing all pair comparisons, total the prices associated with each item and determine a new average price for each item.
6. If desired, use these new item costs to recycle steps 1 through 5 again. This process can be repeated several times. In recycling a second time instead of lowering the price of the least-favored item, *increase* the price of the *most* favored item in each pair to a point where the two items are equal in value. In successive cycles, alternate between decreasing the price of the least-preferred items (on the first, third, fifth, etc. rounds) and increasing the price of the most preferred items (on the second, fourth, sixth, etc. rounds). This process of alternating between high and low is similar to that used in the method of successive ratings.
7. After the process is recycled several times (about five to seven) and it becomes impossible to select preferred and unpreferred items, it produces an interval cost scale that simulates the basic worth of the items analyzed.

Advantages

1. The pair cost comparison method has all of the advantages of regular and scaled pair comparison.
2. The pair cost comparison method establishes a true cost differential between items that is very useful in product pricing.
3. A pair cost comparison can be used to establish a standard for the worth of a product.
4. The method simulates market pricing conditions.

Disadvantages

1. The necessary calculations can be cumbersome without a computer.
2. The method is time-consuming, and people have difficulty determining how much they should reduce a product's price to make it equal in value to a preferred item.
3. A single cost metric analysis is insufficient to establish the worth of an item.
4. Five or six successive cost metric analyses may be required before the worth of each item is established and all items are appropriately priced.

Discussion. Although the pair cost comparison method is cumbersome and difficult to use, it can provide very important product price information. Computer programs have been written that make it easier to use, but even with these programs the method is difficult to use with more than six or seven items.

Bibliography

De Marle, D. J., "Value Measurement," in *Proceedings of the First International Miles Value Foundation Conference,* Honolulu, Hawaii, Spring 1989.

Pessemier, E. A., "Measuring Social, Scientific and Military Benefits in a Dollar Metric," Paper No. 146, Institute for Research in the Behavioral, Economic and Management Sciences, Purdue University, August 1966.

Pessemier, E. A., and Baker, N. R., "Project and Program Decisions in Research and Development," *R & D Management* **2**(1), 3–14, October 1971.

Direct Magnitude Estimation

Description. Direct magnitude estimation (DME) is a method that enables participants to assign numerical indicators as a function of relative merit to a list of proposed items, usually generated from previous discussions. The objective is to obtain ratings from a group. The ratings are then averaged and normalized.

Example. In a task force meeting or where more than two people get together to solve a problem, it becomes extremely difficult to direct everyone's efforts toward a specific aspect of the problem. Each individual is more attentive to problems related to the area in which they work. The application of a DME session in all this chaos will focus task force member's attentions to that portion of the problem that has been established democratically as the majority's concern.

Method

1. Each person randomly assigns any positive number to the first item in a list.
2. A number is assigned to each succeeding item in proportion to the number given the item proceeding it.
3. Calculate the geometric mean:

$$GM \text{ Item } 1 = \sqrt[n]{R_1 \text{ Item } 1 \times R_2 \text{ Item } 1 \times \ldots \times R_n \text{ Item } 1} \qquad (4.1)$$

where

n = the number of raters
R_1 Item 1 = the rating given to item 1 by the first person
R_2 Item 1 = the rating given to item 1 by the second person
R_n Item 1 = the rating given to item 1 by the final person

4. Normalize the geometric means so that their totals add to 100%:

$$GM \% \text{ Item } 1 = \frac{GM \text{ Item } 1 \times 100}{GM \text{ Item } 1 + GM \text{ Item } 2 + \ldots GM \text{ Item } M} \qquad (4.2)$$

where

$$M = \text{the number of items rated}$$

5. Steps 3 and 4 are best done by a computer

Advantages

1. Direct magnitude estimation can be applied to any list of items requiring individual attention by a group.
2. It establishes group preference.
3. It can be used to graphically illustrate group consensus and dissention.
4. It is simple and easy to use.
5. It is an excellent way to show the magnitude of difference between items.
6. It is an effective way to quantify the intangible positive and negative aspects of something.
7. It is an effective vehicle for dialogue and is a good communications device.
8. It can be used with any number of items, but preferably less than 30.
9. It provides a simple way to quantify importance.
10. It leads to a rapid definition of poor value areas.

Disadvantages

1. The necessary calculations can be cumbersome without a computer.
2. It requires a computer computation program to compute geometric means.
3. Waiting for computer output delays feedback unless an on-line or interactive program is used.
4. The open-ended (psychophysical) scaling process tends to confuse people at the outset. Some people have difficulty with all the freedom of expression.

Discussion

1. How does one compare ratings between different people if each person uses a different scale? Normalize all ratings to a percentage. With DME every participant can use their own number scale. In order to compare ratings between different participants, it is necessary to normalize each participant's ratings to a percentage and compare the normalized ratings.

2. Psychophysics shows that people think in ratios, which implies a log-normal scale. Therefore, a geometric mean is used as opposed to an arithmetic mean.

3. The geometric mean also minimizes the effect of extreme ratings, which, if using an arithmetic mean, would considerably skew the results.

4. If computer-generated rating sheets are used, the items can be listed in random order. This may be necessary to minimize order effects or dependencies or subsets. Interaction of items, however, is most prevalent in a list generated by a brainstorming session where there is considerable hitchhiking of ideas.

Bibliography

DeMarle, D. J., "The Nature and Measurement of Value," in *Proceedings, 23rd Annual AIIE Conference,* 1972, pp. 507–512.

Meyer, D. M., "Direct Magnitude Estimation: A Method of Quantifying The Value Index," *Proceedings, Society of American Value Engineers* **6,** 293–298, 1971.

Stevens, S. S., "A Metric For The Social Consensus," *Science* **151,** 1966.

Category Scaling

Description. Whereas DME uses an open-ended scale to allow people to assign numbers to express their perception of merit of ideas or other items, category scaling employs a closed, finite scale with category descriptors.

Background. With DME, each rater uses their own scale, which means that each person has their own reference set. Consequently, consistency in rating and perception across raters is weakened and is not fully compensated by normalizing a respondent's ratings. Having everyone use the same close-ended category scale (developed by the group) alleviates this inconsistency by providing a common reference set.

Method

1. Each person assigns numbers to items using a group-derived category scale with rating descriptors to represent their perception of the importance of the item.

2. Calculate the arithmetic means across all raters for each item rated. In this case, an arithmetic mean is used because a close-ended scale with boundaries (e.g., 1 to 100) is used. Intervals between scales are preset and are presumed to be equal and linear.

3. Arithmetic averages can then be normalized across items to derive percentage merit on a relative basis.

Advantages

1. Category scaling has the same advantages as DME.

2. However, category scaling has further advantages.

(a) It is easier and less confusing to use than an open-ended scale. It requires less explanation.

(b) Category descriptors are more meaningful.

(c) A category scale is more convenient.

(d) A category scale promotes greater consistency among raters in terms of their interpretation, perception, and quantification. Everyone uses the same reference set.

Discussion. There are two types of scales, objective scales and subjective scales. Objective scales are based on or measured by things such as dollars, feet, length, width, weight, and brightness. The level of measurement and scale values are both numerical.

Subjective scales entail things such as risk, morale, importance, difficulty, and seriousness of offenses. The level of measurement is nonnumeric, whereas the scale value is numeric. Levels of measurement in subjective scales are more prone to misinterpretation than levels of measurement in objective scales. With the objective scales, the reference set is more concrete and tangible.

The direction of a scale is also important. Is the scale going in a positive direction, where the higher numbers reflect something good, or in a negative direction, where the higher number reflects something negative? For example, does the highest value equal greatest importance (positive), highest cost (negative), lowest cost (positive), greatest risk (negative), or lowest risk (positive)? A trap that many groups fall into is mixing both positive and negative issues in the same evaluation of items. When rating a group of items, the ratings must be grouped as either all positive or all negative. Otherwise there will be misinterpretation. It is important to be sure that everyone using a scale knows the direction. Volumes have been written about scaling and the psychophysics of scaling. We have abstracted the bare essentials for use with groups. The numbers and scales are employed as a means for better communication among individuals and team members rather than an exact measurement.

Some example category scales that have been used are the following.

RATING SCALE: SEVEN CATEGORIES

Scale Range	Category Descriptor
90–100	Excellent to most desirable; fully satisfies needs in all respects; may even exceed expectations.
70–89	Good to very desirable; satisfies needs very well; definitely better than average.
55–69	Okay to desirable; adequately fulfills needs; could be better, but no significant drawbacks.
45–54	Neutral to really makes no difference one way or another.
30–44	Fair to slightly undesirable; does not completely meet needs; may be some sacrifice to live with it; would not be customer's choice if other options were available.

10–29	Poor to undesirable; does not fulfill needs; considerable inconvenience to live with it.
0–9	Very poor to very undesirable; does not fulfill needs; definitely not a choice.

RATING SCALE: FIVE CATEGORIES

Scale Range	Category Descriptor
90–100	Very significant perceptible improvement; customer may demand the improvement once he or she experiences it; would definitely pay more to continue getting it.
70–89	Considerable contribution to quality; customer would be attracted to it and may pay more to maintain it.
30–69	Moderate contribution to quality; customer notices some difference, but may be neutral to the improvement.
10–29	Very slight contribution to quality; probably not noticeable to customer.
0–9	No change; no contribution to quality; customer probably less than satisfied but will continue to accept as is.

Category scales and descriptors must be tailor-made for each occasion. There is no universal rating scale and descriptor!

Bibliography

Bass, B. M., Cascio, W. F., and O'Connor, E. J., "Standardized Magnitude Estimations of Frequency and Amount for Use in Rating Extensivity," Technical Report 61, Management Research Center, University of Rochester, 1973.

Cooper, M. J., "An Evaluation System For Project Selection," *Research Management,* pp. 29–33, July 1978.

Edwards, A. L., *Techniques of Attitude Scale Construction,* Appleton-Century-Crofts, New York, 1957.

Green, P. E., and Rao, V. R., "Rating Scales and Information Recovery—How Many Scales and Response Categories To Use?" *Journal of Marketing* **34,** 33–39, July 1970.

Guilford, J. P., *Psychometric Methods,* 2nd ed., McGraw-Hill, New York, 1954.

Myers, J. H., and Warner, W. G., "Semantic Properties of Selected Evaluation Adjectives," *Journal of Marketing Research* **5,** 409–412, November 1968.

Torgerson, W. S., *Theory and Methods of Scaling,* Wiley, New York, 1958.

Criteria Analysis

Description. Criteria analysis is a systematic subjective quantification process for evaluating alternatives according to criteria that are determined by the nature of the problem. It is also known as a *decision matrix*. The matrix contains numerical estimates of the contribution of each alternative to each criterion. It is a method for combining the judgments of individuals or groups to obtain quantified measures of both concrete and intangible factors needed for evaluating alternatives. When used in VM projects, alternatives many times are product components and criteria are system functions.

Example. Chapter 3 contains a detailed example of the use of criteria analysis for a decorative candle. In Figure 3.6 the candle's five components are arrayed against its main functions. The numbers inside the matrix cells represent peoples' perception of the contribution of each component to each function. These ratings were multiplied by the importance of each criteria. The numbers were obtained in a Delphi analysis in which individuals used the direct magnitude estimation procedure to quantify their individual preferences.

Calculating the Value of Components. The percentage importance of a component is calculated using the following equation:

$$\% \ Ix = 100 \ [(\% \ Cx_1)(\% \ I_1) + (\% \ Cx_2)(\% \ I_2) + \ldots + (\% \ Cx_n)(\% \ I_n)]$$

where

$\% \ Ix$ = the total importance of component x expressed as a percentage.
$\% \ Cx_n$ = the percentage contribution of component x to function n.
$\% \ I_n$ = the percentage importance of a function n (also known as a
          weighting factor)

Method

1. List the alternatives.
2. List the criteria. It is necessary to have clear descriptions preferably developed by a team.
3. Give weightings to the criteria.
4. Evaluate each alternative using criteria 1.
5. Evaluate each alternative using criteria 2, and so on. Alternatives are quantified by using DME or category scaling.
6. For each alternative, sum the products of the weighting factor times the cell score for each row of the matrix.

Advantages

1. Criteria analysis is a very powerful and discriminating analysis.
2. It offers maximum discrimination and expression.
3. It is an excellent method of achieving a consensus view while preserving individual ratings.

4. Participants are forced to line up information and consider issues (alternatives) one at a time.
5. It reveals how each alternative contributes to or satisfies each criterion.
6. It forces methodical and meticulous consideration of all the factors involved in a decision.
7. It readily reveals areas of agreement and disagreement.
8. Decision rationale is documented and can be repeated in light of new information.
9. The process and discipline required to derive the matrix usually provides as much information and insight as quantifying the matrix itself.
10. It can be used with a team with multiple input or singly with an individual.

Disadvantages

1. It is time-consuming to derive the matrix (mainly criteria).
2. With many raters, it requires a computer program for the computation.
3. If a mainframe computer is used, feedback is delayed.
4. The maximum matrix size is about 15 × 10 (150 discriminations).

Discussion

1. The maximum number of items to be evaluated should be less than 10.
2. Criteria analysis is generally used after the more simpler screening and/or evaluation techniques, such as Pareto voting, have reduced the number of items to be evaluated to a manageable size. Screening techniques are discussed later.
3. With all the other evaluation techniques, there is no common set of evaluation criteria established. That is, each rater used their own reference set and point of view. Therefore, there could be as many different reference sets as there are raters. In this case, how does the decision maker really evaluate the group output produced by the measurement technique(s)? The technique of criteria analysis employing a decision matrix readily addresses this concern. The decision matrix is used to evaluate alternatives using the criteria developed for evaluation.

Criteria Derivation. A criterion or set of criteria can be defined as a standard of reference that is/are used to weight the benefits of an alternative. The basic question used to derive and select criteria is, What are the most important factors in selecting _____?

Guidelines for Defining Criteria:

1. Keep them short. The longer the description, the more likely it is that the statement will contain several standards of reference.

2. They should be unique. The same reference should not be contained or implied in more than one criterion to avoid multiple weighting, confounding, or interaction. For example, "effect on employee morale" and "effect on employee attitude" are really the same. Only one should be used in the matrix. Another example, "energy rationing" and "mileage controls." Mileage controls is merely a subset of energy rationing. Using both would give double weighting to the criterion.

3. They should discriminate among alternatives. That is, if all alternatives are judged equally with reference to the same criterion, the criterion does not help one make a decision. In this case, it really is no criterion at all.

4. They should be relevant.

5. Criteria descriptions should contain measurement words, which, in turn, are stated or implied in the evaluation scale.

6. They should be measurable.

7. There is no one all-purpose set of criteria. Each decision case has its own set.

8. All criteria should go in the same direction. That is, they should either be all positive or all negative within the same decision set. This same principle was discussed in the section on category scaling. It is not a good idea to mix both positive and negative criteria in the same decision set. Doing so can be very misleading and confusing. So, if the highest number in the rating scale represents something positive, then all criteria are to be stated in positive terms and visa versa.

Methods to Change Criteria Direction

1. *Rewording*. For example, cost can be reworded as reduced cost (changed from negative to positive where highest number represents lowest cost).

2. *Reversing the scale*. For example, the highest number represents the lowest cost.

3. *Taking the reciprocal of negative ratings*. For example, 1/cost. Here the highest number represents the highest cost. Inverting ratings is usually only done when it is discovered that a respondent has used the rating scale in reverse or, in some cases, where respondents insist on mixing positive and negative criteria in the same decision set. Some people have difficulty with scales where the highest number represents things like "lowest cost."

Weighting Criteria. After criteria have been established and well defined, they must be weighted. If weighting factors are not used, it is implied that all of the criteria are of equal importance. This, however, is usually not the case. Weighting factors can be obtained from a continuous scale (e.g., 0 to 100) or from a category scale. A category scale might be as follows:

80–100 Most important

60–79   Very important

40–59   Moderately important

20–39   Slightly important

 0–19   Little or no importance

Regardless of the scale or number used, the total of all the weighting factors must total 100 or unity. There are two methods for doing this:

1. Normalizing scores (most likely obtained form DME or scaling).
2. Imposing the restriction on the raters themselves.

The most accurate method is to impose the requirement on the rater's themselves. That is, the total of all weighting factors must equal unity. The respondent has only a finite amount of weight to allocate, so what is awarded to one criterion must be taken away from another and so on. This method is quite different than normalizing individual ratings after they have already been selected from, say, a 100-point scale. When using a scale the respondent does not have to consider any trade-offs or interaction between criteria. Instructions we have used with groups of raters for estimating weighting factors are, "Here are 100 points. Distribute them across the criteria so that the sum totals 100." We ask the participants to do this individually and then have them work as a group to produce a single distribution of the weightings. Alternately, we record each individuals distribution and average them for each criterion. This is followed by a group discussion of the distributions and averages from which the group derives a final distribution.

Bibliography

Bass, B. M., Cascio, W. F., and O'Connor, E. J., "Standardized Magnitude Estimations of Frequency and Amount for Use in Rating Extensivity," Technical Report 61, Management Research Center, University of Rochester, April 1973.

Bick, J. H., "Planning and Forecasting Using a Combined Relevance Analysis and Cross-Impact Matrix Method," General Research Corp., Westgate Research Park, McLean, Va., June 1974.

DeMarle, D. J., "Criteria Analysis of Consumer Products," *Proceedings, Society of American Value Engineers,* **6,** pp. 267–272, May 1971.

Edwards, A. L., *Techniques of Attitude Scale Construction,* Appleton-Century-Crafts, New York, 1957.

Huber, G. P., and Delbecq, A., "Guidelines for Combining The Judgements of Individual Members in Decision Conferences," *Academy of Management Journal,* **15**(2), 161–174, 1972.

Myers, J. H., and Warner, W. G., "Semantic Prospects of Selected Evaluation Adjectives," *Journal of Marketing Research,* pp. 409–412, November 1968.

Schermerhorn, R. S., and Taft, M. I., "Measuring Design Intangibles," *Machine Design,* pp. 108–112, December 1968.

Scoring Models

Description. Scoring models are very similar to criteria analysis, where alternatives are arrayed against a set of judgment criteria. The difference is that a performance measurement scale is established for each criterion. Criteria are weighted and alternatives are scored against the established performance measurement scale instead of the category scale as discussed under Criteria Analysis.

Method

1. List the alternatives.
2. List the criteria. Refer to the guidelines for deriving criteria in the criteria analysis section. One significant difference in labeling criteria for scoring models is that the criteria descriptors should *not* contain measurement words. These will be provided in the performance measurement scale.
3. For each criterion construct a performance measurement scale. See Figure 4.3.
   (a) A unique scale is developed for each individual criterion based upon its characteristics.
   (b) A 1 to 5 scale is often used.
   (c) A scale descriptor is established for each measurement unit and may be based on objective or subjective data. The purpose of the scale descriptor is to convert each measurement space into comparable units to permit comparison across equivalent criterion scale descriptors.
   (d) In all cases 5 is "good" and 1 is "poor."
   (e) It is necessary that the same number of categories be specified for each criterion and that the same numerical scale (i.e., 1 to 5) be utilized.
4. Assign weighting factors to reflect the relative importance of each criterion. Refer to the criteria analysis section for details.

Criteria	Rating Scale Numbers				
	1	2	3	4	5
Project Cost	$1,300,000 to $1,500,000	$1,000,000 to $1,300,000	$800,000 to $1,000,000	$600,000 to $800,000	LESS THAN $600,000
Probability of Achieving Technical Objectives	35%–50%	50%–60%	60%–70%	70%–80%	80%–100%
Availability of Required Resources In-house	Not Available	Partial Availability	Available but in Demand	Available	Readily Available

FIGURE 4.3 Scoring model performance measurement scale.

5. Evaluate each alternative against the criteria. This is done by reading the 1 to 5 scale and selecting the performance measurement scale descriptor that best fits the characteristics of the alternative.

6. For each alternative, sum the products of the weighting factor times the performance rating for each row (alternative) of the matrix. See Figure 4.4.

Advantages. The scoring model has all the advantages of criteria analysis. In addition, because its ratings are calibrated across the criteria, ratings on one criterion are equivalent to ratings on all the others. Thus a score of 2 on one criterion is equal to a score of 2 on another etc.

Discussion

1. One underlying assumption is that there are relatively few criteria needed to discriminate across alternatives.

2. Also see discussion section under Criteria Analysis.

Bibliography

Moore, J. R., and Baker N. R., "An Analytical Approach to Scoring Model Design—Application to R & D Project Selection," *IEEE Transactions On Engineering Management* **EM-16**(3), pp. 90–98, 1969.

Moore, J. R., and Baker, N. R., "Computational Analysis of Scoring Models for R & D Project Selection," *Management Science,* **16**(4), pp. B212–B232, 1969.

| Criteria | Criteria Weighting | ALTERNATIVES | | | | | |
| | | A | | B | | C | |
		Score	Weighted Score	Score	Weighted Score	Score	Weighted Score
Project Cost	0.5	1	0.5	4	2.0	3	1.5
Probability of Achieving Technical Objectives	0.3	3	0.9	2	0.6	3	0.9
Availability of Required Resources In-house	0.2	3	0.6	4	0.8	3	0.6
TOTALS		A = 2.0		B = 3.4		C = 3.0	

FIGURE 4.4 Scoring model decision matrix.

Value Index

Description. Once cost and importance are derived for each item, they can be used to compute a figure of merit, generally called a *value index.* The value index is a numerical description of value, a dimensionless number that furnishes a metric description in which a value ratio greater than 1 represents good value, a ratio equal to 1 represents acceptable value, and a ratio less than 1 indicates areas for potential value improvement.

Method

1. Divide the percentage importance of an item (alternative, component, etc.) by its percentage cost.

$$V = \frac{\% \, I}{\% \, C}$$

where

$$V = \text{average value}$$
$$\% \, I = \text{relative percentage importance}$$
$$\% \, C = \text{relative percentage cost}$$

Percentage importance can be derived using any of the value measurement techniques and then normalizing each individual measure of importance. Percentage cost may be obtained by dividing the incremental cost of an item or component by the total cost of all items or components. If real costs are unknown, percentage cost may also be estimated by the various value measurement techniques and then normalizing each incremental cost. It is important to note that whichever parameters are used to derive the numerator or denominator of the value index, the numbers representing those parameters must be either all relative units or all absolute units. Consistency of measurement units is necessary so that the units of measure cancel out to give a dimensionless index.

2. The value index can be multiplied by the real cost of the component to get a target cost. [See Chapter 3, the section on value targets.]

Advantages

1. The value index indicates which components in a system need attention and serves to define problem areas. It is an excellent method for scoping projects.
2. Because the value index is numerical, it lends itself to graphic portrayal.

Value Graph. A vivid way to illustrate the value index is to plot relative importance versus relative cost for each function or item. There are numerous examples of value graphs in Chapter 3. The 45° line on a value graph has a value index of 1 and represents acceptable value. Areas above this line possess good value, whereas areas below the line represent candidates for value improvement. Value graphs more vividly depict relative value than do tabular listings of numerical value indexes. The graphs display the relative magnitude and relationship of the components that form the structure of the value index for each item rated.

Bibliography

DeMarle, D. J., "The Nature and Measurement of Value," in *Proceedings, 23rd Annual AIIE Conference,* May 1972, pp. 507–512.

DeMarle, D. J., and Shillito, M. L., "Value Engineering," in *Handbook of Industrial Engineering* (G. Salvendy, Ed.), Chapter 7.3, Wiley, New York, 1982.

Meyer, D. M., "Direct Magnitude Estimation: A Method For Quantifying The Value Index," *Proceedings, Society of American Value Engineers,* Vol. 6, pp. 293–298, May 1971.

VALUE SCREENING TECHNIQUES

Pareto Voting

Description. Pareto voting is a formalized preliminary screening to reduce a list of alternatives to a more manageable size. The principle behind the technique is based on Pareto's law of maldistribution, which, when applied to value engineering, states that 80% of the value is invested in 20% of the items. Using this principle, it is possible to narrow down a list of items in order to extract those that possess the greatest amount of attributes and feasibility. Once the list has been reduced by Pareto voting, the items can be further evaluated by more powerful discriminating value measurement techniques described earlier.

Method

1. All items on the list are numbered so that each item has its own unique identification number. This will ease group communication and comparison of items. The team edits the large list of items to eliminate redundancy and similar items. If there is not unanimous agreement on combining certain items, it is best to let the items remain separate as is. The editing promotes team interaction and clarification.

2. Participants vote on the list of items. The total number of votes allowed per person is restricted to 20% of the total number of items generated. Of the allowed number of votes, only one vote is permitted per selected item and the full allotment of votes must be used. Group members vote by secret ballot, recording the identification number of the items they have selected.

3. The team leader collects the ballots and tallies the votes before the group. The number of votes is recorded alongside the respective item on the original chart pad.

4. The entire list is typed and distributed to the team members. Those items receiving no votes are retained on the list for reference but are separated from the rest of the items and are excluded from further consideration. The items are typed in descending order of the number of votes each item received.

5. If the list needs to be further reduced, the procedure is repeated with the already shortened list. Refer to Figure 4.5.

The total number of votes is 16 (20% × 8 = 2 × 8 people = 16 total votes).

Usage

1. This is a quick priority setting method.
2. It reduces a very large list of items to a manageable size.
3. It isolates topics for further discussion/clarification.

Advantages

1. There is fast, simple, immediate feedback.
2. Quick feedback allows momentum of meeting to be retained.
3. No forms are needed.
4. It minimizes conflict and bickering at meetings.

Disadvantages

1. It does not allow discrimination between all items.
2. It does not produce a real scale.

List of Eight Items Voted on by Eight Participants in a VM Project

Item	Total Group Votes
Use a mechanical sensor	6
Use slip sheets	4
Use fiberglass pallets	3
Use fiber drum	2
Paint a "limit" line	1
Use photo sensor	0
Use a drum pallet	0
Use a top sensing device	0

FIGURE 4.5 Regular Pareto voting.

Limitations

1. Pareto voting may not work with small list of items (e.g., less than 10), where there may not be a wide dispersion in votes.
2. It will not work if there are only a few raters (less than three or four).
3. It may not be valid where items are not listed in random order, such as a list generated from a brainstorm session, where there is much hitchhiking of ideas and items are recorded in thought clusters.

Bibliography. Shillito, M. L., "Pareto Voting," *Proceedings, Society of American Value Engineers,* Vol. 8, pp. 131–135, 1973.

Ranked Pareto Voting

Description. Ranked Pareto voting is performed in the same manner as regular Pareto except that after the participants have picked their 20% of the most important items, they are also asked to rank the chosen items where the highest number represents the best. Then, in addition to tallying votes, one tallies the ranks and sums the rank scores for an overall total score by item. See Figure 4.6.

Advantages

1. Using rank totals gives more separation between items. This can be important especially where several items receive the same frequency of votes. In Figure 4.6, items 5 and 8 both received five votes, but item 8 clearly is more important when viewing the rank totals.
2. The distribution of ranks within each item shows the spread in participant perceptions, and gives a good picture of group consensus. The distribution of ranks can be used to trigger further dialogue.

Disadvantages. They are the same as regular Pareto.

Limitations. They are the same as regular Pareto.

Q-Sort

Description. Q-Sort is a screening technique that is very useful in determining qualitative differences in the value of a series of items. It can be useful as a starting point from which to assign quantitative values to the individual items. The result of a Q-sort is a qualitative ranking of categories of items in which items of similar value are grouped together.

Method

1. Write the name of each item on a separate 3 × 5-in. card.
2. Separate the items into two piles, those of high value and those of low value.

Ranked Pareto Exercise

Total number of items = 15
Total participants = 5
Number of votes per participant = 3 (20%)
Total number of votes = 15
Rank scale = 3 (best), 2 (next best), 1 (least best)
Rank total = 30

Item	Distribution of Ranks by Participant	Vote Frequency	Rank Totals
1			
2	2,3	2	5
3			
4			
5	3,1,1,1,2	5	8
6			
7			
8	3,2,3,2,2	5	12
9			
10	1,1	2	2
11			
12	3	1	3
13			
14			
15			
		15	30

FIGURE 4.6 Ranked Pareto voting.

3. Separate the high-value items into "high" and "very high" piles. Separate the low-value items into "low" and "very low" piles.
4. Separate the high items into "high" and "medium" and the low items into "low" and "medium," putting all the mediums into one pile. The result will be five distinct categories: very high, high, medium, low, and very low. See Figure 4.7.
5. Examine each pile to make sure it contains only the items desired.

Advantages/Disadvantages

1. This method becomes very useful with a large number of items *if* only a qualitative ranking is desired. The categories, however, may not be equally spaced and only a hierarchy of categories is obtained, not a continuous scale of values.
2. It is quick and "very dirty."
3. It is good for grouping like items into sets for further analysis. For example, "let's work on all the L items." Also those items in the very high group may

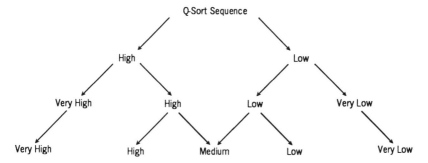

FIGURE 4.7 Q-sort procedure.

be candidates for further discriminating value measurement techniques, such as criteria analysis or pair comparison.

4. Q-Sort is very useful with a large number of items if only a qualitative ranking is desired.

5. Categories may not be equally spaced; only a hierarchy of categories is obtained.

Bibliography

Melin, A. F., and Souder, W. E., "Experimental Test of a Q-Sort Procedure for Prioritizing R&D Projects," *IEEE Transactions On Engineering Management,* **EM-21**(4), 159–164, 1974.

Souder, W. E., "Field Studies with a Q-Sort/Nominal Group Process for Selecting R & D Projects," *Research Policy,* **4,** 172–188, 1975.

Souder, W. E., "A System for Using R & D Project Evaluation Methods," *Research Management,* **21**(5), 29–37, 1978.

WHAT DOES IT ALL MEAN?

Measurement in the context of value measurement has been shown as a numerical procedure to obtain subjective opinions and perceptions from participants about items or attributes of items. The numbers represent the relative strength of a respondent's perception. It is assumed that the magnitude of this perception, in turn, is indicative of the objective properties, events, states, and traits of the items in question and the individuals subjective estimation of their relative value. The numbers collected and the subsequent manipulation produce four different types of scales.

Nominal Scale

This is the lowest form of measurement. It basically classifies or clusters items together by name or similarity. For example, items in a room that are large or small. Pareto voting and Q-sort produce nominal scales.

Ordinal Scale

An ordinal scale arranges items by position in an order. It allows us to place names in some sort of order. Ranking is commonly used to arrange items in an increasing or decreasing order. Pareto vote tallies produce an order as does Q-sort. FAST diagrams and flow charts order items by causality and/or time.

Interval Scales

Interval scales show name, order, and the distance between items. Unlike ordinal scales where the distance between items is uniform, such as one unit, as in a ranking of items 1,2,3,4, . . . , n, an interval scale indicates the distance between adjacent items. Interval scales contain much more information about the relative value of items. However, they are situational and only represent the relative attributes of the items evaluated, because no comparisons or standards exist in these scales. Criteria analysis, DME, regular and scaled pair comparisons, successive ratings, and category scaling produce interval scales.

Ratio Scales

Ratio scales give us the name, order, distance, and an absolute standard for reference and comparison. Because they contain a reference, items in a ratio scale can be compared to items originally not included in the set of items. A ruler is a common example of a ratio scale and can be used to measure the relative length of items. Converting value to energy units produces a ratio scale for value. Pair cost comparisons and scoring models produce ratio scales. Notice each scale's relationship with the other scales. We described them in order, with each successive scale containing the attributes of the prior scales.

$$
\begin{aligned}
\text{Nominal} &= \text{name} \\
\text{Ordinal} &= \text{name} + \text{order} \\
\text{Interval} &= \text{name} + \text{order} + \text{interval} \\
\text{Ratio} &= \text{name} + \text{order} + \text{interval} + \text{reference}
\end{aligned}
$$

How you use value measurement techniques depends on the relative discrimination you need. Nominal and ordinal scales are involved in value analysis. Value measurement involves interval and ratio scales.

SUMMARY

As we mentioned earlier, the value measurement and screening techniques are used as a numerical language to enhance communications, so that decision makers and value teams can make better and more informed decisions. All of these techniques have been used by many people many times with success. They have helped people and groups make many important decisions that have resulted in many improvements. Finally the value index and value graphs are used to focus the efforts of a VM team on the most effective areas for value improvement. They can be used not only on a macro scale to select a VM project candidate, but also on a micro scale to select parts of a product for further value improvement.

5

MODELING THE
DYNAMICS OF VALUE

David J. De Marle

EVALUATIVE MODELING

In recent years powerful computer programs have been developed that allow dynamic mathematical models to be constructed and run on personal computers. These models can simulate the operation of complex systems and can evaluate the operation of products and services long before they are actually constructed. They are especially useful when the system being modeled is expensive or difficult to construct and test or when the system is dynamic and changes over time. In this chapter I will describe their use in assessing the value of seasonal products which change with time.

DYNAMIC MODELING

Classical decision making uses mathematical models such as criteria analysis to analyze the relative value of different approaches to a problem. Simple spreadsheet criteria analysis models help clarify many decision-making situations. However, these models fail to describe systems that are interactive and dynamic. Here system dynamic models are appropriate. System dynamic modeling was developed at Massachusetts Institute of Technology by Professor J. Forester. In developing system dynamics, Forester applied control theory feedback concepts to the study of closed loop systems. Figure 5.1 compares classical decision-making models (5.1a) with system dynamic models (5.1b) that contain feedback loops. System dynamic models allow us to simulate the effects that interactions and feedback have on systems as they act through time.

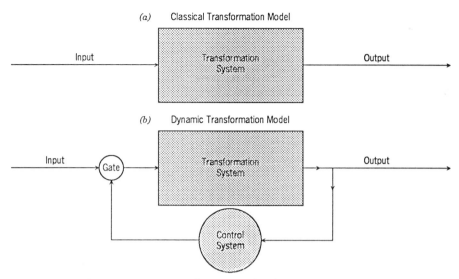

FIGURE 5.1 Comparison of conventional & dynamic feedback models.

SYSTEM DYNAMIC MODELING

Initially system dynamic models were written in languages like Fortran by skilled programmers who required many hours to construct relatively small dynamic models. These models were run on large mainframe computers in universities and large corporations and were expensive to build and run. System dynamics programs were next written in Dynamo, an extremely powerful programming language developed specifically for system dynamics modeling. Although this language simplified the task of programming and made it easier to model the behavior of complex systems, it still required the services of a knowledgeable programmer.

Recently a new language called Stella™ was developed by High Performance Systems Inc., which greatly simplifies the construction of dynamic models.[1] Stella evolved from system dynamics studies at Dartmouth and uses a series of icons to represent elements present in any system. Stella runs on Macintosh computers and allows people with little or no programming to easily construct interactive models.

GRAPHICAL REPRESENTATION OF SYSTEMS

Stella uses graphical icons to represent the stocks and flows Forester used to describe a dynamic system. A *stock* represents a state, such as the amount of money in a bank, that describes an important condition of the system under study. A *flow* represents a rate, such as the interest rate paid on a deposit, that controls a dynamic property of the system under study. In system dynamics, rectangles represent stocks and arrows represent flows in a system. Converters, represented by small circles, transform operations in the system, for example, a converter could convert dollars into foreign currency. Converter arrows connect stocks or other converters into a dynamic model drawing.

Figure 5.2 shows the icons used to build simulation models in Stella. Flow direction is shown by a large hollow *flow pipe arrow* in this diagram. A converter, shown as a circle below this arrow, regulates flow through the faucet in the flow pipe arrow. The flow collects in a rectangular stock box on the right-hand side of the diagram. A small black arrow connects the stock to the converter. This arrow shows that the converter uses information about the stock to regulate flow into it. The small arrows do not depict flows, they provide information about the state of the system needed by converters to regulate flows. A cloud is drawn to the left of the large hollow flow arrow. Clouds are used to depict parameters that are exogenous to the model and are not part of its closed architecture.

An example may help explain the system concept portrayed by the icons in Figure 5.2. Think of the diagram as a plumbing diagram in which water flows through the hollow flow pipe arrow into the tank depicted by the rectangular stock box. A faucet in this pipe allows water to flow into the rectangular stock box. It is opened or closed by the converter (A) depending on the level of water in the tank. Feedback information on the amount of water in the tank is provided to the converter by the arrow. The converter can be programmed to open the faucet and allow water to flow into the tank at a specified rate when the tank is empty. Similarly it can turn off the faucet when the tank is full. Water flow is regulated by the converter and depends on level of water in the tank. The system described is a closed system that describes how the tank will be filled with water. The cloud is used to indicate that the model does not describe where the water comes from.

In Stella, an individual draws a picture of a system that she uses to represent a system. She selects icons from a window drawn on the Macintosh computer screen. Using the computer's mouse these icons are dragged onto the screen and connected together.[2] As the drawing is constructed, the computer builds a mathematical model to represent the logic of the system being modeled. Stella uses question marks to prompt the programmer to add simple level and rate information to its plumbing. In this way a system dynamic model consisting of a set of sequential equations is created. In this case the model is

1. Water Tank = Water Tank + dt × (Fill rate)[3]
2. INIT(Water Tank) = 0[4]
3. Fill rate = IF Water Tank ≥ 100 THEN 0 Else 10[5]

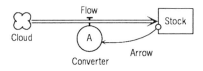

FIGURE 5.2 System dynamic icons used in Stella ©.

After some simple calibration and testing, a Stella model is obtained that can be used to simulate operation of the system. Figure 5.3 is a time plot of the amount of water in the tank.

This example shows how a simple system dynamic model is constructed. Stella is simple to learn and use and has excellent graphical capabilities. Of course the model shown is trivial, anyone can calculate that an empty 100-gallon tank will fill in 10-hr if water is added at a rate of 10 gallons/hr. I used this model to illustrate how a set of simple graphic icons can be used to construct a computer program, not to solve a difficult problem. It is a simple matter to extend the model to a more complex situation.

Consider this problem. A 100-gallon tank is half full of water. Water is added to this tank at a rate of 10 gallons/hr. The tank has a small drain at the bottom that allows water to flow from the tank. The tank can be drained of water at hourly rates of 10%, 20%, or 40% of the water in the tank. Determine the amount of water in the tank under each of these conditions. This problem is more complex, and is easily solved using dynamic modeling. The first model can solve the 0% drainage question, but its structure needs to be modified to solve the other problems. Figure 5.4 shows how the initial model (Figure 5.2) was modified to include water drainage.

The new model contains a flow to drain water from the tank. This flow has a "drain rate" converter that can be used to simulate the effect of various drainage rates. The drain converter controls the flow of water from the tank. It multiplys the stock of water in the tank by the drain rate and subtracts this amount of water from the tank. At the same time new water enters the tank at a rate of 10 gallons/hr. The initial amount of water in the tank was set at 50 gallons. The small arrows connecting the stocks and converters show the linkages used to construct the mathematical model that lies behind this graphic representation. The program listing is:

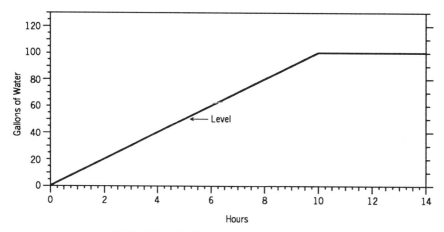

FIGURE 5.3 Stella © graph showing water tank level.

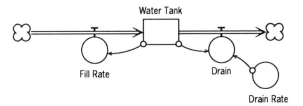

FIGURE 5.4 Input output dynamic model.

```
Water Tank = Water Tank + dt × (Fill rate - drain)
INIT(Water Tank) = 50
   {tank is half full initially}
drain = Water Tank × drain rate
   {calculates amount by multiplying the water stock by the drainage rate}
drain rate = .10
   {ten percent drainage rate, Note that 0, .20, and .40 were used to simulate
   other drainage rates}
Fill rate = IF Water Tank ≥ 100 THEN 0 Else 10
   {Tank capacity is 100 gallons and is being filled at a rate of 10 gallons per
   hour}
```

Figure 5.5 shows the simulation results of this model for the conditions studied.

A RESERVOIR MODEL

Figure 5.6 shows a system dynamic model for a reservoir. Although considerably more complex than those described above, this model is similar in structure. It illustrates how system dynamics can be used to model the operation of an expensive

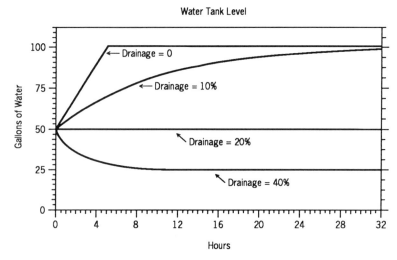

FIGURE 5.5 Plot of water tank levels.

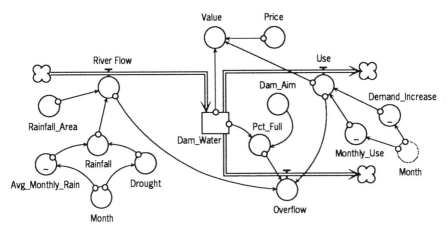

FIGURE 5.6 Stella model for a reservoir near Capetown South Africa.

and complex project. The model simulates how a dam will operate under different conditions of water supply and demand. The model uses monthly rainfall, seasonal water demands, and annual demand increases of 7% to simulate operating conditions for a large reservoir near Capetown, South Africa.[6] The effect of droughts of different intensity and duration can be simulated. Based on water usage, water stock, and sale price, the model calculates a value index for water in the dam.

Water Input

The left-hand side of Figure 5.6 shows the converters that regulate the flow of river water into the dam. The average monthly rainfall at Capetown is an input that is modified by a converter for droughts. A drought can happen for several months or several years and is represented as a percentage decrease in normal rainfall. Rainfall is calculated by multiplying the drought percentage by the average rainfall conditions for the area. Each month, the drainage basin collects rainfall and converts it into river water that flows to the dam.

Water Flow from the Dam

Two water flows can take water from the dam. Water use by the community takes water constantly from the dam, and overflow through the dam's sluice ways remove water from the dam when the dam is full. Water use varies from month to month on a regular seasonal basis and increases at a constant rate as population and consumption increase at a 7% annual rate.

Water Value

The value of water in the dam is a function of the available water supply and demand. A value index is calculated as the ratio of the actual use times the price of water divided by the stock of water available in the dam. The natural logarithm of the index is used to restrict the numerical value of this ratio to a reasonable range. Because the natural logarithm of 1 equals 0, zero represents a value index of 1. Negative values offer poor value while positive values represent good value.

Program Listing

```
Dam Water = Dam Water + dt × (− Use + River Flow − Overflow)
INIT(Dam Water) = 100,000,000            {Cubic Meters}
Dam Aim = 100000000                      {Cubic Meters}
Drought = IF (Month ≥ 12) AND (Month ≤ 24) THEN .4 ELSE 1.0
   {A one year long drought starts at the beginning of the second year. 40% of
   the normal rainfall occurs in the second year. Normal rainfall occurs in
   the first and third years.}
Month = TIME
Overflow = IF (River Flow ≥ Use) AND (Pct full ≥ .98) THEN River Flow − Use ELSE
   0
   {When the dam is 98% full, water is diverted to sluice ways to prevent
   overfilling of the dam}
Pct full = Dam Water/Dam Aim
   {Calculates how full the dam is as a percentage of its capacity aim point.}
Price = 7
   {Water in the dam is priced at this amount per cubic meter}⁷
Rainfall = Avg Monthly Rain × Drought
   {Actual rainfall is equal to the effects of a drought on average seasonal
   amounts}
Rainfall Area = 4000000
   {Conversion factor for drainage basin}
River Flow = Rainfall Area × Rainfall
Use = Monthly Use × 15000000 × (1 + Demand increase)
Value = LOGN((Use × Price/Dam Water))
   {The value of water in the dam is equal to the amount used times its price.
   The natural logarithm of this value index is used to restrict the ratio to
   a reasonable range}
Avg Monthly Rain = graph(Month)
(1.00,3.00),(2.00,2.00),(3.00,3.00),(4.00,6.00),(5.00,8.00),
(6.00,9.00),(7.00,10.00),(8.00,9.00),(9.00,7.00),(10.00,5.00),
(11.00,3.00), (12.00,3.00),(13.00,3.00),(14.00,2.00),(15.00,3.00),
(16.00,6.00),(17.00,9.00),(18.00,9.00),(19.00,10.00),(20.00,9.00),
(21.00,7.00),(22.00,5.00),(23.00,3.00),(24.00,3.00),(25.00,3.00),
(26.00,2.00), (27.00,3.00),(28.00,6.00),(29.00,9.00),(30.00,9.00),
(31.00,10.00),(32.00,9.00),(33.00,7.00),(34.00,5.00),(35.00,3.00),
(36.00,3.00)
Demand increase = graph(Month)
(1.00,0.0),(2.00,0.00375),(3.00,0.00875),(4.00,0.0138),(5.00,0.0187),
(6.00,0.0238),(7.00,0.0300),(8.00,0.0350),(9.00,0.0400),
(10.00,0.0450),(11.00,0.0512),(12.00,0.0575),(13.00,0.0625),
(14.00,0.0688),(15.00,0.0750),(16.00,0.0813),(17.00,0.0875),
(18.00,0.0950),(19.00,0.101),(20.00,0.109),(21.00,0.115),
(22.00,0.122),(23.00,0.130),(24.00,0.136), (25.00,0.142),
(26.00,0.149),(27.00,0.154),(28.00,0.159),(29.00,0.165),
(30.00,0.170),(31.00,0.175),(32.00,0.180),(33.00,0.185),
(34.00,0.189),(35.00,0.194),(36.00,0.200)
```

```
Monthly Use = graph(Month)
(1.00,1.00),(2.00,0.900),(3.00,0.800),(4.00,0.700),(5.00,0.600),
(6.00,0.500),(7.00,0.600),(8.00,0.700),(9.00,0.800),(10.00,0.900),
(11.00,1.00),(12.00,1.00),(13.00,0.900),(14.00,0.800),(15.00,0.0700),
(16.00,0.600),(17.00,0.500),(18.00,0.600),(19.00,0.700),(20.00,0.800),
(21.00,0.900),(22.00,1.00),(23.00,1.00),(24.00,0.900),(25.00,0.800),
(26.00,0.700),(27.00,0.600),(28.00,0.495),(29.00,0.600),
(30.00,0.700),(31.00,0.800),(32.00,0.900), (33.00,1.00),(34.00,1.00),
(35.00,0.900), (36.00,0.800)
```

Simulation Results

Several simulations were run to ascertain the operation of the reservoir over a 3-year time span. I wanted to test the effect of a drought on the reservoir's capacity. To do this I introduce a drought in which the normal rainfall was cut by 60% for 1 year. The drought starts in the second year and runs till month 24.[8] Figure 5.7 shows the effect of the drought on the dam's water. Line 1 on this graph shows the amount of water in the dam by month. The water drops as water is used. The dam is finally filled to capacity in August of the third year. Rainfall and overflow are plotted at the bottom of this graph. Note that rainfall declines in the second year. In the first year an overflow started in May and ended in December. The drought was so severe that no overflow occurred in the second year and was delayed until July in the third year.

Value of Water

Water is a commodity that is vital to the operation of any community. In most communities, water is supplied by a government public utility at cost. A fundamental purpose of government is to provide pure water to its citizens at a reasonable cost. Water projects usually involve the expenditure of large amounts of public money on

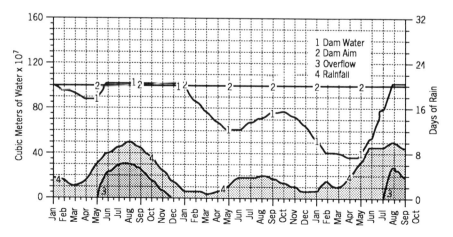

FIGURE 5.7 Effect of a year long drought on reservoir water.

large capital projects. Accordingly, water prices often remain constant for years, slowly increasing as population demands and use create the need for further capital projects. Yet water, like everything else, actually varies in value depending on supply and demand. To show this I calculate a value index for water in this reservoir.[9] The value of the water in the dam is a function of its use (demand) and its availability (supply). Figure 5.8 plots the value of water. For comparison, I plot the use and the supply of water in the dam. As the supply declines, the remaining water increases in value, and as use increases, the value of water increases.

The value index is expressed as a natural logarithm of the ratio of water supply and demand. The natural logarithm is used to restrict the ratio to a range from −1.5 to +1.5. Because the natural logarithm of 1 is 0, the 0 in Figure 5.8 represents an aim point for value. Note that in this model I express demand in economic terms by multiplying the amount of water used by the cost of water. I set the cost of water taken from the dam at seven times the cost of water in the dam.[10]

The relationship of value to the water supply and demand is shown in Figure 5.9. Note the large scatter. Traditional economics uses general supply and demand curves to depict value. This graph was produced by the Stella model as these simulations were run. While an overall correlation is obvious the individual correlations are situation specific.

MODELING THE VALUE OF TWO DIFFERENT TYPES OF HOME AIR CONDITIONERS

In recent years off-peak air-conditioning systems have been developed to decrease the cost of air-conditioning. Called *thermal storage systems,* they use ice or chilled water made at night to air-condition a building the following day. These cooling

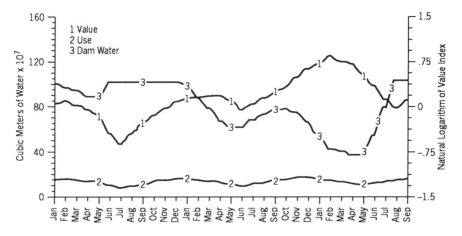

FIGURE 5.8 Value of water in a reservoir.

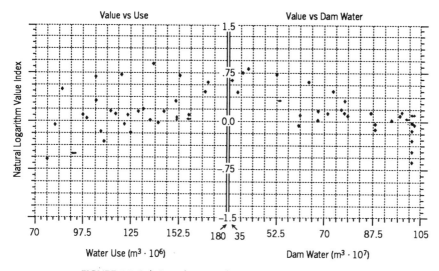

FIGURE 5.9 Relation of water value to water supply and demand.

storage systems have been well publicized and have been used effectively at numerous industrial and commercial sites.[11-13] These systems are also being investigated for home air-conditioning.

An air conditioner is a good example of a product that is valuable for only part of the year. It is used during the summer when it is hot. In the winter and through much of the spring and fall, it is not needed. Money invested in the system when it is not in use is really wasted; it cannot be re-invested or used.

Dynamic models can be used to calculate the life cycle costs of seasonal products. I used Stella to construct a model that compares the life cycle cost of a conventional home air-conditioning system to an ice-making home air-conditioning system. Because the climate has a profound effect on air-conditioning needs, I used this model to simulate the relative value of these air conditioners at several different locations in the United States.

Model Description

In the previous examples I studied water flows, because people can readily understand the dynamics of plumbing systems. In this model, we model the flow of money into and out of a budget. We set aside a sum of money into a cooling budget that we will use to cool a home for a five-year period. Money left in this account grows as the unspent principal earns interest at a specified rate. Money withdrawn from the account is spent and can no longer earn interest. Money is initially withdrawn from the cooling budget to purchase equipment, either a conventional central home air conditioner, or an ice-making thermal storage system. Subsequently, money is withdrawn to pay for the electricity used to operate the air conditioner. Climate affects the length of time an air conditioner operates. In the model, the average monthly temperature of different cities in the United States is compared to a comfort temperature of 78°F to determine the need for air-conditioning. When appropriate, the air conditioner is turned on and begins to consume electricity. It continues to use electricity until the outside temperature is low enough to turn the unit off. Although the thermal storage equipment is more expensive than conventional air-conditioning, it uses low-cost off-peak electricity and costs less to operate. The model evaluates the trade-off between equipment cost and energy use in three cities: Phoenix, Arizona, Kansas City, Missouri, and Rochester, New York. Figure 5.10 shows the structure of the model.

Program Listing

```
COOLING BUDGET = COOLING BUDGET + dt × (INTEREST INCOME - ELECTRICAL COST -
EQUIPMENT COST)
INIT(COOLING BUDGET) = 12000
COOLING BUDGET 2 = COOLING BUDGET 2 + dt × (INTEREST INCOME 2 - ELECTRICAL COST
2 - EQUIPMENT COST 2)
INIT(COOLING BUDGET 2) = 16000
   {Amount necessary for Phoenix for 8 years}
electric bill = electric bill + dt × (ELECTRICAL COST)
INIT(electric bill) = 0
electric bill 2 = electric bill 2 + dt × (ELECTRICAL COST 2)
INIT(electric bill 2) = 0
MONTH 3 = MONTH 3 + dt × (TICK TOCK)
INIT(MONTH 3) = 0
bill difference = electric bill - electric bill 2
```

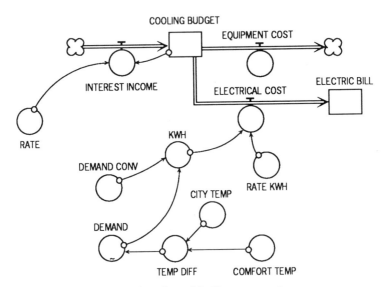

FIGURE 5.10 Life cycle model of home air conditioners.

```
budget difference = COOLING BUDGET - COOLING BUDGET 2
COMFORT TEMP = 78
demand conv = 6000
ELECTRICAL COST = RATE KWH × KWH
ELECTRICAL COST 2 = RATE KWH 2 × KWH 2
EQUIPMENT COST = PULSE(2500,3,1000)
EQUIPMENT COST 2 = PULSE(4000,3,1000)
INTEREST INCOME = RATE × COOLING BUDGET
INTEREST INCOME 2 = COOLING BUDGET 2 × RATE
Kansas City = 0
KWH = demand conv × DEMAND 2
KWH 2 = DEMAND × demand conv
Phoenix = 1
RATE = .005
   {6% per year}
RATE KWH = .066
   {Current normal electrical rate.}
RATE KWH 2 = .036
   {Current off peak electrical rate.}
Rochester = 0
TEMP DIFF = -COMFORT TEMP + Kansas City Temp × Kansas City + Rochester TEMP ×
Rochester + Phoenix Temp × Phoenix
TICK = 1
TOCK = PULSE(MONTH 3,12,12)
DEMAND = graph(TEMP DIFF)
(-18.00,0.0),(-16.20,0.0350),(-14.40,0.0700),(-12.60,0.110),
(-10.80,0.165),(-9.00,0.225),(-7.20,0.315),(-5.40,0.410),
(-3.60,0.540),(-1.80,0.715),(0.0,1.00)
DEMAND 2 = graph(TEMP DIFF)
(-18.00,0.0),(-16.20,0.0350),
(-14.40,0.0700),(-12.60,0.110),(-10.80,0.165),(-9.00,0.225),
(-7.20,0.315),(-5.40,0.410), (-3.60,0.540),(-1.80,0.715),(0.0,1.00)
Kansas City Temp = graph(MONTH 3)
(0.0,0.0),(1.00,30.00),(2.00,32.50),(3.00,43.50),(4.00,55.50),
(5.00,65.00),(6.00,74.00),(7.00,79.50),(8.00,77.50),(9.00,70.00),
(10.00,58.50),(11.00,44.50),(12.00,33.50)
Phoenix Temp = graph(MONTH 3)
(0.0,0.0),(1.00,52.00),(2.00,56.00),(3.00,61.00),(4.00,67.50),
(5.00,75.50),(6.00,85.00),(7.00,90.50),(8.00,88.50),(9.00,83.00),
(10.00,71.00),(11.00,60.00),(12.00,53.00)
Rochester TEMP = graph(MONTH 3)
(0.0,0.0),(1.00,24.80),(2.00,24.00),(3.00,33.00),(4.00,44.00),
(5.00,57.00),(6.00,67.00),(7.00,71.00),(8.00,69.00),(9.00,62.00),
(10.00,51.00),(11.00,39.00),(12.00,28.00)
```

Air-Conditioning Needs

The average monthly temperature for Phoenix, Kansas City, and Rochester help determine the demand for air-conditioning. The demand curves for these cities is plotted in Figure 5.11. A scale from 0 to 1 indicates the fraction of the month that air-conditioning is needed. The area below each demand curve is shaded to allow visual comparisons of the different air-conditioning needs of each city.

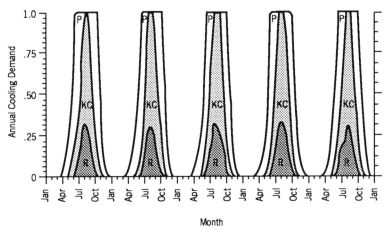

FIGURE 5.11 Air-conditioning needs for Phoenix, Kansas City and Rochester.

Electrical Utility Costs

These different needs are reflected by the electrical bills for each system in each city. Over a 5-year period a home owner in Phoenix with a conventional central air conditioner would pay over $10,000 in electrical bills. In this same period of time, a thermal storage air-conditioning system in Phoenix would have electrical costs equal to about $6000. In comparison, a home in Rochester would require from $900 to $1500 to pay its electrical bill depending on its air-conditioning system. Figure 5.12 shows the electrical costs for these cities. The upper line in each shaded area represents the electrical operating costs of a conventional central air conditioner, whereas the lower line represents the operating cost of an ice-making air conditioner.

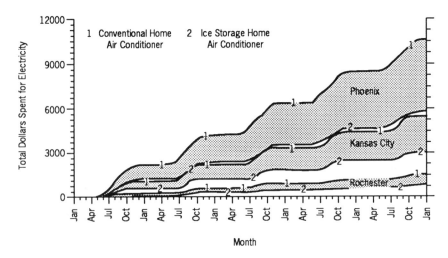

FIGURE 5.12 Cost of electricity used by air-conditioners in different cities.

Air-Conditioning Budget Returns

One way to examine the effectiveness of an investment is to examine its effect on a budgeted amount of money. Accordingly, I estimated the amount of money a home owner in Phoenix would need to budget to cover air-conditioning expenses for 8 years, and $16,000 was used to initialize the Phoenix cooling budget in January. For comparison purposes, this same amount of money was put in the cooling budgets for Kansas City and Rochester even though they require less. In this example, this money will grow at a rate of 6% per year if left on deposit.[14] In April, central air-conditioning equipment was purchased and installed: $2500 for a conventional system; $4000 for an ice-making air-conditioning system. Immediately the units in Phoenix began to operate, within a month the Kansas City units were operating, and the Rochester units were turned on part-time a month later. These units consumed electricity as they operated, and as the electrical bills were paid, the money was removed from the cooling budget. Any unused money continued to grow at a rate of 6% per year and added to the cooling budget. The net effect on the cooling budget is plotted in Figure 5.13, which shows 5 years of operation of both systems in each city.

Note that the thermal storage system is cost-effective in both Phoenix and Kansas City, but is not cost-effective in Rochester. In Phoenix, the financial breakeven point occurs in a little over one year, whereas in Kansas City, it occurs in three and a half years.

This simple model was used to illustrate how value, which changes dramatically with time, can be modeled. Here I compared the value of a seasonal product to that of a constant investment. This simple model can be used to simulate the effect of a wide variety of factors on an investment over time. If desired the model can be modified to show the value of air-conditioning investments in other cities.

All of the models described in this chapter are relatively simple. They can easily be expanded to include more dynamic situations, i.e. where the interest rate, or the

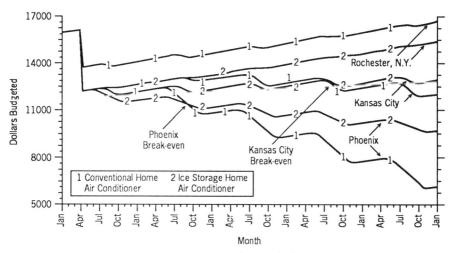

FIGURE 5.13 Air-conditioning budgets.

electrical rates change over time. Also, maintenance costs can be added to the model that change as the equipment ages etc. The point is that system dynamic modeling allows us to measure the dynamics of value. We can do this before investing in a system or developing a product or service.

MODELING THE VALUE OF POOL CHLORINATION PROCEDURES

In much of the United States, swimming pools are used during the summer and then closed during the winter and through much of the spring and fall. Accordingly, a dynamic model like that for the air-conditioning systems just described could be used to calculate the life cycle costs of swimming pools in different areas of the country. Instead, in order to illustrate the rapid change in value that can occur daily in an interactive system, I chose to model fluctuations in the value of a system that is prone to rapid change, in this case, the swimming pool's chlorination system.

Model Description

This model simulates the growth and death of algae in a swimming pool that is periodically chlorinated. Figure 5.14 shows the structure of this interactive model. Two algae "flows" constantly fill the swimming pool with algae. The first flow comes from seed algae that enter the pool constantly from the air. A second flow of algae comes from the rapid growth of algae in the pool. These two flows create a stock of algae. Algae are withdrawn from this pool when killed by chlorine, which is fed into

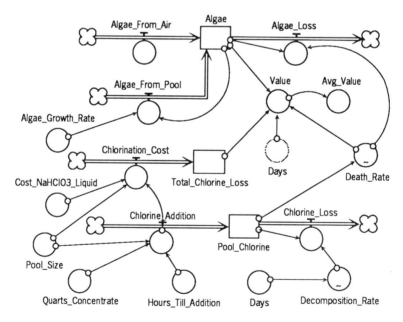

FIGURE 5.14 Swimming pool chlorination model.

the pool periodically by adding liquid sodium hypochlorite to it. The amount of chlorine determines the effectiveness of algae kill. Once in the pool, the chlorine begins to decompose, and this decomposition is accelerated during daylight.

This model is very interactive and has several strong positive and negative feedback loops. An example of a positive feedback loop is the algae growth cycle, where the level of algae in the pool can cause algae to grow at an explosive rate as more algae accumulates in the pool. Because chlorine kills algae, the chlorine level in the pool provides a strong negative feedback that moderates and controls algae growth. If a large concentration of chlorine is added to the pool,[15] all the algae will die, and new growth will begin only after this chlorine dissipates and new algae seed enters the pool from the air.

Chlorine is added to the pool as a liquid concentrate. The chlorine is unstable and decomposes rapidly when diluted and exposed to sunlight. The chlorine is stable at night but decomposes rapidly during the day and is normally added to the pool once a day. The model can be used to test the effect of adding chlorine at different rates and times. Because algae will grow rapidly when chlorine levels drop, it is important to add chlorine to control this growth. A value index is calculated as the ratio of the importance of chlorinating the pool to the cost of chlorination. The importance of chlorination is a product of the need to chlorinate times the effectiveness of chlorination in removing the algae. The value index in this model changes rapidly. Value indices in dynamics models cannot be normalized by converting cost and importance values to percentages during a simulation. Accordingly, the value index is situation specific, in this case representing a ratio between algae counts and chlorine costs. Figure 5.15 shows the way in which this value index fluctuates with time. These variations are so pronounced that it is advantageous to calculate an average value to facilitate comparisons of different chlorination procedures. A 24-hr smoothed value is calculated in the model, which is also plotted on Figure 5.15.

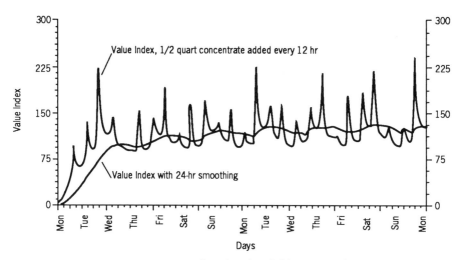

FIGURE 5.15 Variation in the value of pool chlorination with time.

Simulation of Various Chlorination Procedures

This model was used to simulate the value of several different chlorination procedures. The goal is to optimize value in a way that ensures that a sanitary pool is
maintained at minimum cost. Figure 5.16 compares the value of adding large con-
centrations of chlorine to the pool once a day to the value of adding smaller con-
centrations twice a day. Optimum value occurs when small quantities of chlorine are
added to the pool more frequently. The simulations support the old adage that "an
ounce of prevention is worth a pound of cure."

Figure 5.17 shows the algae and chlorine concentration in the pool over a 2-week
period. Note how the level of chlorine controls the algae level and how this level can
drop over 24 hr.

Program Listing

```
Algae = Algae + dt × (Algae From Pool + Algae From Air - Algae Loss)
INIT(Algae) = 100
    {initial value is 100}
Pool Chlorine = Pool Chlorine + dt × (-Chlorine Loss + Chlorine Addition)
INIT(Pool Chlorine) = Quarts Concentrate/Pool Size
    {initial chlorine equals the quarts of concentrate divided by the pool size
    in gallons}
Total Chlorine Cost = Total Chlorine Cost + dt × (Chlorination Cost)
INIT(Total Chlorine Cost) = Cost NaHClO3 Liquid × Quarts Concentrate
Algae From Air = 100
    {same initial value as algae in pool}
Algae From Pool = Algae × Algae Growth Rate {positive feedback loop}
Algae Growth Rate = .1
    {Algae grows at a rate of 10% per hour}
```

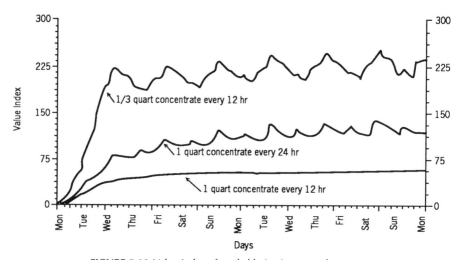

FIGURE 5.16 Value index of pool chlorination procedures.

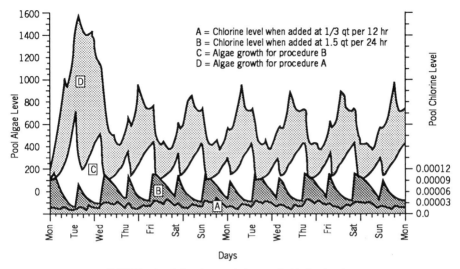

FIGURE 5.17 Swimming pool algae and chlorine levels.

```
Algae Loss = Death Rate × Algae
    {Algae in pool are lost proportional to the death rate}
Avg Value = SMTH1(Value,24)
    {SMTH 1(Value,24) is a smoothing function which averages Value for the
    previous 24 hours}
Chlorination Cost = Chlorine Addition × (Cost NaHClO3 Liquid × Pool Size)
    {have to multiply by pool size to eliminate dilution}
Chlorine Loss = Pool Chlorine × Decomposition Rate
    {Chlorine is lost depending on the numerical value of the decomposition
    rate}
Chlorine Addition = PULSE(Quarts Concentrate,0,Hours Till Addition)/Pool
    Size
    {Chlorine is added all at once at time 0 and then repeated as a pulse every
    12 hours}
Cost NaHClO3 Liquid = 1.00
    {Chlorine concentrate costs $2 gallon.}
Days = TIME/24
    {A day consists of 24 hours}
Hours Till Addition = 12
    {Concentrate is added every 12 hours as a pulse}
Pool Size = 20000
    {Swimming pool size in gallons.}
Quarts Concentrate = .5
    {A half of a quart is added}
Value = (Algae × Death Rate)/(Total Chlorine Cost/Days)
    {Value is calculated as the ratio of importance to cost. In this case impor-
    tance is a product of the need to chlorinate (The Algae in the pool at this
    time) times the effectiveness of removing the algae (Death Rate at this
    time). Since the Total Chlorine cost continues to grow, the average cost
    per day is obtained by dividing the total cost by the number of days that
    have passed.}
```

```
Death Rate = graph(Pool Chlorine)
   {This data string shows the effect of chlorine concentration on the algae
   death rate. When the concentration of chlorine in the pool is equal to or
   greater than .0001 quarts of chlorine concentrate all of the algae die.}
(0.0,0.0),(0.0000100,0.100),(0.0000200,0.200),(0.0000300,0.300),
(0.0000400,0.400),(0.0000500,0.500),(0.0000600,0.600),
(0.0000700,0.700),(0.0000800,0.800),(0.0000900,0.900),(0.000100,1.00)
Decomposition Rate = graph(Days)
   {Chlorine decomposes at a rate of 10% per hour for twelve hours when the sun
   shines, and does not decompose for 12 hours during darkness}
(1.00,0.100),(2.00,0.0),(3.00,0.100),(4.00,0.0),(5.00,0.100),
(6.00,0.0),(7.00,0.100),(8.00,0.0),(9.00,0.100),(10.00,0.0),
(11.00,0.100), (12.00,0.0),(13.00,0.100),(14.00,0.0)
```

SUMMARY

In this chapter I described how system dynamic models can be created to model changes that occur in value over time. Value indices were used to monitor changes in value that occur as supply and demand change. The value of almost all products and services change over time. Some products change dramatically within a day, others within a week, and others annually. It is possible to simulate the value of all of these situations using dynamic models.

Dynamic modeling allows value management studies to be made of complex systems and extends the appplicability of VA and VE to many new fields. I have developed value models for a wide variety of products and services. Products studied ranged from space platforms to doughnut production and sale. Dynamic modeling is very useful when applied to manufacturing and distribution procedures and inventory analysis and production control.

Competitive product analysis and strategic planning are other areas where modeling value is most beneficial. In these areas dynamic models have been constructed that predict sales and market share based on the dynamics of market factors, such as product value, advertising, distribution, and research and development.

REFERENCES AND ENDNOTES

1. Stella is a copywrited computer software program. It was used to construct the modes in this chapter. It and an expanded program called Ithink are available from High Performance Systems, Inc., 45 Lyme Road, Suite 300, Hanover, NH 03755.

2. In a manner analogous to that used in a drawing or paint program, a graphical representation of the system is drawn on the Macintosh screen.

3. The level of water in the tank equals the previous level of the tank plus water added in the period of time *dt*.

4. The initial amount of water in the tank is zero.

5. The tank capacity is 100 gallons, and it is being filled at a rate of 10 gallons/hr.

6. Stella allows data to be introduced as a data string. The Stella icon for a graphical converter is indicated by an ampersand placed inside of the converters circular icon. (Note Avg-Monthly-rain converter in Figure 5.6.) Data strings for this model are listed at the end of the program listing on page 108.

7. This figure was selected since the capacity of the dam is about seven times its average usage rate. Although seldom done with water, the price of water can be used to reduce usage. To do this the value index would be used to modulate price. When water is readily available, its price would be lower than during scarce periods such as a drought. This model can be easily modified to simulate such a situation. This is the way a private business would manage a reservoir. It would reduce overflow waste and optimize the economic return on investment.

8. Drought = IF (Month $\geq$ 12) AND (Month $\leq$ 24) THEN .4 ELSE 1.0.

9. Value = LOGN((Use $\times$ Price/Dam Water)) {The value of water in the dam is equal to the amount used times its price.}

10. This allows the sale price to reflect the ratio between use and the normal capacity of the dam.

11. "Thermal Storage, the State of the Market." *American Society of Heating, Refrigerating and Air-Conditioning Engineers (ASHRAE) Journal,* pp. 20–25, May 1986.

12. "Thermal Storage, A Showcase on Cost Savings," *ASHRAE Journal,* pp. 28–31, May 1986.

13. *Thermal Storage,* ASHRAE Technical Data Bulletin, pp. 1–145, 1985.

14. By changing the monthly interest rate in the program, it is easy to simulate the effect of other interest rates.

15. This is referred to as superchlorination, and is a necessary practice if the pool is left unattended for several days.

6

VALUE AND
DECISION MAKING

David J. De Marle
M. Larry Shillito

INFORMATION EXPLOSION

It has been said that the 19th century was the age of power, and the 20th century is the age of information. Since 1950, the world has seen tremendous growth in the sciences dealing with information.[1] Value management practitioners and decision makers are concerned with both the type and amount of information available for decision making.

According to Dalkey[2] there are roughly three types of information that can play a role in decision making:

1. *Knowledge.* Assertions that are highly confirmed and for which there is considerable evidence to support them.
2. *Speculation.* Material that has little or no evidence to back it up.
3. *Opinion.* The broad area of material that has some foundation for credence, yet is not adequately validated to justify being classified as knowledge. Some people may even consider this an area of intuition.

The distinction between these three areas is not well defined. Within what area of the data spectrum do most decision inputs lie? We can't always hold decisions until we have 100% knowledge, nor can we quickly base all decisions on speculation. Instead, a significant number of decisions are based on opinion. Using value measurement and management methodology, we can begin to structure data such that we can base decisions on better, more informed opinion.

The amount of information available for decision makers has been growing exponentially. People have been quickly thrust into information overload.[3] Coupling information overload with the readily available personal computer power and software programs available to structure and process information also inflicts man with yet another malady known as *analysis paralysis.* How can we begin to cope with the bewildering assortment of technical and electronic wizardry?

Information overload suggests that there is a limit to how much information one can process. Studies by Alderson and Sproull[4] indicate that the capability of the human mind to interrelate and correlate a large number of variables falls off at about 20 alternatives. Beyond twenty, the probability of making a knowledgeable decision is greatly reduced. This limit also correlates with the conclusion of Miller,[5] which indicates that the maximum number of bits of information that can be handled at one time is roughly "seven plus or minus two."

The actual numbers of 20 alternatives or seven bits are guidelines, but more important they indicate the need for assistance in data/information processing and decision making.

First it is helpful to try and understand a person's decision and selection process. We can then relate this to the rationale for using the various subjective value measurement techniques (discussed in Chapter 4) as aides for decision making. Information overload is analogous to the problem of detecting signals from noise. We will equate signals to value stimuli and then discuss a person's reaction to and processing of these stimuli.

A Person as an Information Processor

A person receives data from the environment through some form of energy. These data are coded by the sense organs, be it the eyes, ears, or just "gut feeling." It is stored and processed to make a decision based on some perception of value. Both value and information processing involve energy. That is, information and value can stimulate the senses just as heat, light, or electricity. Energy is always present to transmit information. Value as a form of energy has already been discussed in Chapter 1.

VALUE AS STIMULUS

The task involved in value measurement embodies the recognition, comparison, and judgment of stimuli. *Stimuli* is a much used term in psychology, but we will use it here to apply to value measurement as well. Referring to the energy theory in Chapter 1, the perception of value is proportional to the magnitude of a value stimulus, positive or negative. Work by such pioneers in the field of psychophysics as Thurstone[6,7] and Stevens[8,9] have laid the foundations to support this concept. One must be able to detect a stimulus, and unfortunately a stimulus generally does not occur in isolation. There is usually a background of sensory noise from which a stimulus is sensed.

Value Threshold

One way of viewing a person's ability to detect value stimuli would be to determine the degree of value intensity above which a person will always detect a stimulus and below which he would never experience it. However, a fixed-point value threshold most likely cannot be established. Such a point would vary according to the nature of the problem, as well as with time, and would really not be that helpful even if it could be established. Rather, a smooth curve, such as the one in Figure 6.1, represents the probability of correctly detecting a value stimulus relative to its intensity. This function can vary not only with the type of alternative but also with the level of the person's motivation, the quality of instructions he receives, the phrasing of the questions asked, the clarity and type of judgment criteria, and personal paradigms that filter incoming information. In complex decision tasks, conditions are rarely optimal.

Value thresholds are not nearly so important as how the subject varies the criteria for deciding if a given alternative constitutes a signal. This is one reason why decision matrix techniques like criteria analysis[10] discussed in Chapter 4, and combinex[11] play an important part in the selection process. They require the decision maker to define requirements by assigning numerical measures of relative importance; they force the decision maker to face, one at a time, the major factors leading to a decision. Because there is a continuum of value stimuli and associated noise, a statistical model of selection seems to be more useful to explain value measurement and selection than a single sensory threshold.

SIGNAL DETECTION THEORY

The statistical decision viewpoint, known as the theory of signal detection[1,12] emerged in the 1960s. Science has made much progress in sorting signals from noise especially in electronics. We will examine the theory of signal detection and relate it to the task of separating signals from noise when making subjective opinions and decisions regarding value.

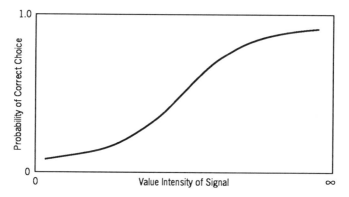

FIGURE 6.1 Probability of correct selection of a value alternative.

Fitts and Posner[13] describe signal detection very nicely. According to signal detection theory, every variation in energy at a sense organ gives rise to some change within the organism. Every new stimulus occurs against a background of stimulation already present. That is, a decision maker is confronted with two distributions: One distribution is that of a background of noise plus a signal (N+S), and the other is that of the background noise alone (N).

The probability of detecting a value-type signal depends upon the value intensity relative to the background. Neither the background noise nor the stimulus (signal) is constant in it's effects. Both can be represented by distributions as shown in Figure 6.2. These curves represent the magnitude of assumed value intensity for the occurrence of the background noise alone (N) and for the background noise and the signal together (N+S).

Referring to Figure 6.2, the decision maker makes a decision, x, which because of noise can vary randomly in magnitude. The noise can either be external due to the overwhelming amount of combinations of data, or it can be internal due to the random activity of the neural pathways and the brain due to confusion.

The decision x, for those instances in which a signal is present is taken as having a similar distribution to the noise but with each observation increased by the amount of noise present. We will assume in using this model that both distributions are normal, although they most likely are highly skewed such as the binomial or the Poisson.

The decision maker must establish a cutoff point, X_c, based on his/her judgment of value and/or opportunity loss. Any level (decision) of x above this point is treated as a signal (a choice), and any point below it is considered as noise alone.

If the signals are strong enough for the two distributions to be well separated, discrimination is virtually complete and a decision can easily be made. If, however,

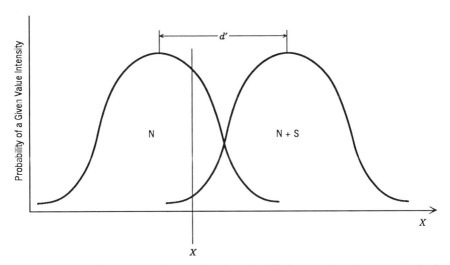

FIGURE 6.2 Hypothetical curve showing the theoretical distribution of varying amounts of value intensity for noise and signal-plus noise.

the distinction between alternatives is weak and nebulous so that there is considerable overlap between the two distributions, discrimination (decision making) is more difficult and cannot always be accurate. Part of one or the other distributions, or both, will inevitably be on the wrong side of X_c so that errors (less optimum selections) will be made.

The two parameters of this model are d' and β, where d' is the distance between the means of the two distributions as measured in standard deviation units (ϕ). Thus,

$$d' = \frac{X_{S+N} - X_N}{\phi}$$

β is the *likelihood ratio* or the ratio of the frequencies (the heights of the ordinates) at the cutoff point X_c,

$$\beta = \frac{f_{S+N}}{f_N} \text{ at point } X_c$$

For example, from Figure 6.3, $f_{S+N} = DE$ and $f_N = CE$ so that $\beta = DE/CE$.

The value of d' is the measure of the true discrimination in terms of the ratio (in the comparative sense) between signal strengths and the variability of the noise level. In other words, d' is a measure of the amount of distinction between the two distributions. This measure is important in decision making and decision-making techniques. It is obvious that all effort should be directed toward maximizing d'. That is, the design of a decision mechanism or value measurement technique should be such that the more optimum choices can be more easily detected. This is done by minimizing the number of redundant and undesirable combinations of criteria and

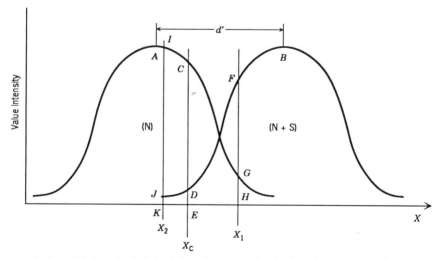

FIGURE 6.3 Hypothetical signal detection curves showing basis for parameters d' and β.

alternatives and is the premise behind value measurement methodology discussed in Chapter 4.

Alderson and Sproull[4] state that criteria must be understandable, applicable, independent, and complete. Fallon[11] goes a step further by setting upper and lower bounds to the ranges of choices within a criterion to maximize the relevancy of information going into the construction of a decision matrix. That is, he adjusts for utility. The use of decision criteria is a very important point. They should always be documented where a decision process is involved. This then limits the number of combinations of noise and signals. The more noise and the more combinations a decision maker must cope with, the less will be the resulting magnitude of d'. This is where value screening techniques like Q-sort and Pareto voting (discussed in Chapter 4) are helpful; they are used in the beginning stages of evaluation to filter through and eliminate the most obvious and recognizable forms of noise. The highest-scoring items are then used to construct a decision matrix consisting of the most pertinent items of concern. So, Q-sort or Pareto voting in the initial stages helps to maximize d' by surfacing those items with stronger value intensity. Once this is accomplished, criteria are established so a decision algorithm such as criteria analysis can be constructed in order to fine tune your perception of the remaining items for judgment.

On the other hand, β as a measure of the cutoff point X_c can vary independently of d'. It can be thought of as a measure of the caution exercised by the decision maker or of the confidence with which his judgments are made. It is also a measure of whether he clearly understands the objective or the purpose for which a decision or selection is being made. This can also be the realm of intuition.

Referring to Figure 6.3, a large value of β (where $\beta = FH/GH$) at point X_1 implying a high cutoff well to the right of X_c means that the decision maker is being cautious and is demanding a high degree of confidence before making a decision. It is also a measure of the decision maker's perceived value intensity of a particular choice. The result will be that he will probably select only near-perfect choices to the possible exclusion of other possible acceptable choices that should be considered or selected as well. Conversely, a cutoff point X_2 to the left of X_c (where $\beta = IK/JK$) implies a loose decision where the decision maker selects a less optimum choice.

The decision maker is able to vary the cutoff point at will and can make judgments in terms of more than one cutoff point by defining different levels of value. This is analogous to using a rating scale to judge alternatives against one criterion at a time. Each different cutoff point corresponds to a different criterion and a different rating.

Another factor that determines the location of the cutoff point X_c is the objective of the decision at hand as well as the criteria. Criteria and value measurement techniques can also influence the shape of the distributions. Objectives are derived from the results expected to come from a decision. Fitts and Posner[13] offer the example of a lookout trying to spot forest fires.

> It might be good strategy for him to set his criterion far to the left and report a large number of false alarms, but lessen the chance of a fire starting without being detected. Or he might set it far to the right, minimizing the number of false alarms, but increasing the possibility of undetected fires. Clearly his instructions including the rewards for

being correct and the punishments for error can be thought of as causing variation in the criterion which he uses.[13 p. 46.]

A clear statement of the objective cannot be overemphasized. The rater's instructions, including the rewards for being correct and the punishments or the opportunity loss or error can be thought of as causing variation in the placement of X_c. If the rater can use different cutoff points implying different degrees of value perception, it is reasonable to suppose that he can somehow scale the whole X axis in the same way. Such a scale must presumably be correlated with the level of value perception.

The magnitude of the β ratio can be affected in two ways: (1) the placement of the cutoff point X_c by the decision maker as mentioned before or (2) the shape of the two distributions, that is, the amount of skewness and kurtosis. The shape of the distributions are largely dependent on the design of the decision matrix itself or other value measurement techniques and how much confusion, noise, or redundancy is contained therein. Obviously, the goal is to have the distribution as leptokurtic as possible. This can be accomplished by establishing judgment criteria and constructing a decision algorithm. A decision matrix helps accomplish systematic shrinking of the two distributions.

Value measurement techniques are also geared to increasing the distance d' between the two distributions N and S+N, as well as shrinking the range of the distributions. This is done by structuring a problem to disaggregate a decision and its complex interrelated variables into a series of one-dimensional comparisons or separate subdecisions that are more manageable for evaluation. According to Morris[14] these one-dimensional comparisons are easy for the unaided intuition. Shepard[15] also notes that the subtle weighing and combining of the factors required for subjective decisions can only be accomplished by the mysterious intuitive deliberations of the human brain. However, difficulty arises when the decision maker attempts to take into account and combine all of the subdecisions simultaneously to reaggregate them to make an absolute judgment of the overall decision. This integration of the many one-dimensional subdecisions is performed much better by a computer and should be left for it. If the individual chooses not to use the computer for joining the subelements but chooses to mentally integrate them, the more the person will rely on intuition to make the final integrated decision. Integration tends to shrink the distance d' between the two distributions. Intuition will always play some role in decision making. However, by breaking a complex issue into smaller, less complex interrelated variables, the greater the chance of increasing the distance d' between the two distributions.

We use signal detection theory as a new approach to describing decision making and value measurement. It is also an attempt to show that there is foundation in using the various subjective value measurement techniques.

A Numerical Language To Amplify Signals

Value measurement techniques are very useful for separating signals from noise. Numbers, even though they are subjective, allow us to begin plotting distributions.

We can compare various raters numbers and begin to see differences. This, in turn, allows us to engage in more dialogue to reevaluate an issue that brings about yet new distributions that may be more convergent or divergent.

It is not the premise here to blindly apply subjective measures under the pretext that they can be used for a final decision. It is instead an attempt to communicate with numbers, for such a technique gives a rich supply of answers in an information-seeking mode. It is the use of numbers to replace semantics as a way of communicating perceptions concerning sometimes vague concepts, that is, numbers give us a common language by making opinions measurable.

A decision matrix, in turn, makes people think by forcing them to line up information and face issues one at a time in order to get the information in a form to make a better decision. These techniques will assist one in arriving at a better decision by giving one a better source for asking further more pertinent questions to seek more information.

We are aware that these subjective quantification techniques may have weaknesses. It should be stressed again that they are intended not to produce decisions, but rather to produce information that will facilitate decisions. Indeed, these techniques are merely thinking structures to force a methodical and meticulous consideration of all the factors involved in making a decision. Hopefully, this in turn will reduce the dangerous practice of making seat-of-the-pants decisions.

Schermerhorn and Taft[16] illustrate this very nicely with their term *systematic subjectivity*. It accentuates who believes what is important so we can test who is correct.

Objective Subjectivity

Just because we use subjective numbers in a structured decision process does not guarantee that we have and act on truth. We admit that evaluation is not as simple as averaging subjective numbers and concentrating on the highest ratings. Opinions and opinion distributions have to be tested against standards that are fact. People in VE, the behavioral sciences, market research, and so on are finding it increasingly important to introduce a measurable standard. Such a standard helps to make subjective opinions more objective by providing an absolute measurement. The more objective standards are used, the less we will have to be concerned about the signal-to-noise ratio.

The closest we can come to measurable standards are the development and use of value standards. Value standards help reduce subjectivity by providing an objective standard for comparison of alternatives.

Value Standards

Value standards have been used for many years. They evolved as a result of many VE studies where value engineers painstakingly strived to produce an objective basis for cost comparison and cost measurement. Such objective external cost comparisons were used as the basis for determining function worth. Fallon[11] was one of the pioneers in this area who refined the decision matrix. His method, *combinex*, utilizes utility theory for developing "commensurable units" and a "standard scale" to produce benchmarks for comparison of alternatives. He establishes "a range of choice" by setting limits using real measurable attributes like weight, time, dimensions, dollars, tolerances, and speed. He bounds his limits with "the best practical" and "the least favorable but acceptable" measures. This is a commendable attempt at developing objective measures for comparison and decision making.

The Department of Defense's VE course book[17] describes the use of value standards in cost estimating. They recognize two standards, historical and mathematical. Historical standards are

> . . . based on historical cost data for products that are the same or similar to the product in question. They are . . . based on products that have been in existence for some time, are highly competitive, are produced efficiently, and sold at a reasonable price. This standard assumes that such cost is a reasonable indication of value.[17, p. 3–41.]

Mathematical standards are "based on the scientific or physical equations that define the function of the item."[17, p. 3–43.] Minimum possible cost is then calculated for the function. Functions such as transmit torque, support weight, conduct current, and contain liquid can be expressed in the fundamental variables which can be measured using the laws of physics.

Figure 6.4 illustrates the use of mathematical cost standards. The theoretical cost of several different types of containers was calculated and is plotted versus their capacity. The cost of any container is equal to the weight of material used in making

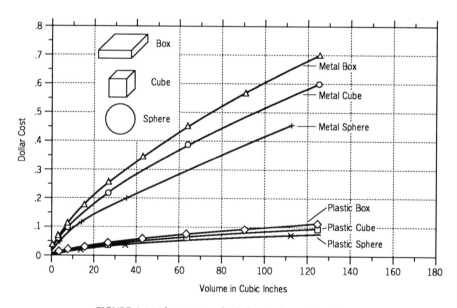

FIGURE 6.4 Value cost standards for simple small containers.

the container times the cost per unit weight of its construction material. The three equations shown below were used to calculate the costs in Figure 6.4.

$$\text{For a cubic box: } C = 6 \times L^2 \times t \times d \times P$$

where

C = cost of cube in cents
L = length of a side in inches
d = density of material in pounds per cubic inch
P = price per pound
t = thickness of container wall in inches

$$\text{For a sphere: } C = \pi \times D^2 \times t \times d \times P$$

where

C = cost of sphere in cents
D = diameter of the sphere in inches
d = density of material in pounds per cubic inch
P = price per pound
t = thickness of container wall in inches
π = pi 3.14159 . . .

$$\text{For a rectangular box: } C = 2 \times [(L \times W) + (W \times H) + (L \times H)] \times t \times d \times P$$

where

C = cost of box in cents
L = length of the box in inches
W = width of the box in inches
H = height of the box in inches
d = density of material in pounds per cubic inch
P = price per pound
t = thickness of container wall in inches

The weight of the material used is equal to the product of the volume of material used times the density of the material. The volume of material used can be calculated by multiplying the thickness times the surface area of the container. In the case illustrated in Figure 6.4, we derived standard cost data from the surface area equations for a cube, a sphere, and a box in which the length was twice the width of the box and four times its height. A $\frac{1}{16}$-in. thickness was assumed, and the volume of material used to construct each container was obtained by multiplying this thickness by the total surface area of each container. Costs are plotted for a plastic container with a specific gravity of 0.8 and a cost of 35¢ per pound and a metallic container with a cost of 65¢ per pound and a specific gravity of 2.7.

Value standards like this reduce subjectivity. We can calculate the advantages of different design shapes. Notice how a sphere is a more efficient container than a cube or a rectangular box. Notice also how the efficiency of these containers improves as they increase in size. Value standards allow design comparisons to be developed based on knowledge rather than opinion. Based on the physical and chemical properties of materials, they are more precise and meaningful than historical standards. Value standards allow objective comparisons to be made among alternatives and reduce the signal to noise problem. The VM literature is replete with examples of value standards. Some of the most recent are Bartlett[18] and Caldwell.[19]

VALUE ENERGY STANDARDS

Chapter 1 described value in terms of energy and briefly mentioned how energy accounting could be used to develop an objective measure for value. Let us now return to that example and describe how the amount of energy used to manufacture these products was calculated. The amount of energy required to make a product can be calculated from mathematical equations and used as a value standard to judge the efficiency of value creation. The process is analogous to that described above but substitutes energy units such as kilocalories or British thermal units for costs.

In determining the energy content of a product, you need to calculate the total energy required to make a product. To do this you need to calculate the energy used to produce the raw materials used in the product as well as the energy used to fabricate and assemble the product from its component parts. This is analogous to calculating the cost of a product from the cost of its component parts and their assembly.

Energy Used to Make the Raw Materials Used in an Ice Cube Tray

The ice cube trays described in Chapter 1 were made from aluminum and polyethylene. These materials are produced from raw materials found in nature. The energy used to produce metals, plastics, and so on can be calculated from chemical equations. The amount of energy varies considerably from material to material and is a function of the heat of formation of the compounds involved. To calculate the amount of energy used to produce a specified weight of product, you must examine the amount of energy consumed in the chemical reactions used to prepare them.

Energy Used to Make Aluminum Ice Cube Trays   Aluminum is made from bauxite (Al_2O_3) by electrolysis. The chemical formula for producing aluminum is

$$2Al_2O_3 \rightarrow 4Al + 3O_2$$

Two moles of bauxite produce 4 moles of aluminum plus 3 moles of oxygen. A mole of aluminum is equal to its molecular weight expressed in grams (26.98 g). The energy required to decompose 1 mole of Al_2O_3 is equal to 364.84 kg. cal. This is the amount of heat produced when 1 mole of aluminum oxide is formed from its elements. To calculate a value energy standard for aluminum, we need to calculate the minimum energy required to produce 1 g of aluminum. To create 1 g of aluminum from bauxite we must theoretically expend at least

$$(2 \times 384.84)/(4 \times 26.98) \text{ kcal of energy} = 7.1319 \text{ kg cal/g}$$

Thus, at least 7.13 kg cal of energy is required to produce 1 g of aluminum. We can calculate the energy required to produce any part made of aluminum by multiplying 7.13 kg cal times the weight of the part. Product A is composed of two units, a tray and a lever arm assembly. The weight and energy content of these units are shown in the following.

Unit	Weight (g)	Energy to Produce (kg cal)
Tray	135.5	966.4
Arm assembly	157.0	1119.7
Total	292.5	2086.1

These calculations show the *minimum* amount of energy required to form the aluminum used in this product. This assumes that all of the electrical energy was used to produce aluminum. The minimum amount of energy is used as a standard. If 3000 kg cal is actually used (due to the production of heat), you calculate the efficiency of the reaction. In this case it would be 69.5% efficient. In the example in Chapter 1, we used the minimum energy required, because we were unable to find data on the actual energy consumed.

Before continuing with this example we want to point out that we can substitute energy values for costs in the value standard equations we described previously. Thus,

For an aluminum rectangular box:
$$E = 2 \times [(L \times W) + (W \times H) + (L \times H)] \times t \times d \times 116.84$$

where

E = energy of box in kilogram calories
L = length of the box in inches
W = width of the box in inches
H = height of the box in inches
d = density of material in pounds per cubic inch
t = thickness of container wall in inches
116.64 = kilogram calories per cubic inch of aluminum

(Since 7.13 kg cal of energy is required to produce 1 g of aluminum, 116.84 kg cal of energy is used to produce a cubic inch of aluminum.)

Energy Used to Fabricate the Parts in an Ice Cube Tray You can calculate the amount of energy required to fabricate the parts used in this product by calculating the minimum force required to overcome the resistance of aluminum to cutting. Aluminum has a specific resistance to cutting of about 50 kg/mm². The total aluminum surface parted in stamping out the various parts used in this product equals 9600 mm². The minimum force required to part this aluminum surface is 34.55 kg cal. We estimate that an additional 17.2 kg cal is needed to shape aluminum, and 4.2 kg cal is required to assemble the part. Thus a total of 56 kg cal is used to manufacture each aluminum ice cube tray. Hence the minimum amount of energy used to fabricate the aluminum ice cube tray (product A in Chapter 1) is 2142 kg cal. Over half of this energy is used to make the lever arm assembly that separates and removes ice from the tray. In the other designs, this function is provided much more efficiently.

Energy Required to Produce Polyethylene Ice-Cooling Products Polyethylene is manufactured commercially from ethane, a natural gas product. Manufacture requires two separate steps. In the first step, energy is required to convert ethane into ethylene. In the second step, energy is produced as ethylene is converted into polyethylene in a catalytic operation. The reactions involved are

1. C_2H_6 (ethane) → C_2H_4 (ethylene) + H_2 (hydrogen)
2. nC_2H_4 (ethylene) → $(CH_2CH_2)_n$ (polyethylene)

The heat of formation of ethane equals −20.24 kg cal/mole and the heat of formation of ethylene equals +12.5 kg cal/mole. The negative sign indicates that 20.24 kg cal is required to decompose a mole of ethane, whereas the positive sign means that a mole of ethylene will give off 12.5 kg cal of heat during polymerization to polyethylene. Theoretically, if all of this heat is recovered, the net energy required to produce a mole of ethylene from a mole of ethane is 7.74 kg cal. To calculate the energy required to produce 1 g of polyethylene, you divide this quantity by 28, the grams in 1 mole of ethylene, and find that theoretically only 0.276 kg cal is required to produce 1 g of polyethylene. Since little of this energy is recovered, we estimate that about 0.45 kg cal is required to produce a gram of polyethylene.

If you multiply the 0.45 kg cal required to produce a gram of polyethylene by the total weight in grams of each product, you can find the energy required to produce the amount of polyethylene in each product. Thus,

Product	Weight (g)	Energy Required (kg cal)
B	230	103.5
C	236	106.2
D	20.2	9.1
E	167.5	75.4

These values were used to calculate the energy content of the plastic ice coolers described in Chapter 1.

THE INFLUENCE AND POWER OF PARADIGMS

A paradigm is a mental model that we use in decision making. So far we have treated decision making analysis and the design and use of value measurement without examining the influence of paradigms. Where a decision maker places the cutoff point X_c (Figure 6.3) is highly influenced by paradigms. Joel Baker, the futurist, describes paradigms as "a set of rules and regulations that: 1) define boundaries; and 2) tells you what to do to be successful within those boundaries. (Success is measured by the problems you solve using these rules and regulations.)"[20, p. 14]

Paradigms constantly filter incoming information. They influence the way we do things and the way we make decisions. Based on past experience, they can block our vision of the future. If we are aware of our paradigms we can take measures to reduce their effect.

Paradigms can be surfaced in group awareness sessions. We call them *paradigm workshops*. Such a session, using a third-party facilitator, involves answering paradigm surfacing questions like, What are some paradigms that can influence the way we make decisions about _____? Both positive and negative statements are produced that, depending on their number, can be aggregated into themes using the affinity diagrams discussed in Chapter 8. Surfacing paradigms and their effects on the decision process can help increase the distance of d' and make it clearer where to set the decision point X_c. As discussed in Chapter 13, the paradigm surfacing meeting can be introduced by showing the video *Discovering the Future: The Business of Paradigms.*[21] Discussion of this video gets people to start thinking in paradigms.

A Behavioral Science Perspective

Conflict is another source of noise. For example, when person A talks to person B, it is assumed that B has correctly received and decoded the message that person A has encoded and sent. For many reasons this does not always happen. However, if A communicates with B through a structured dialogue like that required to evaluate a decision matrix, noise is minimized. The value measurement techniques described in Chapter 4 serve as a vehicle for dialogue, which in turn helps to increase the distance d' between the two signal/noise distributions. These value measurement techniques force clarification and dialogue through a common frame of reference based on structure. Argyris[22] elaborates further by segregating noise related to individual factors and noise generated from organizational factors. He also introduces another dimension that ranges from cognitive noise to emotional noise.

Fraser[23] relates both of these dimensions in a two-dimensional matrix as in Figure 6.5. The individual cognitive noise source has already been discussed in terms of information overload and an individual's processing capacity. Another source of individual noise is jargon and, as just mentioned, paradigms, which filter perception. Individual emotional noise comes from ego defense mechanisms used to satisfy

	Individual	Organization
Cognitive	Information Processing Capabilities	Management Systems
Emotional	Ego Defense Mechanisms	Organizational Defense Mechanisms

FIGURE 6.5 Causes of information noise (Fraser).
Reprinted by permission of the Society of American Value Engineers.

personal needs and maintain a good self-image. Argyris[22] discusses this and other defense mechanisms well. For example, organizational defense mechanisms exist in the form of policies and actions designed to "protect turf." The longer these policies exist, the more entrenched and legitimate they become. Organizational cognitive noise, on the other hand, is generated by management information systems that can generate excess data and create information overload. These factors are all at work when evaluating new ideas.

Evaluating New Ideas (Finding the Needle in the Haystack)

Although the brainstorming technique has been widely used, very little emphasis has been placed on the evaluation of ideas generated by these sessions. Idea lists, no matter how long, how original, or how elegant, are worthless until they are used to solve existing problems.

Ideation techniques such as brainstorming are based on the use of suspended judgment. An environment conducive to divergent thought is created when evaluation is delayed. In such situations, fluency and originality prosper, and a large number of ideas are obtained. Every idea is duly recorded, and a master list of all ideas is prepared. This list contains poor as well as good ideas. Because judgment was suspended, the problem of evaluating these ideas is complex and difficult.

Most authors writing in this field have neglected the problem of evaluation. Osborn, one of the original authors in the field, has been criticized on the ground that his book *Applied Imagination* overrated ideation and underrated judgment.[24] In defense of this, Osborn states: "Everyone is fairly strong in judgment, and ever eager to apply it at every turn. As a result of the pressures of experience and the disciplines of education, the critical faculty of the average person is relatively well trained; and, instinctively, it is the first function to be applied to almost any question."[24] Osborn is quoted here because his feelings are typical of most authors in this field. They assume that one needs little or no training in evaluation and that anyone can easily select good ideas from bad.

It has been our experience that the problem of evaluation is of such magnitude that unless it is properly met, the ideation techniques may hinder more than help a development effort. Faced with job pressures, product deadlines, and economic restrictions, ideation lists can confound a pragmatic supervisor.

Rating by Experts Over the last 30 years we have conducted a large number of creativity sessions. These meetings were attended by a wide variety of technical and managerial people with considerable job experience. The sessions often dealt with technical problems related to the design and manufacture of new products. These brainstorming sessions produced many ideas, having an average yield of between 100 to 150. The ideas produced were common and unique, simple and complex, fallacious and good. Originally, with considerable naïveté, we compiled idea lists and submitted them to various experts for evaluation. Screening sessions were held in which professional people, experts in the field, were invited to evaluate the ideas. Often evaluation sessions were conducted in group meetings in which each idea was discussed and then rated. These meetings proved to be long and arduous. Extroverts tended to dominate the meetings and influence voting. Screening was also conducted on an individual basis where rating sheets were mailed to each judge. In this situation our experts either refused to rate ideas they were unfamiliar with or gave them average ratings. This evaluation technique is commonly used and was suggested by Osborn and others. It is based on the theory that people can screen good from bad ideas.

Case History of a Failure Let us now use a case history to show how paradigms prevented a group of experts from selecting the "right answer" from a large list of ideas developed in a creativity session. Some years ago we used brainstorming to create a list of ideas designed to correct a quality problem that had developed unexpectedly in a product development project. A total of 107 ideas was recorded in the session and subsequently evaluated in a screening session in which seven professional "experts" rated each idea using a numerical scale from 1 to 5. A rating of 1 designated poor; 2, fair; 3, average; 4, good; and 5, excellent. No evaluation criteria other than the scale was used, and judgment was subjective. At the end of the session, we compiled the total score for each idea by summing individual ratings and produced a list of all the ideas ordered according to score. Figure 6.6 is a histogram of the ratings for these ideas.

Four ideas received a total vote of 34, one vote short of a perfect total score. With considerable enthusiasm we tested these four "best" ideas, only to find that they didn't solve the problem. Undaunted, we then tested all 17 ideas that had been rated 27 or higher. Unfortunately, none of these ideas solved the problem. The best we could say at the end of our tests was that we knew 17 things that would not solve the problem. Unfortunately, although our experts had little difficulty rating the brainstorming ideas, they were unable to select ideas that could solve this problem!

After nearly a year of effort, the problem was solved. Reinspection of the original idea list revealed four ideas that would have solved the problem. Had we tested any one of these ideas, we would have been led directly to the final solution of the

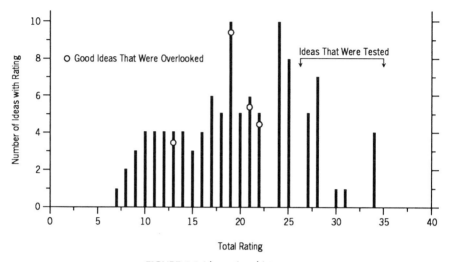

FIGURE 6.6 Idea ratings histogram.

problem. The four ideas are plotted on the histogram and were originally rated 13, 19, 21, and 22, respectively. The experts had considered them either average or below average. Table 6.1 shows how the experts rated these ideas. For comparison, each judge's average ratings of these four ideas are given, along with their average rating for all of the ideas.

Judge B was the only judge who rated these ideas as "good" ideas. However, judge B rated many ideas "good." He had difficulty differentiating these ideas from others that he rated highly. Clearly our experts were unable to recognize the solution to this problem even when confronted with it! Yet the experts were highly trained scientific personnel well versed in the problem area. Unfortunately their "expertness" biased their evaluation. They were working under a false paradigm, in this case a set of

TABLE 6.1 Experts' ratings of brainstorming ideas

	Individual Ratings				Average Rating for These Ideas	Average Rating for All Ideas
Judge	Idea 1	Idea 2	Idea 3	Idea 4		
A	3	2	3	3	2.75	2.63
B	5	4	3	4	4.00	3.25
C	4	1	4	2	2.75	2.76
D	2	2	2	1	1.75	2.85
E	4	1	4	4	3.25	2.96
F	4	2	5	1	3.00	2.78
G	1	1	1	4	1.75	2.00

assumptions that caused them to concentrate on a specific area that they believed had caused the problem.

Fortunately, this example is an exception to a general rule. Collecting and screening brainstorming ideas to select the best ideas usually works. We need not abandon the screening process if it sometimes fails. However, we need to take steps to improve the screening process. Simply allowing experts to screen ideas as poor, fair, and good without examining bias can eliminate some good ideas. Through the years we have found that this example is not an isolated case. In this and other cases the experts selected the wrong ideas. When experts are biased, they often fail to recognize good ideas.

The problem lies in the paradigms that experts use to rate ideas and in a screening process that fails to challenge these paradigms. The failures occur frequently enough to contradict Osborn's statements relative to judgment and have serious implications regarding the use of brainstorming to help solve many technical problems. In retrospect we might ask ourselves this logical question: If the experts had not solved the problem before a brainstorming session was held, why should they be expected to recognize the solution from a brainstorming list? Brainstorming and other creative techniques allow us to "suspend judgment" long enough to voice all kinds of foreign ideas, but what safeguards exist to prevent "expert bias" from discarding good ideas when judgment is resumed?

OBJECTIVE TECHNIQUES FOR SCREENING IDEAS

We have experimented with several promising ways to improve the objectivity of the screening process. Listed below are several techniques that are easy to use and help make the evaluation more objective.

Paradigm Analysis

Paradigm analysis is a term that describes a method for objectively screening ideas. Paradigm analysis is a scientific method in which basic hypotheses are drawn about the problem and then tested. In paradigm analyses, we categorize a number of ideas, draw premises about each category, and test the hypotheses. Evaluation is analytical and objective.

The ideas are first arranged in categories of related ideas. In the example mentioned previously, the original 107 ideas were divided into various groups. Twenty-eight ideas were related to a mechanical support, 17 ideas involved an amplifier, 9 ideas related to a switch, 22 ideas pertained to a storage device, 23 ideas involved a recorder, and 11 ideas were of a miscellaneous nature. Categorization is the first step in premise analysis.

After the ideas are categorized, one or more paradigms are drawn for each category. General premises such as "the support is the cause of this defect" are obtained. Each premise is then tested in the laboratory. Experiments can be designed that test several premises at once by use of statistical factorial designs. Paradigm analysis substitutes objective testing for subjective rating.

After testing the category premises, a considerable reduction can usually be made in the number of ideas to be tested. In the example cited, paradigm analysis would reduce the testing of all 107 ideas to testing 22 ideas related to the storage device. Had we asked the experts to judge the contribution of the storage device to the defect they would have said it had little to do with the defect. Eventually when they got around to testing this premise, they found that it was deeply involved. By formulating premises and testing them, the evaluation is rendered objective.

In actual use we find that the premises drawn seem trite because we usually have firm opinions regarding them. No matter how obvious the answer to the premise may seem, it is important to test its validity. Experts can be wrong, and breakthroughs usually result from the disruption of established opinion.

Evolutionary Evaluation

The solution to many complex problems is accomplished via an evolutionary process. Problem solving is dynamic, not static. A complex problem is difficult to solve quickly. Often you need to break it down into a set of problems that you can study. This takes time, and you need to develop a strategy for evaluating complex problems.

Problems vary considerably in complexity. Consider the following list of problems that are ranked in increasing complexity.

1. Naming a new product.
2. Developing an advertising slogan for a product.
3. Designing a packaging unit for a product.
4. Increasing production capacity of a product.
5. Troubleshooting a product defect that suddenly appears.
6. Improving an established product.
7. Inventing a new product.
8. Inventing a family of new products.

Alex Osborn developed and used brainstorming successfully on problems such as items 1 and 2 above. Getting and evaluating advertising ideas are relatively simple, and many different and imaginative ideas can be used. As you go down the list, the problems are more technical and complex, and many factors exist that constrain a solution. In these problems you should not expect to find a solution clearly spelled out in brainstorming lists. Here people should look to brainstorming lists for approaches rather than solutions. More time and effort are needed to define and attack these problems. People need time to study problems as they arise, to gather data, and to develop and test ideas. They often need to redefine the problem and expand or narrow its scope before reaching a solution. The evaluation process is evolutionary, often depending on the acquisition of new data and insights. Only as you learn more about the problem can you expect to reach a solution.

In evolutionary evaluation we establish a sequence of meetings on a given problem. These meetings follow the VA/VE job plan. The value team collects and organizes ideas and screens them as the program progresses. A feedback of information is effected, and following a period of discussion, more ideation is allowed. As time progresses, the problem is redefined, and the sequence is repeated. Information, such as reports, meeting notes, laboratory results, cost data, and so on is continually circulated to members of this group. The sessions thus make use of alternative periods of discussion and ideation. This feedback of information aids evaluation and provides a stimulus for further ideation. As more information is gathered on a problem, the team can either proceed as planned or redefine the problem and open up new areas for ideation. Evaluation does not limit ideas, it shifts their focus and emphasis. Note that this process is different from that used in many VE workshops. Value analysis and value engineering studies are often conducted in a single 40-hr workshop conducted in a week-long intensive cost-reduction process. We believe that it is much better to conduct a VE study over a 2- or 3-month period where the VA/VE team meets on a regular basis several times a week. This allows sufficient time for the team to apply VA/VE effectively to most industrial problems.

Simplicity Screening

Evolutionary evaluation depends on gathering information. It is important to test ideas quickly and feedback the test results. The faster this feedback occurs, the more efficient the evaluation process. For this reason, we recommend evaluating simple ideas and premises first. They can be quickly tested before evaluating more complex ideas. Often the test results may obviate the need to test more complex ideas.

Screening ideas for simplicity is a good way to do this. A Pareto vote can be used to establish a list of ideas that can be quickly tested. Alternately, a Q-sort or a categorical rating can be used to screen out difficult ideas. Things that can be easily done are rated 1, difficult things rated 3, and a 2 rating is used for items of moderate difficulty. Simplicity screening provides a maximum feedback of information in minimum time, is economical, and gives a value team a sense of accomplishment.

Frequency Analysis

Another method for evaluation that we have experimented with is based on examining the frequency with which individuals suggest an idea. This method yields information on the uniqueness or originality of ideas. In this method, several persons are asked to individually brainstorm a given problem. After a problem is described, each individual lists his or her ideas on a piece of paper. In a creative session, no communication or discussion is permitted until all of the individuals have listed all of their ideas. Now the ideas that all of the individuals had are compiled and compared. A frequency distribution of ideas is obtained that separates commonly occurring ideas from more unique ideas. In conducting this type of analysis, we have often timed the occurrence of different ideas. This can be done in a meeting by asking the people to mark their papers with a check mark upon a signal from a person who is timing the meeting. Now an average time of occurrence can be calculated for each idea. We have found that the time of occurrence is a good measure of uniqueness since common ideas occur early in these brainstorming sessions. The average time of occurrence provides an objective measure of the uniqueness of brainstorming ideas.

Using the frequency with which an idea is suggested or the time it took for an idea to be listed can be used to help screen ideas. Common ideas are applied to new problems, whereas unique ideas are used where the problem is an "old chestnut." This technique is free of expert bias. However, it is time-consuming and experimental.

Emotional Impact Analysis

Gordon describes a method in which ideas are evaluated on the basis of their emotional impact. In his book *Synectics,* he describes the use of this method.[25] Gordon believes that a pleasurable excitement or hedonic response accompanies an inventor's valid intuitions. He believes that creative people have trained themselves to feel these responses and use them to direct their work. In ideation sessions, a good idea is signaled by a sudden burst of enthusiasm and excitement in the group. Gordon suggests recording these sessions for subsequent analysis. Videotapes can then be examined to find portions of the meetings where considerable enthusiasm was shown.

Concept Selection

Stewart Pugh, professor of design engineering at the Loughborough University of Technology in Leicester, Scotland, has devised a simple nonnumeric comparison process for evaluating design concepts.[26] His concept selection process is based on the premise that engineers and designers are often blinded by their preference for their favorite engineering and design approaches. Strong adherence to a conceptual paradigm makes it difficult to select alternate concepts. When a poor concept is chosen from a set of design candidates, the resulting design, according to Pugh, suffers from "conceptual weakness."

Pugh has developed a progressive and disciplined concept formulation and evaluation process to minimize conceptual weakness. A matrix is generated that arrays sketches of alternative design approaches against a set of product specification criteria. Figure 6.7 illustrates his concept selection matrix.

One of the concepts is selected as a *datum* reference against which all of the other concepts are compared. Often this datum is the existing product. An evaluation scheme is established for comparing alternative concepts. If an alternate design is better than the datum for a particular criterion, it is rated as a plus (+). If it is worse than the datum, it is rated as a minus (−). When doubt exists about the advantage or disadvantage of an alternative, it is given an S to signify it is the same as the datum. After all of the alternatives have been evaluated and the entire matrix is completed, the + and − ratings in each column of the matrix are summed. Column totals are used to identify the best concept. Pugh recommends repeating the process several times with other reference datum as a consistency check. The process is usually repeated

Criteria	Concept Sketches 1	2	3	4	5
A	+	Reference	+	−	S
B	+		+	−	+
C	S		+	S	−
D	S		+	−	−
E	−		S	+	S
Totals	2 + 1 −		4 +	1 + 3 −	1 + 2 −

FIGURE 6.7 Pugh concept selection matrix.

with more detailed criteria for the few strong concepts that emerge from the first round of ratings.

INTUITION: A CONTRADICTION?

In spite of our emphasis and insistence on value measurement to improve the information and communication needed for a decision and the need for more objective data and value standards, many decision makers will still be influenced by and in some cases dependent on a gut feeling to make a final decision. Just as much development is being done in the field of decision making, so too, considerable attention is being given to intuition. Weston Agor has contributed much to this subject and has written several books on intuition. He encourages interdisciplinary research on the use of intuition in leadership and management. His book, *Intuition in Organizations,* is a collection of works from numerous authors and experts on the subject.[27] The book contains background information on the subject and provides a foundation for developing and using intuition in decision making to increase productivity.

Intuition will be used more. The closer we get to the 21st century, the more we will rely on what Lonnie Helgeson, director of the Intuitive Leadership Project at the Hubert II. Humphry Institute of Public Affairs at the University of Minnesota, describes as "intuitive leadership." Value measurement will provide the numerical language for intuitive leaders to translate their thoughts into actionable targets for managers. Dr. Alden Lank, director of studies of the International Management Institute in Divonne, France, remarked in April 1988 that "we have identified intuition as one of the key learnings for managers in the 21st Century."[28, p. 23]

We believe there can be unique integration of intuition and value measurement. Over 20 years ago, Morris[29] stressed the importance of intuition and the fact that intuition plays a major part in management decisions. He describes four ways intuition can be developed: (1) "the school of hard knocks," (2) "experience," (3) "teaching the analytical aspects of management science," and (4) "through education by encouraging the development of the personal characteristics and organizational conditions associated with the flourishing of intuitive skill."[29, pp. B163–B164]

Harian Cleveland of the Minneapolis Star and Tribune vividly describes the practical aspects of intuitive leadership.[30] He, too, states that intuition is coming into vogue as the practical uses of visioning and reasoning are being discovered, used, and accepted. More corporations and businesses are spending more bottom-line money and time for visioning. Visioning leads to empowerment of individuals, groups, and departments. Empowerment will require maximum use of intuitive brainpower. However, intuition is not and will not become a substitute for reason and rational problem solving.

SUMMARY

In this chapter we used signal detection theory to explain how humans separate valuable information signals from noise. We discussed the deleterious effect of too little and too much information on value practitioners and decision makers. Signals are akin to stimuli and people process value stimuli as signals. There is an analogy between people's processing of data and an instrument's separation of signals from noise. We used signal-to-noise theory to describe a cutoff point for a decision. The clarity of the objectives and decision criteria used to make a decision determine the placement of the cutoff point.

Decisions are improved by the use of value measurement and value screening techniques, such as a decision matrix. However, expert opinions should be verified by factual data. Value standards allow objective comparisons to be made among alternatives and improve decision making. Value standards based on energy accounting were developed to illustrate this process.

Confounding the signal-to-noise dilemma in decision making are the effects of paradigms. We showed how paradigms prevented experts from selecting the right ideas from a list of ideas generated in a brainstorming session. We described how several new decision enhancing techniques, paradigm analysis, evolutionary evaluation, simplicity screening, frequency analysis, emotional impact analysis, and a concept selection process, can help isolate the best workable ideas from a large composite of ideas.

What then of intuition? Intuition will be used in spite of our best efforts to develop and improve upon rational decision-making methods. Yet it should be used in conjunction with these processes. We believe that intuition and these methods can be synergistic. Intuition offers an imaginative counterpoint to rational analytic processes that should be used to evaluate intuitive ideas and plans.

REFERENCES

1. Welford, A. T., *Fundamentals of Skill,* Methuen & Co., London, 1968.
2. Dalkey, N. C., *The Delphi Method: An Experimental Study of Group Opinion,* Rand Corp., Santa Monica, Calif., 1969.

3. Knox, W. T., "The Pathology of Information," *Journal of Technical Writing And Communication,* **2**(1), 85–92, 1972.

4. Alderson, R. C., and Sproull, W. C., "Requirement Analysis, Need Forecasting, and Technology Planning, Using the Honeywell PATTERN Technique," in *Industrial Applications of Technological Forecasting,* (M. J. Cetron, and C. A. Ralph, Eds.), Wiley–Interscience, New York, pp. 428–443, 1971.

5. Miller, G., "The Magical Number Seven Plus or Minus Two," *Psychological Review,* **63,** 81–97, 1956.

6. Thurstone, L. L., "Psychophysical Analysis," *American Journal of Psychology* **38,** 368–389, 1927.

7. Thurstone, L. L., "The Measurement of Values," *Psychological Review,* **61**(1), 47–58, 1954.

8. Stevens, S. S., and Galanter, E. H., "Ratio Scales and Category Scales for a Dozen Perceptual Continua," *Journal of Experimental Psychology,* **54**(6), 377–411, 1957.

9. Stevens, S. S., "A Metric for the Social Consensus," *Science,* **151,** 530–541, 1966.

10. DeMarle, D. J., "Criteria Analysis of Consumer Products," *Proceedings, Society of American Value Engineers* **6,** 267–272, 1971.

11. Fallon, C., *Value Analysis to Improve Productivity,* Wiley–Interscience, New York, 1971.

12. Green, D. M., and Swets, J. A., *Signal Detection Theory and Psychophysics,* Wiley, New York, 1966.

13. Fitts, P. M., and Posner, M. I., *Human Performance,* Brooks/Cole, Belmont, Calif., 1967.

14. Morris, W. T., "Matching Decision Aids with Intuitive Styles," in *Decision Making* (M. S. Brinkers, Ed.), Chapter 1, Ohio State Univ. Press, 1972, Columbus, Ohio.

15. Shepard, R. N., "On Subjectively Optimum Selection Among Multiattribute Alternatives," in *Human Judgements and Optimality* (M. W. Shelley, and G. L. Bryan, Eds.), Chapter 14, Wiley, New York. 1964.

16. Schermerhorn, R. S., and Taft, M. I., "Measuring Design Intangibles," *Machine Design* **40,** 108–112, December 19, 1968.

17. U.S. Department of Defense, *"Principles and Applications of Value Engineering,"* Vol. 1, U.S. Gov. Printing Office, Washington, D.C., 1968.

18. Bartlett, R. L., "Functional Value Development and Synthesis," *Proceedings, Society of American Value Engineers* **24,** 9–25, 1989.

19. Caldwell, R. D., "Value Engineering—A New Systematic Way of Determining Cost Reduction Candidates for Manufacturable Products," *Proceedings, Society of American Value Engineers* **24,** 26–35, 1989.

20. Barker, J. A., *Discovering the Future: The Business of Paradigms,* ILI Press, St. Paul, Minn., 1985.

21. Barker, J. A., *Discovering the Future: The Business of Paradigms,* video, Chart House Learning Corp., Burnesville, Minn., 1984.

22. Argyris, C., *Strategy Change and Defensive Routines,* Pittman, Boston, 1985.

23. Fraser, R. A., "Decision Making And The Generation of Information," *Value World,* pp. 11–14, October/November/December 1988.

24. Osborn, A. F., *Applied Imagination,* p. 248, revised edition, Charles Scribner's & Sons, New York, 1957.

25. Gordon, W. J. J., *Synectics,* Harper & Bros., New York, 1961.

26. Pugh, S., "Concept Selection—A Method That Works," paper presented at the International Conference on Engineering Design, Rome, Italy, March 9–13, 1981.

27. Agor, W. H., *Intuition in Organizations,* Sage, Newbury, Calif., 1989.

28. Helgeson, L., "Intuitive Leadership: A Comprehensive Model of 7 Perspectives," in: The Intuitive Leadership Project Report, The Reflective Leadership Project, Hubert H. Humphry Institute for Public Affairs, Univ. of Minnesota, Minneapolis, Minn., pp. 23–28, 1988.

29. Morris, W. T., "Intuition and Relevance," *Management Science* **14**(4), B157–165, 1967.

30. Cleveland, H., "Training Intuitive Leaders," *Minneapolis Star and Tribune,* July 9, 1989.

BIBLIOGRAPHY

1. Pugh, S., *Total Design*, Addison-Wesley, Reading, Mass,. 1990.

THE DESIGN OF VALUE

This section is devoted to a description of function analysis and function matrices that form a basis for design. Patterned on the old engineering adage that form follows function, this section describes different ways to map functions and create system designs of improved value. A total value concept based on satisfying customer, retailer, and producer needs is described. Function analysis diagrams and indentured function matrices are used to portray how, why, and at what cost these needs are met in every element of a design. Detailed "how-to" descriptions of three new leading-edge function design matrix systems, quality function deployment, technology road-maps, and customer-oriented product concepting, are provided. Value measurement techniques are used to quantify the value of new system designs.

7

FUNCTION ANALYSIS

David J. De Marle
M. Larry Shillito

FUNCTION ANALYSIS

Function analysis is a technique for analyzing the performance and usefulness of a product or service. It assumes needs exist for a product and specifies how these needs are met in a design. It distinguishes the value disciplines of VA and VE from other design or cost reduction processes. It is the heart of VA/VE and was conceptualized by Larry Miles, the father of VA/VE.[1] According to Miles, poor value is a people problem.[2] His experience with many cost-reduction projects revealed that by merely determining what things cost and documenting cost breakdown, you can reduce costs 5%; improving the choice of material and methods can reduce cost yet another 10%. However, by developing a better way of performing what a product was originally intended to do, you could save upward of 30%. Miles used short two-word statements to describe the functions of components in a product. He invented the technique of function analysis (FA) while working at General Electric during World War II. Since then, it has had a long history of application and a considerable track record.

A function is a generic statement of what needs to be accomplished without specifying the means. Consequently, all things, products, processes, services, procedures, and so on can be described by functions. As such, FA facilitates communication across disciplines and technologies by providing a universal or common language. Function analysis consists of definitional and structural techniques that employ a semantic clarification of function. It provides the basis for defining the performance and utility of an object in detail. Function analysis also furnishes a basis for altering a design by creating alternative ways for accomplishing a product's functions.

In FA, a product and all its components are converted into functions. The method requires functions to be described with only two words, a verb and a noun. By so restricting function specifications, clear descriptions of the functions are possible, descriptions that are not confounded by unnecessary modifying phrases, adjectives, and adverbs. Lengthy function descriptions increase ambiguity and inhibit creative approaches, making it difficult to develop a rational approach to design change.

The rules of function description are the following.

1. Determine the user's needs for a product or service. What are the qualities, traits, or characteristics that define what the product must be able to do? Why is the product needed?

2. Use only one verb and one noun to describe a function. The verb should answer the question, What does it do? The noun should answer, What does it do it to or with? Where possible, nouns should be measurable, and verbs should be demonstrable or action oriented.

3. Avoid passive or indirect verbs such as *provides, supplies, gives, furnishes, is,* and *prepares.* Such verbs contain very little information.

4. Avoid goallike words or phrases, such as *improve, maximize, optimize, prevent, least, most,* and *100%.*

5. List a large number of two-word combinations and then select the best pair. Teams can be used to derive a group definition of function. Examples are

   (a) Pencil: make marks.

   (b) Coffee cup: hold liquid.

   (c) Screwdriver: transfer torque or—if a painter uses it to open cans—transfer linear force. Function depends on the actual use of a product.

Function descriptions should be derived for a product and for all of its components. Occasionally items may legitimately have more than one function. For example, a military tank is multifunctional; it delivers armor, mobility, and firepower. When listing components and their functions, keep them at the same level of abstraction. For example, during the analysis of a wooden pencil, the components listed as wood, graphite, eraser, eraser holder, paint, glue, and so on would be kept at a single level of assembly. Components in the paint (pigment, solvent, carrier, etc.) are constituents of the paint component and exist at a second lower level of design indenture. They should not be considered with components at the first level of assembly.

Defining functions can be a difficult task and requires a lot of skill and practice. A helpful checklist and memory jogger has been published by Dick Park.[3] He lists over 250 verbs and 450 nouns that have been aggregated into 10 categories covering a wide range of applications over the past 20 years.

He also provides a clear explanation about defining functions. According to Park, a function is not an action but rather an objective of an action. His example "file papers" is an action. The objective of the action may be "store information." Using this as the function, you will quickly realize that there can be numerous actions that can fulfill the function "store information." His categories of functions were developed from his realization and experience that the same functions have appeared over and over in many different situations and cases.

Miles originated a simple system in which functions are classified as either basic or secondary.[1] A basic function is the prime reason for the existence of the product. It describes the output of the product in terms of the user's need. A good question to determine the basic function is, If you take this function from the product, will the purpose of the product still be fulfilled? Secondary functions, on the other hand, support the basic function(s) and allow them to occur. They generally are present as a result of choosing a specific design approach. Secondary functions may improve dependability or convenience or may serve solely sensory or aesthetic functions. The primary function of a fuel-ignited cigarette lighter is to "ignite cigarette." A secondary function of the cigarette lighter is to "prevent evaporation," which it does when the lid is closed. This secondary function is necessary only because a fuel ignition design was chosen. Such secondary functions are prime candidates for elimination, improvement, and innovation. Classifying functions as basic or secondary is very helpful for identifying redundant or unnecessary functions and for identifying the cost of functions within a product. For example, in many products more dollars are spent on secondary functions than on basic or required functions. Because some secondary functions may be unnecessary and others may be modified, this often signals a potential for cost and performance improvement.

Originally FA was performed on a product by listing the components and then determining the function for each component. Functions were then classified as basic or secondary (required, unwanted, aesthetic, etc.) to the performance of the product. In this chapter, we will illustrate function analysis by analyzing the composition of an ordinary incandescent lightbulb.

Figure 7.1 shows the composition of a light bulb, which consists of an assembly of 10 components. A tungsten filament is mounted at the center of an air-tight glass bulb containing inert argon gas. Light and heat are produced when electricity flows through the filament. A central glass stem supports a wire frame assembly from which the filament and a heat-deflecting disc are suspended. The stem contains lead-in wires, that conduct electricity from a metal base to the filament and back to a center contact at the base of the bulb. A glass stem press, formed during assembly, anchors and supports the wire assembly. Lead-in wires are attached to the filament and are insulated and supported by the glass stem. The base rim is threaded to screw into a standard lamp outlet. An insulator in the base prevents electrical shorting between the base rim and the central contact and also supports the glass stem. The bulb confines argon gas, which prevents oxidation, reduces vaporization of the tungsten, and transmits light and heat from the unit. The functions of each component in the light bulb are shown in Figure 7.2.

This method of defining and classifying functions was used extensively for about 20 years. Functions were described for components in a product and classified as primary or secondary. The classification was based on experience and educated guess work. As Kaufman notes,[4] this function classification did not show the relationship of one function to another. In addition, because functions were generated only from

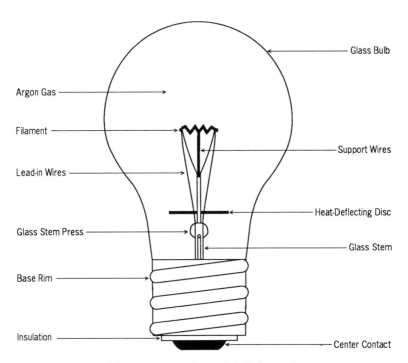

FIGURE 7.1 Incandescent light bulb assembly.

PARTS	FUNCTIONS	BASIC	SECONDARY
Filament	Produce Light	X	
	Convert Energy		X
Support Wires	Support Filament		X
Lead-in Wires	Conduct Current		X
Stem	Support Wires		X
	Insulate Conductors		X
	Mount Disc		X
Heat Deflector Disc	Deflect Heat		X
Center Contact	Conduct Current		X
Insulator	Insulate Conductors		X
Base	Conduct Current		X
	Mount Lamp		X
Glass Bulb	Exclude Oxygen		X
	Retain Argon		X
	Transmit Light and Heat		X
Argon Gas	Prevent Oxidation		X
	Retard Vaporization		X
	Transmit Light and Heat		X

FIGURE 7.2 Functions of components in a light bulb.

parts, many operational functions were missing. Note that the function of the stem press is not listed above. Although the Miles identification of functions as basic or secondary proved to be a useful technique, there really should be a way to include product and operational functions which illustrates their relationship. Fortunately, a method to do this has been developed.

FAST Diagrams

A questioning logic was developed by Charles W. Bytheway at Univac Salt Lake City in 1964[5] and presented at the 1965 Society of American Value Engineers'National Conference. Bytheway's *basic function determination logic*[6] allows functions to be ordered in a hierarchy based on cause and consequence. This questioning logic was the missing link needed to show the relationship of functions. It also identified important operational functions and included these with the functions provided by components. The function determination logic was named the function analysis system technique, or FAST.

The original set of nine questions in the Bytheway basic function determination logic are[5]

1. What subject or problem would you like to discuss _____?
2. What are you really trying to do when you _____?
3. What higher level function has caused _____ to come into being?
4. Why is it necessary to _____?
5. How is _____ actually accomplished or how is it proposed to be accomplished?
6. Does the method selected to _____ cause any supporting functions to come into being?
7. If you didn't have to _____, would you still have to perform the other functions?
8. When you _____, do apparent dependent functions come into existence as a result of the system conceived?
9. What or who actually _____ ?

The blanks are filled in with a function (verb–noun). The questions were designed to isolate the basic function, as well as identify higher-level functions. For example, questions 1, 2, 3 and 4 are used to identify higher-level logic or the functions that appear to the left of another function. Questions 4, 5, and 6 are used to develop the critical path, and questions 7, 8, and 9 are the basic function determination logic that identifies the dependence or independence of a list of functions with respect to each other.[7]

The questioning logic produced a major breakthrough in FA in that you no longer had to guess at the basic and secondary classifications. For the first time we can explain and diagram the interrelationship of the functions. Bytheway's original FAST diagram of a light bulb is illustrated in Figure 7.3. The light bulb diagram is in Figure 7.1.

Compare Bytheway's FAST diagram with the random function determination and classification in Figure 7.2. In the FAST diagram, we have structure, relationship, and a critical path. In the Miles function determination, there is some guess work in identifying basic and secondary functions. The classification can be argued and changed depending on a person's point of view. The traditional classification produced 14 functions, whereas the FAST diagram generated 21. Notice, too, that operational functions such as "provide air tight seals," "connect lead-in wires," and so on are included in this FAST diagram. FAST diagrams can be drawn in different ways. Another FAST diagram of a light bulb is shown in Figure 7.4.

Frank Wojciechowski[8] has done an excellent job tracing the evolution of variations of the FAST diagram. These variations resulted from specific needs for a particular application. According to Wojciechowski the first variation was developed by Wayne Ruggles in 1968. Ruggles' version of FAST is shown in Figure 7.4. Because of a tight schedule in teaching FAST diagraming in a VE course, he needed to shorten the process. So he took the Bytheway critical path logic questions (questions 4, 5, and 6) to form a simple critical path consisting of four or five functions. All other noncritical path functions were positioned above and below the critical path. Such functions were "all-the-time functions" and "at-the-same-time functions." Requirements and specifications such as "provide 1000 lumens" were also included (inside dashed boxes). Ruggles used two dashed scope lines to bound the study. Any

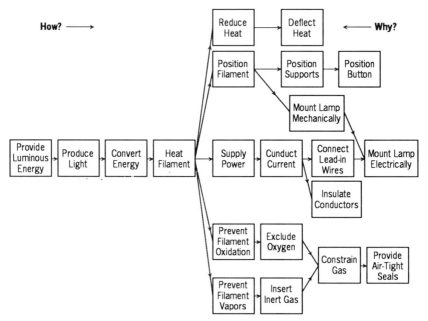

FIGURE 7.3 FAST diagram of light bulb (Bytheway). Reprinted by permission of the Society of American Value Engineers.

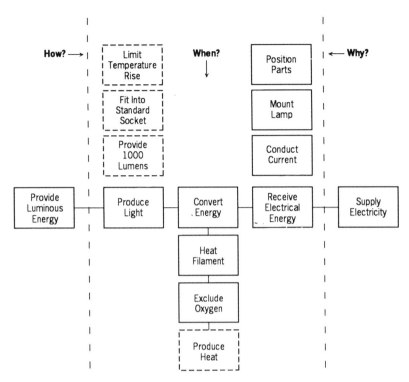

FIGURE 7.4 FAST diagram of light bulb (Ruggles). Reprinted by permission of the Society of American Value Engineers.

functions located outside the scope lines were not included as part of the VA/VE study. Scope lines soon became a permanent part of FAST diagrams.

In Figure 7.5 Wojciechowski[8] shows another form of FAST that was developed by Richard J. Park (then at Chrysler Corporation) and Frank Wojciechowski (then at Xerox Corporation). They, too, started with the how–why logic and the scope lines but put additional emphasis on branching if more than one function was needed to answer the how and/or why questions. The how–why logic was used to drive all paths from one scope line to the other.

Expanding the Park–Wojciechowski type of FAST diagram far enough to the right will eventually contain parts and components within the function description. Wojciechowski[8] illustrates how Ted Fowler and Tom Snodgrass produced yet another version that concentrated on the analysis of an entire product from the customer's or user's point of view. See Figure 7.6. For the first time, the customer and marketing were considered. Fowler and Snodgrass immediately branch out from a customer need, located outside the left scope line, to five generic customer-oriented functions: the basic function, assure convenience, assure dependability, attract user, and satisfy user. These are also expanded using the how–why logic. Notice that the Fowler–Snodgrass diagram considers the entire product and the customer interface

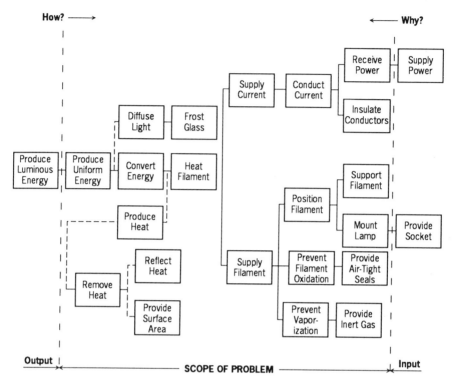

FIGURE 7.5 FAST diagram of light bulb (Park–Wojciechowski). Reprinted by permission of the Society of American Value Engineers.

where as the previous versions of FAST apply to a portion of the whole. Their interrelationship is shown in Figure 7.6.

Which version of FAST should you use? According to Wojciechowski this depends upon the product and the point of view of the client. The purpose of a VA/VE study could range from designing a new light bulb, designing a new filament-type light bulb, reducing the cost of the current light bulb design, to designing a new light bulb to improve market position. The original purpose of a FAST diagram was to stimulate creativity in a logical fashion. Today, they help facilitate design and manufacturing while providing insight into customer features and costs. It is interesting to note that the size and complexity of a FAST diagram is independent of the size of the object under study. FAST is used to understand a product not redesign it. We suggest the reader refer to the Wojciechowski reference for a detailed and enlightening account of the evolution of the FAST diagram.

We use the following process to create a FAST diagram:

1. Use verb–noun function descriptions to define the functions performed by the product and its components. Limit each description to one verb and one noun. Never use phrases! Write each function on a separate small card to facilitate construction of

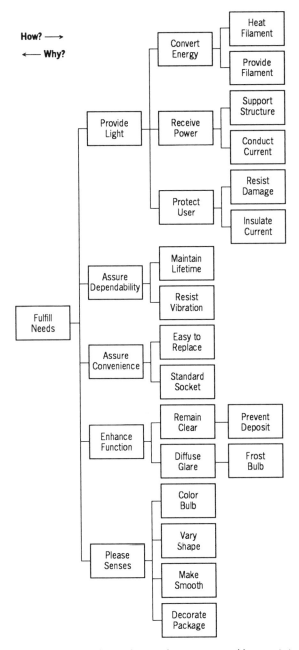

FIGURE 7.6 FAST diagram of light bulb (Fowler–Snodgrass). Reprinted by permission of the Society of American Value Engineers.

the diagram (We find Post-its work well). Place the cards on a surface where they are visible, easily accessible, and can be moved about. It is convenient to place a large sheet of paper on a wall and attach Post-its to the paper.

2. Create a logical relationship across the diagram by placing functions that answer how something is done to the right of a specific function and other functions that tell why something is done to the left of a specific function. Select from among the function descriptions the one that best describes the basic function of the product. Place this description to the far left of the diagram and create a branching tree structure from this point across the diagram to the right. Answering the how question results in branching and is repeated until branching has stopped and the function order is in a logical sequence. Post-its provide a convenient aid to arranging and laying out the order of the diagram.

3. The logic structure is verified in the reverse direction by asking why a function is performed. The how–why questions test the logic of the entire diagram. Answers to the how and why questions must make sense in both directions. Frequently, new function descriptions need to be constructed to accommodate the logic. Write these new descriptions on Post-its or cards and place them on the diagram.

4. After a diagram has been constructed, a *critical function path* should be portrayed on the diagram. This is done by selecting a function path composed of only essential functions. Many of the functions listed on the diagram may be of questionable value and may add unnecessary cost. They are prime candidates for improvement and innovation and should be placed outside the critical function path. The critical function path should contain only functions that must be performed to accomplish the basic function.

5. The limits of the diagram are specified by scope lines, which delineate the limits of the study. Functions outside the right scope lines are performed by the user. For example, when analyzing an ice cube tray, the FAST diagram might include a "freeze water" function but would not include how a freezer does this. A line entered on the right side of the diagram defines the limit of this and other how responsibilities. Similarly, the answer to why ice cubes are created may lie to the left of a scope line that limits the why logic of the diagram.

A FAST diagram may appear confusing and formidable to the beginner. An important advantage gained from constructing the diagram is the intensive questioning and penetrating analysis required for its development. In this respect, the construction process serves as an excellent learning and communications tool. A FAST diagram is not a sequential flow diagram related to time. It is a diagram that depicts the interrelationship of functions in a hierarchical order. The FAST diagram becomes more valuable and useful when the cost and importance of the functions are displayed on the diagram. For more detailed information on the construction and use of FAST diagrams, the reader is referred to References 9 to 15.

FAST diagrams can be used to describe a large variety of complex machines and procedures. The trick is to keep the diagram at an appropriate level of detail—not too large and not too small. It is wise to start with a general function description. This will prevent excessive detail and help prevent nit-picking. More detailed FAST diagrams can be created to describe sections of the general FAST diagram.

Figure 7.7 is a FAST diagram that includes cost data. It describes the function costs of a unique service provided by the U.S. Coast Guard that safeguards shipping in the North Atlantic Ocean.

After the ocean liner Titanic collided with an iceberg and sank, the International Ice Patrol was formed to monitor the position of icebergs that enter the North Atlantic shipping lanes. From late winter to mid summer a Coast Guard unit located in St. John's Newfoundland flies daily over the Davis Straight to locate and track icebergs entering the vital shipping lane between Europe and America. After locating icebergs the Coast Guard reports their positions by radio to any ships in the North Atlantic. Figures 7.7 and 7.8 show the cost of this operation in 1983.

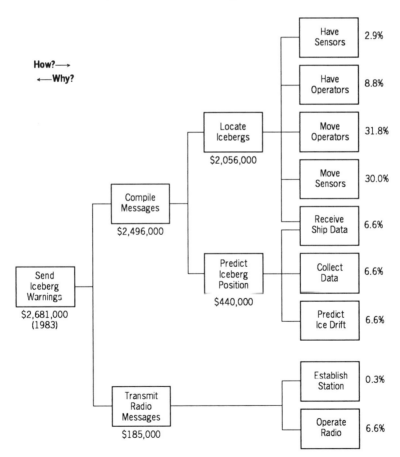

International Ice Patrol Cost by Function–1983

FIGURE 7.7 FAST cost diagram of International Ice Patrol Operation.

		Aircraft Personnel	Fuel	Aircraft Maintenance	Aircraft Operational Support	Office Personnel	Travel & Lodging	Leasing	Buoys	Radar Film	Misc.	Administration
Have Sensors	77							1	63	13		
Have Operators	235	235										
Move Operators	853	235	247	245	76		50					
Move Sensors	803	235	247	245	76							
Receive Ship Data	176					56						120
Collect Data	176					56						120
Predict Drift	176					56						120
Establish Stations	9							4		5		
Operate Transmitter	176					56						120
$2,681,000 ↳		705	494	490	152	224	50	5	63	13	5	480

FIGURE 7.8 Function cost matrix, International Ice Patrol.

Function Matrices

Function matrices can often be used as substitutes for part of a FAST diagram. A function matrix is a spreadsheet in which components, such as parts, subassemblies or assemblies, or process steps such as activities, tasks, and procedures, are arrayed against the functions they perform. Depending on the level of analysis, a matrix may array assemblies against their functions or parts against their functions. However, the matrix should show only one component level. A function matrix lists the components on one side of the matrix and functions on the other. These matrices are often used to collect, distribute, and display function costs. Component costs are distributed on the matrix at the intersection of a component and a function to indicate how much of a component's cost is allocated to a function's cost. A function matrix is a convenient device for displaying design duplication and redundancy. An example of a function cost matrix is shown in Figure 7.8. This matrix details the source and distribution of the costs shown in Figure 7.7.

Function Road Maps

A function road map differs in composition from a function matrix. A road map is created by listing a basic function flow down one side of the matrix and arraying against these functions alternative ways of accomplishing them. A road map of this type shows many potential ways in a which a product or service might be provided.

Road maps are extremely useful in the design phase of projects and can be used to plan for future products. Function road maps are discussed further in Chapter 9.

An example of a function road map is shown in Figure 7.9. The matrix in this figure shows alternative devices for constructing an optical ice detector. Matrices such as these are very useful in VE studies.

FUNCTION COST ANALYSIS

Function cost analysis answers the basic VE question, What does it cost? Function costs are derived by determining the cost of the items, components, or labor necessary to provide respective functions. Costs may consist of actual or "hard" costs, such as material and labor dollars. If available, hard costs should be used. They can be obtained by customary cost accounting methods. Figures 7.7 and 7.8 show how function costs are calculated and displayed from actual cost data.

In many cases, however, actual costs may not be available, and "soft" costs may be all that can be used. Soft costs are estimates of actual costs. Estimates of material and assembly costs can be derived by comparing the relative cost of different components using paired comparisons or other numerical rating techniques. Soft costs are often established for the components in a competitive product by comparing and rating them with the cost of vendor-supplied items. When soft costs are estimated, an individual assigns costs to items in proportion to their perceived relative costs. These ratings are then normalized to individual percentages by dividing each rating by the sum of all the items' ratings. When a group of people estimate cost, an average cost can be obtained for an item by averaging across all of the participants' ratings.

It should be noted that hard costs may also be normalized to percentages. Normalized costs are often used since they show individual costs as a porportion of the total overall cost of the product or system. This is especially useful when studying components of systems, procedures, or services.

Functions	Ice Detectors Possible Devices			
Provide Light	White Light	Infrared	Chromatic	Laser
Pulse Light	Oscillator	Chopper	Nicol Prism	Strobe
Transmit Light	Fiber Optics	Beam	Mirrors	
Filter Light	Colored Filters	Diffraction	Polarizing	No Filters
Focus Light	Normal Lens	Fiber Lens	Fresnel Lens	
Detect Light	Photocell	Diode		
Sense Signal	Light Absent	Light Detected		
Amplify Signal	Transistor	Amplifier	Operational Amp.	

FIGURE 7.9 Function road map for optical ice detector.

Occasionally, individual function or component percentages can be converted to dollars by multiplying them by the total dollar cost of the entire system. This is very useful for establishing target costs or goals for components and tasks of projects at the outset of a VE project. All costs, real or relative, may be posted to the respective functions on FAST diagrams, explosion diagrams, or similar function listings. This allocation process translates the traditional item–cost viewpoint to a systems' view of cost relationships, which is often quite revealing. The total function cost of basic or secondary functions can now be obtained by summing costs on the function diagram.

To illustrate this process, consider the compass shown in Figure 7.10. This compass consists of an assembly of nine components. The relative cost and importance of each component was determined using the pair comparison method. A value index (VI) was calculated for each part. These values are posted on the drawing. Note that the pin has the highest value index and the handle has the lowest. The pencil (25%)

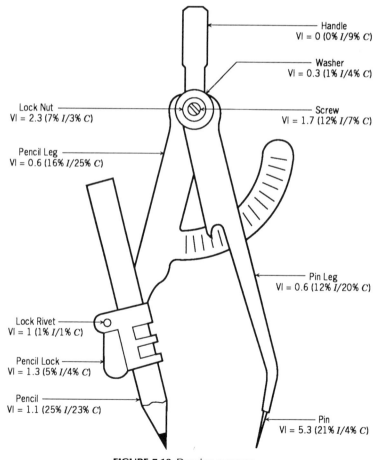

FIGURE 7.10 Drawing compass.

and the pin (21%) were judged most important, while the pencil (23%), pin leg (20%), and pencil leg (25%) were the most expensive components.

A FAST diagram was constructed and is shown in Figure 7.11. Scope lines are used to separate user functions from those of the compass itself. Note that on the right-hand side of Figure 7.11, individual compass components are listed directly to the right of the functions they provide. Because several components provide more than one function, they are posted more than once on the diagram. The contribution of these multifunctional parts to individual functions is indicated for each part. These estimates are expressed as fractions next to the name of each part. Each estimate is based on a subjective appraisal of the contribution of a part to a function, and these estimates are used to distribute cost and importance data on the drawing.

We have positioned the letters A to E in Figure 7.11 to help explain the distribution of cost and importance estimates on this diagram. Inside each function descriptor box, we have printed a fraction. The numerator represents the percentage importance and the denominator the percentage cost. This fraction is 25/23 at A, which denotes that when a pencil is used as the marker in this compass, it holds 25% of the importance and 23% of the cost of the compass (see estimates in Figure 7.10). This 25/23 fraction is posted in both the "use pencil" and the "have marker" function boxes. If a ballpoint pen were used in place of the pencil, its importance and cost would be posted inside of a "use pen" box and its fraction would also be posted inside of the "have marker" function box. Because the pencil's only function in this compass is to serve as a marker, all of its cost and importance data are posted here.

B designates the "have legs" function, which contains the fraction 18.8/30.5. This fraction contains data for both the pencil and pin legs of the compass. Referring to Figure 7.10, we find that the pencil leg has a 16/25 ratio of importance to cost, while the pin leg has a 12/20 value ratio. Both of these legs also provide other functions in the compass. The end of the pencil leg is bent and molded so it can "contain marker." Because we estimate that half of the cost of the leg contributes to this function [as indicated by the (1/2) posted after the pencil leg's name in Figure 7.11], we estimate that $(16 \times 0.5/25 \times 0.5)$, or 8/12.5 represents its contribution to the "have legs" and "contain marker" functions.[16] The pin leg has been modified slightly so it can "hold pin." We estimate that only 10% of its cost is involved in the "hold pin" function and 90% in the "have legs" function, so we post $(12 \times 0.9/20 \times 0.9)$ or 10.8/18 to the "have legs" function.[17] Now we add the cost contribution from each leg together to obtain a denominator of $(12.5 + 18) = 30.5$. In the same way, we add the importance contribution of each part together to obtain a numerator of 18.8.

Cost and importance is aggregated when posted to a FAST diagram. At C the set radius fraction (29.3/39.5) was obtained by adding together the importance and cost data from the "have legs," "angle legs," and "fasten legs" functions. Here the cost denominator of 39.5 is equal to 30.5 + 4 + 5. The importance numerator of 29.3 is equal to 18.8 + 1 + 9.5. Note that the set radius function, with a value index of 0.74, is a low value area in the design.

Sometimes the cost and importance of a single function contributes to several higher-order functions. This occurs when the "anchor center" function at D supports both the "rotate marker" and the "position marker" functions. In Figure 7.11, its

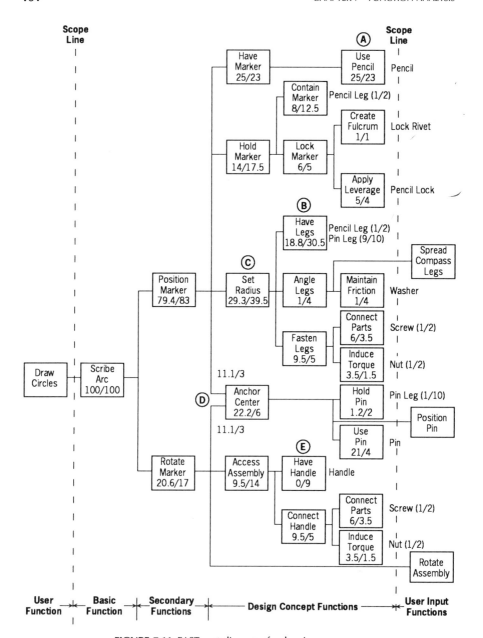

FIGURE 7.11 FAST cost diagram of a drawing compass.

favorable high value fraction of 22.2/6 is divided equally among these functions, contributing 11.1 units of importance and 3 units of cost to each of these higher functions. Note that while the "rotate marker" function has a favorable value ratio of 20.6/17, its value would be even higher if the handle was judged to be important. At E we find that the handle contributes 9% to the cost of the design while contributing 0% importance. This rating represents an insight that the handle is hard to use and could be replaced by a pencil in the design. A longer pencil is easy to hold and makes using the compass easier.

In a new design, a larger pencil is used to replace both the handle and the pencil leg. Two variations of this design are shown in Figure 7.12. Additional performance features are provided by including an eraser and changing the aesthetics of these designs. A study of school curricula indicates that compasses are first used by 8 to 10-year-olds. The compass on the right has been designed with them in mind.

DYSFUNCTION ANALYSIS

Dysfunction analysis is a technique for charting the ways in which problems are overcome in a machine or procedure. It is a new technique combining some of the

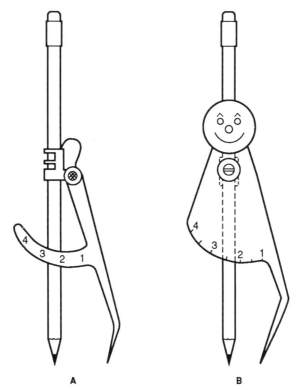

A   B

FIGURE 7.12 New compass designs.

features of function analysis with network analysis systems such as Program Evaluation Review Technique to allow designers to ascertain the cost of dysfunctions. In this procedure, the primary component in an item is identified along with its dysfunction.

Figure 7.13 is a dysfunction diagram for an incandescent light bulb.

Dysfunction Analysis Example

In this light bulb, the primary component is the filament that "produces light." However, the filament only "emits light" when it is heated. This dysfunction is overcome by passing an electric current through the filament. Lead-in wires connected to the filament carry electricity to the filament. Insulation prevents shorting of this current. When the filament is heated, it will oxidize and burn in air. This dysfunction is overcome by surrounding the filament with an inert gas. This gas is confined by a glass bulb and helps keep the filament from vaporizing. The gas transmits light and conducts heat away from the hot filament.

Notice that as components are added to a design to correct dysfunctions they often bring in new dysfunctions. These new dysfunctions must then be eliminated. Eventually a design is achieved where all of the most serious dysfunctions have been eliminated. Any remaining dysfunctions are accepted. These dysfunctions may subsequently be overcome by other products, for example, a light bulb manufacturer sells light bulbs with dysfunctions that are overcome by a lamp manufacturer.

To conduct a dysfunction analysis use the following procedure.

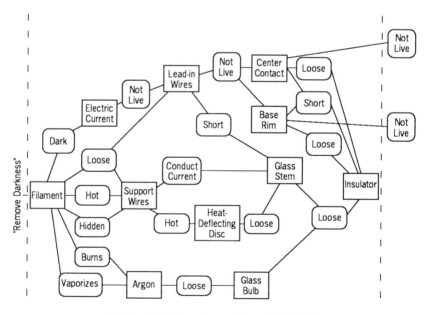

FIGURE 7.13 Dysfunction analysis of a light bulb.

1. Identify the primary dysfunction that a product or service needs to overcome.

A light bulb "removes darkness."

2. Identify a primary component in the product that overcomes this dysfunction. Use a noun to describe this component. Write the name of this component on a card or Post-it. Place this name to the left of a large piece of paper.

FILAMENT (light bulb example)

3. Identify any dysfunctions of the primary component. Describe them in one or two words (usually adjectives), and write the name of each dysfunction on a card or Post-it. Identify these dysfunctions by filling in the following statement. A dysfunction of the _____ (primary component) is that it is _____.

A dysfunction of the filament is that it is dark.

Place the names of these primary dysfunctions next to the name of the primary component. Draw lines from the primary component to each dysfunction (like spokes on a wheel).

Dark

Hot

Loose

4. For each dysfunction, write the name of a component or operation that will overcome it. Place these names next to each dysfunction and draw lines to connect the primary dysfunctions with the appropriate components or operations. Now a network of lines will start to branch outward from the original lines on the diagram.

Electric current

Wire supports

Inert gas (argon)

5. Identify any new dysfunctions these new components or operations bring into the design. Describe them in one or two words, and write the name of each dysfunction on a card or Post-it. Arrange these new dysfunctions next to the name of the appropriate component and draw lines from the components to the new dysfunctions.

6. For each new dysfunction, write the name of a component or operation required to directly overcome it. Add these new names on the outside of the diagram next to the appropriate dysfunctions. Draw lines to connect the components to their dysfunctions.

7. Repeat this process until the cause and cure of all of the dysfunctions in the product have been identified. A scope line can be drawn on the right-hand side of the diagram to indicate any dysfunctions that have not been eliminated by components in the design.

8. Add the cost of components and their assembly to the dysfunction diagram. If a component overcomes more than one dysfunction, distribute its cost proportional to your perception of its specific contribution to each dysfunction.

9. Sum costs along each dysfunction chain you are interested in. Do this by adding together the costs of all components and operations in the network paths that branch from the dysfunction.

ADJACENCY DIAGRAMS

The assembly of components into manufactured devices costs a great deal of money. It is common to find that costs involved with "holding parts" contribute 20% to 40% of product cost. Parts that are close together are often held together. A technique that examines the adjacency of parts can help identify parts that can be combined together. Parts consolidation will lower assembly and material costs and lower manufacturing costs. Products with fewer parts are often more reliable and easier to maintain. An adjacency diagram shows how parts are positioned in a product. Adjacency diagrams are a specialized form of dysfunctional analysis that examine the way parts are assembled in a product. They are relatively easy to create and can lead to simplified and less complex designs. The following procedure can be used to create an adjacency diagram. [Refer to Figure 7.14.]

1. Identify a primary component in an item or service. Write its name on a small card or Post-it.

In a compass, this would be the pencil.

2. Identify any parts in the design that come in contact with this part. Write these names on small cards or Post-its, and place them to the right of the primary component's name.

Pencil leg

Pencil lock

3. Now determine the component or components that physically contact these new components. Write these names on small cards or Post-its, and place them to the right of the secondary components' names.

Lock rivet (contacts the pencil lock and pencil arm)

Handle, pin arm, screw (contact the pencil arm)

4. Repeat this operation successively until all of the components involved in the design under study have been included. Enter the material cost of each component on its card.

5. Draw lines between the components that are in contact. Determine the assembly operations involved in joining each pair of parts, and enter the assembly costs on the lines connecting each component. You may want to indicate different kinds of assembly operations on the diagram.

Insert

Position

Close

6. Now sum up the assembly costs involved in the design. Post these costs on the diagram. Note that savings would accrue if several adjacent parts could be formed into a single piece.

7. Examine the diagram to see what parts might be combined. Determine how these parts might be combined (plastic molding, powder metallurgy, etc.). Prepare a list of adjacent parts that could be combined.

An adjacency diagram is not an assembly diagram. It does not show an actual sequence of manufacturing operations. Instead it shows the material placement of parts and the costs involved in joining parts together. Adjacency diagrams are a specialized form of dysfunction analysis devoted to analyzing the general dysfunction of assembly. Adjacency diagrams can be used to provide input to, or used in combination with, the design for assembly process devised by Boothroyd and Dewhurst.[18] Figure 7.14 is an adjacency diagram for the compass shown in Figure 7.10. For comparison, we include a PERT diagram that shows the assembly of the compass (Figure 7.15).

SUMMARY

In this chapter we described FA. Developed by Miles in the 1940s, it is the heart of VA/VE and was the breakthrough that broke the sound barrier of cost reduction. In the original version of FA a product and its components are described in two-word verb–noun abridgements. These verb–noun descriptions are further classified as basic or secondary functions. Creativity was applied to the functions to improve design and reduce cost.

This model worked well for a number of years. However, FA and classification does not show the relationship between functions. Also, because functions were generated from parts and components, many operational functions were missing. Another breakthrough, FAST diagramming, was developed that provided the missing link to identify and order functions in a hierarchy of cause and consequence. FAST removed the guess work in classification and provided an ordered, structured self-correcting procedure instead of a random function identification and classification. Value practitioners refined and varied the FAST process over the years. There

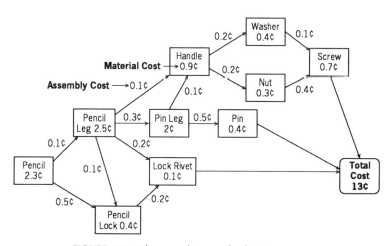

FIGURE 7.14 Adjacency diagram of a drawing compass.

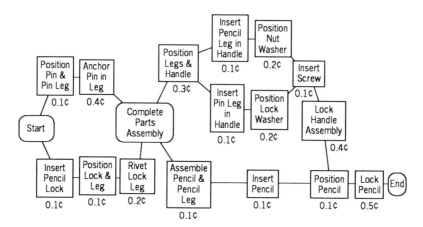

FIGURE 7.15 Drawing compass assembly diagram.

basically are four types of FAST diagrams, each appropriate for a particular application. FAST is used extensively today.

Another FA technique, the function matrix is a spread sheet that arrays parts, pieces, activities, and tasks against the functions they perform. This method is useful for collecting function costs and for displaying design duplication and redundancy. In addition, we also discussed the function road map that arrays operational functions against alternative ways of accomplishing them. It is useful in the design phase of projects as well as product planning.

Dysfunction analysis is a hybrid of FA and PERT. It is used to surface problem areas in products. It allows designers to determine the cost of dysfunctions and provides the structure for charting possible solutions to those problems.

Adjacency diagrams are described that allow a team to identify redundancy in a design and provides the basis for part and/or function consolidation. This methodology is a variation of dysfunction analysis that examines the way products are assembled with a view toward cost reduction, increased reliability, and less maintenance.

In summary, FA is a technique for analyzing the performance and usefulness of a product. It is a backbone of VA/VE. Functional analysis is used to understand a product by providing a language and a structure to increase communication across multiple disciplines. It is vehicle for dialogue that provides a common understanding and a template for improving the design and value of a product.

REFERENCES

1. Miles, L. D., *Techniques of Value Analysis and Engineering*, McGraw-Hill Book Company Inc., New York, 1961.
2. Jarboe, R. R., and Ferguson, J. E., *Function Analysis*, Department of Engineering, University of Wisconsin-Extension, Madison, 1977.

3. Park, R. J., *Park's Catalogue of Frequently Used Functions in Value Engineering*, R. J. Park and Associates, Birmingham, Mich., 1982.

4. Kaufman, J. J., *Value Engineering for the Practitioner*, North Carolina State University, School of Engineering, Raleigh, 1985.

5. Bytheway, C. W., "FAST Diagrams for Creative Analysis," *Journal of Value Engineering*, pp. 6–10, March 1971.

6. Bytheway, C. W., "Basic Function Determination Logic," *Proceedings, Society of American Value Engineers*, pp. 38.1–38.21, 1965.

7. Bytheway, "FAST Diagrams for Creative Analysis."

8. Wojciechowski, F. X., "The Various Types and Uses of the FAST Diagram," *Proceedings, Society of American Value Engineers*, **13,** 153–159, 1978.

9. Bytheway, "FAST Diagrams for Creative Analysis."

10. Bytheway, C. W., "Innovation to FAST," *Proceedings, SAVE Regional Conference*, Detroit, 1972, pp. 6.1–6.7.

11. Cook, T. F., "Function Analysis Systems Technique, Task Oriented FAST Diagram," *Value World* **3**(2), 24–28, 1979.

12. Creasey, R., *FAST Manual*, Value Design Press, Ft. Worth, Tex., July 1973.

13. Ronen, E., "Functional Analysis of Procedures and Organizational Structures," *Performance* **5**(6), 12–17, 1975.

14. Fowlkes, J. K., Ruggles, W. F., and Groothuis, J. D., "Advanced FAST Diagramming," *Proceedings, Society of American Value Engineers*, 7, 45–52, June 1972.

15. Wojcieshowski, F. X., "FAST Diagram—Its Many Uses,"*Proceedings, SAVE Regional Conference*, Detroit, 1972, pp. 10.1–10.4

16. Note that we posted 8/12.5 to the "contain marker" function.

17. The remaining 1.2/2 is posted to the "hold pin" function.

18. Boothroyd, G., and Dewhurst, P., *Design for Assembly*, Department of Mechanical Engineering, University of Massachusetts, Amherst, 1983.

8

QUALITY FUNCTION DEPLOYMENT: THE TOTAL PRODUCT CONCEPT

M. Larry Shillito

APPLYING VM TO TOTAL PRODUCT DESIGN

Historically and traditionally, VM applications have focused on use value and less on exchange and esteem value.[1] A review of the literature and past applications reveals that the thrust of a VM study was/is to reduce costs primarily in manufacturing. Regarding the product itself, VM effort was/is applied to parts and assemblies within the product by engineering and manufacturing personnel. Little, if ever, was VM originally applied to the entire product.

Function analysis including FAST diagrams also concentrated primarily on use functions. Use functions are known as work functions when used to refer to the entire product in a value improvement capacity. They are functions that are essential to the customer performing a desired task. They are the essence of how a product works. However, the primary thrust of VM applications was cost reduction in product design and manufacturing. With the work function, part-piece orientation, the application of VM quickly evolved into the now popular and still highly successful 40-hour Value Analysis Workshop and Training Seminar. This format was developed in 1952 by Larry Miles and Roy Fountain while at General Electric.[2]

The work function cost-reduction approach worked nicely, because it fit into a 40-hr time frame and could serve as both training and real-time application. Consequently, this approach was used most widely and blindly progressed until today. However, there has been a conspicuous absence of customer input and attention to customer/user needs and wants.

Because the applications were internal in nature and dealt primarily with manufacturing, little or no attention was given to the sell functions pertaining to exchange and esteem value. *Sell functions* are defined and categorized by Tom Snodgrass of the

172

University of Wisconsin, Madison and President of Value Standards Inc.[3] as "attract user," "satisfy user," "assure convenience," and "assure dependability."

The sell functions were traditionally left to marketing personnel because they were highly subjective and pertained to customer wants and delights.[1] Manufacturing and design people talk in terms of measurable needs, whereas marketing people talk in terms of subjective needs, which are nonmeasurable. Engineers and manufacturing people are eager to reduce costs in design and manufacturing. After all, their performance appraisals are based on cost reduction and problem solving. They receive little recognition for working on marketing issues and customer delights. This encourages polarization of manufacturing and marketing personnel, which further weakens communications. Unfortunately, this results in a paradox. According to Tom Cook,[4] "VA Teams remove much of the manufacturing cost while sales and marketing add features to the product. Thus total product cost could rise despite the best of VA results."

The traditional work function approach works well for cost reduction and for working or studying components or subassemblies within a product. However, it does not necessarily always improve overall product value. A customer purchases and equates value to an entire product. The goal of a VM study should be not only to reduce cost but to increase product value. When the emphasis is on improving value, the sell functions readily become very important. Cook[5] has found repeatedly that there is no direct relationship between cost to manufacture and the selling price. If a product has high user value the customer is usually willing to pay a higher price. Wasserman[6] also notes that "buyer needs are expressed as use value and are satisfied by work functions. Buyer wants are expressed by sell functions." A customer/user considers both the work functions and the sell functions when buying a product to accomplish a task. It is work functions that allow the customer to accomplish the task. These functions are taken for granted by the customer. All competitors' products must have these functions. There is some performance variation across competitors' products for the work functions. The real competition in products lies in the sell functions. The producer who has a performance edge in work functions and who also provides the most attractive sell functions will have a distinct advantage and will attract more customers.

Work functions are generally considered the price of admission to the market. They have to be there, and they better work well. The sell functions, on the other hand, become the perks that attract the user and give the producer market leverage. We sometimes refer to these as *perk functions*. Some exceptions are safety and reliability, where, again, the user assumes the product is safe and reliable. The producer gets no credit for these but the customer is unhappy and concerned when they are not there.

The work function–sell function balance is further compounded by the fact that it is a natural tendency for producers to improve those functions and features that *they* believe are important to the customer and that *they* are especially good at in manufacturing. This can result in overdesigned products and unfulfilled user needs. Many products on the market support this honest, but wrong, belief of the producer.

So what is the best balance between work functions and sell functions? It is this very situation that gave birth to customer oriented value engineering (COVE) described by Bryant[1] and the process developed by Snodgrass and Kasi called customer-oriented product design (COPD).[3] Although most of the tools are the same in traditional VA/VE and COVE, the scoping and emphasis are different. With COVE, management generally selects the product to be studied. An interdiscipline team determines the scope of the project based on considering the entire product and using life cycle costs. The customer emphasis also requires adding marketing, sales, and customer service personnel to the team.

Next, the team determines the buyer, the user and/or the chief buying influence. That is, who is it that we are trying to please when designing and selling the product? Features and characteristics are documented using surveys, panels, focus groups, and internal company personnel. This activity sounds easier than it is. How are customer likes and dislikes measured? The VM team could determine what is needed in the marketplace, but they have only limited knowledge. Market research can be used, but caution must be exercised in sample size and population selection. Here, too, the respondents when answering questions about product features many times have never used the product and, worse yet, most likely have not gone through the anxiety of the buy decision. The best way to get customer feedback, for those products already on the market, is to ask customers who have already purchased the product and who have used it for some time. This part of the COVE process elicits what the customer likes as well as faults and complaints. Customer-oriented brainstorming using internal company personnel has also been used. Caution should be used in this approach because it does not always truly represent the customer, which in turn, can weaken credibility of the study. Internal Guru's have a greater chance of being wrong because there is a spectrum of customers whose preferences change with time and because they are also producer biased.

The importance of the features/characteristics are evaluated and their performance range, or limits of acceptability, documented where appropriate. The actual definition of the features/characteristics is just as important as their quantification and are many times more difficult to establish. Benchmarking with competitor's products is also appropriate. Competitor cost comparisons are often performed. The approach is to determine what it would cost to produce the competitor's product in your plant with your labor, material, and manufacturing methods and not what it would cost the competitor to build it in his plant. Finally, the current cost of the features and characteristics are posted to the functions to which they pertain. The task-oriented FAST diagram was designed for COVE and COPD applications.[7,8] The function costs are used to locate areas of value mismatch. A value mismatch is an imbalance of cost or performance between work functions and sell functions. The design and specifications are not matched to the needs of the marketplace. The result is usually overdesign or underdesign. These value mismatches become the objectives that enter into the speculative phase of a more traditional VA/VE study to reduce costs and improve product performance.

There are some excellent resources on COVE such as Cook,[4,5] Bryant,[1] Snodgrass and Kasi,[3] and Jarboe and Ferguson.[9] Courses and a diploma program developed by Tom Snodgrass are also available at the University of Wisconsin—Extension, Department of Engineering and Applied Science.[10]

We will not dwell on COVE and COPD, because they have been around for some time and the previous references have covered the subject well.

We chose to go beyond these methods. For example, the customer-oriented application of VM has been expanded and applied in many different capacities such as product concepting,[11] competitive analysis,[12] forecasting future technology,[13] as well as to sales.[6] The most recent and probably the most significant customer-oriented value-type methodology to appear has been Quality Function Deployment, usually referred to as QFD.[14-21]

QUALITY FUNCTION DEPLOYMENT

History

Quality Function Deployment (QFD) originated in Japan and was first introduced at the Kobe shipyards of Mitsubishi Heavy Industries Ltd. around 1972. Its application has since proliferated in Japan, but its use is still not universal. The majority of QFD applications are centered in the high-tech and transportation industries and are applied primarily in those companies whose products represent a significant part of Japan's export business.[22]

Because quality circles were already well established in companies and because employees were well versed in statistical quality techniques, the groundwork was laid nicely for introducing QFD for developing a competitive advantage in quality, cost, development cycle time, and delivery.

The U.S. exposure to QFD was in 1983 through an article in *Quality Progress* by Kogure and Akao[17] and through Ford Motor Company and the Cambridge Corporation, an international management consulting firm.[13] The interest and application of QFD in the United States is growing at an incredible rate, despite its brief history.

The two major training sources in QFD are GOAL/QPC in Methuen, Massachusetts, and the American Supplier Institute in Dearborn, Michigan. Both sources developed their own but similar QFD models. The ASI uses a basic four-matrix method developed by Macabe, a Japanese reliability engineer. GOAL/QPC advocates a multiple matrix method developed by Akao and incorporates many disciplines (including VE) in a less structured format consisting of a matrix of matrices. Akao has collected the multiple matrix applications from many Japanese practitioners and assembled them into a new book, *Quality Function Deployment*.[19]

Overview

Before describing the QFD process in detail it is useful to briefly describe QFD. Quality function deployment is an interdiscipline team process to plan and design new or improved products or services in a way that:

1. Focuses on customer requirements,
2. Uses competitive environment and marketing potential to prioritize design goals,

3. Uses and strengthens interfunctional team work,
4. Provides flexible easy-to-assimilate documentation, and
5. Translates soft customer requirements into measurable goals, so that the right products and services are introduced to market faster and correctly the first time.

The QFD methodology consists of a structured multiple matrix-driven process to:

1. Translate customer requirements into engineering or design requirements,
2. Translate engineering or design requirements into product or part characteristics,
3. Translate product or part characteristics into manufacturing operations,
4. Translate manufacturing operations into specific operations and controls.

Based on the American Supplier Institute model,[16] the translation mechanism is a series of four connected matrices (see Figure 8.1).

The QFD process starts with the needs of the customer and applies them to the entire product life cycle of conceiving, developing, planning, and producing a product or service. Customer requirements are most often referred to as *the voice of the customer*[16] because the requirements are expressed in the customer's own words and, in many cases, are written as direct quotes. The process is an attempt to preserve the customer demands by "insulating them against constant reinterpretation throughout the product development cycle and providing a frame work against which future changes can be measured."[22,p. 2]

Preparation

Before describing the construction and use of the QFD matrices, it is necessary to describe three basic structural techniques used to analyze and structure qualitative data. These tools are used to build a matrix of customer information and product features and measures. They are part of a set of seven QC tools developed in Japan

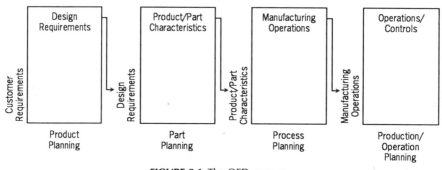

FIGURE 8.1 The QFD process.

by the Society of QC Techniques Development and are known as the *seven new tools.*[23]

The first tool is called an *affinity diagram.* It is used to gather large amounts of qualitative data to organize it into subgroupings based on similarities between items. For example, the QFD team collects and/or generates the needs of the customer for product X. Each individual item (customer need) is written on an index card or Post-it. All cards are laid at random on a table or the Post-its are stuck on a wall or chart board. At this time, the items are unstructured because the data were generated in a random stream-of-consciousness manner. Based on intuition and gut feeling the items are first paired together based on similar attributes and then further aggregated by the team into larger clusters that represent a common theme. Generally items group into five to ten main clusters that each contain 1 to 15 items. The 5 to 10 clusters can be further aggregated many times to form a three- or four-level hierarchy.

To illustrate the affinity diagram we will apply the process to the customer needs of a candle. Based on market surveys, interviews, and/or internal brainstorming, individual customer needs are generated by team members and written on cards (Figure 8.2). The number of items is considerably reduced for this example. Normally there may be 20 to 80 items. Cards were then rearranged by team members (Figure 8.3). For example, the team decided that "be visually attractive" and "be fragrant" were related to a general theme of aesthetics. In like manner, the other four customers needs were grouped. Note that the team did not develop categories first and then assign the items to the categories. It is more creative and more information-rich to cluster items first and then assign a heading to the cluster. The affinity diagram forces organization, fosters a general level of understanding, and surfaces hidden relationships.

The clusters of items from the affinity diagram can then be arranged horizontally into a second tool called a *tree diagram* (Figure 8.4). That is, the affinity diagram, which is based on intuition and gut feeling, is used to construct a tree diagram based more on logic and analytical skills. The various branching levels of the tree diagram are used to search for gaps and omissions in the affinity diagram. For example, teams often discover new items that were missed during the first generation of needs, as well as uncover new branches and regroupings. The tree diagram allows the team to add, expand, and elaborate on the voice of the customer to form a more complete structure based on interrelationships.

The same identical procedures were used to generate product features and measures also called *the voice of the company.* These measures were also arranged into a tree diagram. The two trees were then arrayed at a right angle to each other, so that a *matrix diagram* could be formed based on the third (or lowest) level items of the two trees (Figure 8.5). The matrix, the third tool, very conveniently allows items in

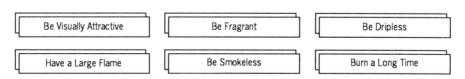

FIGURE 8.2 Candle user needs.

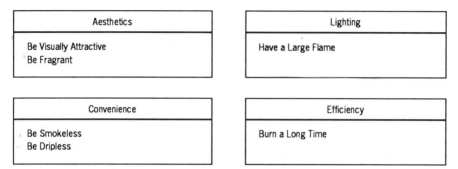

FIGURE 8.3 Affinity diagram of customer needs.

one dimension (e.g., voice of the customer) to be mapped against items in another dimension (voice of the company). The matrix provides a structure to systematically evaluate the relationship between the items in both dimensions. Each cell of the matrix provides the basis for asking questions about the relationship between customer needs and product characteristics that may not have been obvious before.

Any relationship between rows and columns may be coded by symbols such as those used by the Japanese. That is, a double circle indicates a strong relationship usually given a score of 9, a single circle indicates a moderate relationship, given a score of 3, a triangle indicates a possible or low relationship, given a score of one, and a blank cell represents no relationship.

The intersection of the two trees to form a matrix serves as the basis for constructing the first QFD matrix, termed the house of quality.

The House of Quality

The house of quality (HOQ) is the nerve center and the engine that drives the entire QFD process. It is, according to Hauser, "a kind of conceptual map that provides the

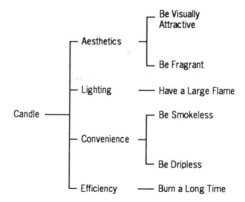

FIGURE 8.4 Tree diagram of customer needs.

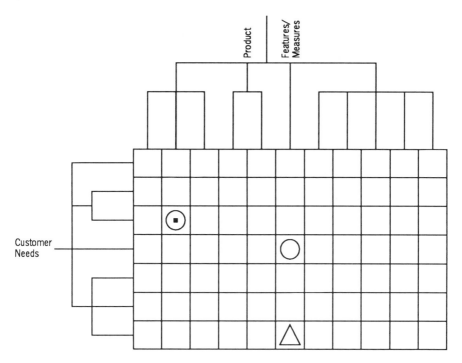

FIGURE 8.5 Matrix diagram.

means for interfunctional planning and communications"[18,p. 63]. The HOQ is a large matrix that contains seven different elements. (See Figure 8.6)
The seven elements in the HOQ are the following.

1. *Customer Needs.* These are the voice of the customer. They are also known as *customer attributes,* or *customer requirements.* Very often they are generated by the affinity diagram and the tree diagram. Examples are for a car door, "easy to open;" for a bank, "no waiting in lines;" for a lawnmower, "easy to start."

2. *Product Features.* These are also called *design requirements, engineering attributes,* or *characteristics.* They, too, can be developed using the affinity diagram and tree diagram. The product features will become the measure to determine how well we satisfy the customer needs. That is, marketing tells us what to do and the engineers/designers tell us how to do it. Examples are for a car door, energy (in ft/lbs) to close door; for a bank, computer downtime frequency; for a lawnmower, pulling force to rotate shaft. Product features must be stated in measurable and benchmarkable terms. Many teams use the customer needs to generate the measurable characteristics.

3. *Importance of Customer Needs.* Not only do we need to know what the customer wants, but also how important those needs are.

4. *Planning Matrix.* This portion of the HOQ contains a competitive analysis of the company's product with major competitor's products for each customer need.

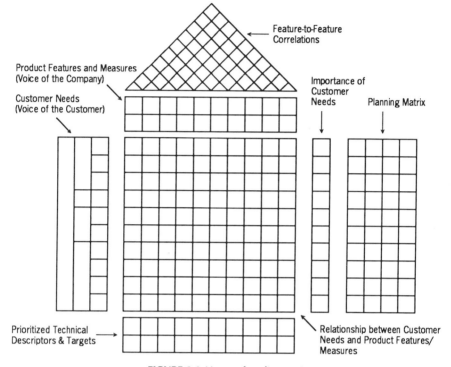

FIGURE 8.6 House of quality matrix.

There are columns to judge how much improvement is needed in the current product, how much sales leverage may result from the improvements, and a final overall score for each customer need. Each score is factored by customer importance, needed improvement and sales leverage.

5. *Relationship Between Customer Needs and Product Features.* The interfunctional team uses the body of the matrix to indicate how much each product feature (engineering characteristic) affects each customer need.

6. *Feature-to-Feature Correlation.* To what extent will a change in one feature affect other features? Too often a magnitude change in one feature results in a negative affect on another feature. This correlation allows one to identify such impacts. It is in effect a potential problem analysis.

7. *Prioritized Technical Descriptors or Targets.* This section is a summation of the affects of all prior variables on each product feature. It may also contain target measures for engineering characteristics as well as a competitive analysis of other manufacturer's measures for the same variables.

Using these seven elements, the HOQ becomes a repository of information that can be used as a mechanism for applying common sense engineering.

Starting the QFD Process—Building the House of Quality

There is nothing magical about the HOQ. It is not difficult to construct. It does require a fair amount of team time, which may induce anxiety on part of the team members and be perceived as a heavy front-end load. As the team gets into the process they become committed to completing it. Let us now work through the steps in building the HOQ using Figure 8.7 as a reference. The example product will again be the candle used earlier.

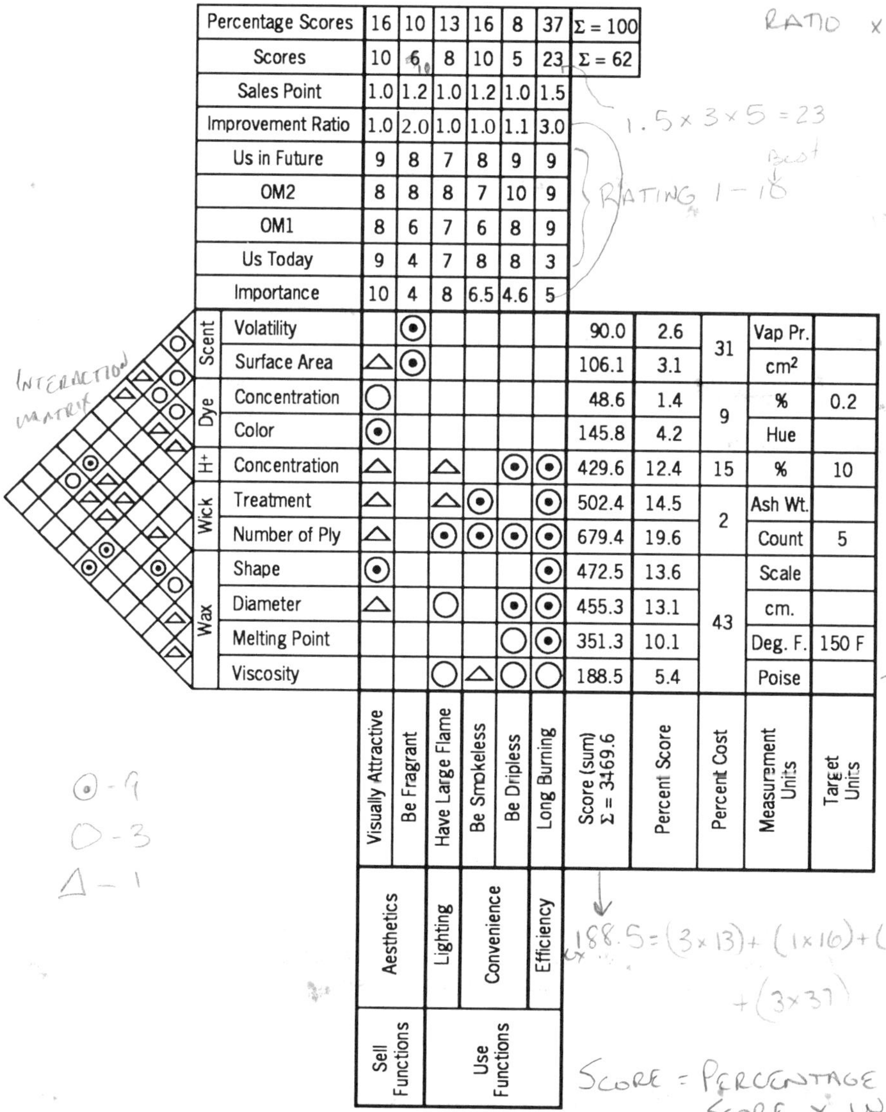

(Handwritten annotations around figure:)

SCORE – IMPORTANCE × IMPROVEMENT RATIO × SALES POINT

$1.5 \times 3 \times 5 = 23$

Best ↓ RATING 1 – 10

Percentage Scores	16	10	13	16	8	37	Σ = 100
Scores	10	6	8	10	5	23	Σ = 62
Sales Point	1.0	1.2	1.0	1.2	1.0	1.5	
Improvement Ratio	1.0	2.0	1.0	1.0	1.1	3.0	
Us in Future	9	8	7	8	9	9	
OM2	8	8	8	7	10	9	
OM1	8	6	7	6	8	9	
Us Today	9	4	7	8	8	3	
Importance	10	4	8	6.5	4.6	5	

INTERACTION MATRIX

Category	Characteristic	Score (sum)	Percent Score	Percent Cost	Measurement Units	Target Units
Scent	Volatility	90.0	2.6	31	Vap Pr.	
Scent	Surface Area	106.1	3.1		cm²	
Dye	Concentration	48.6	1.4	9	%	0.2
Dye	Color	145.8	4.2		Hue	
Wick	Concentration	429.6	12.4	15	%	10
Wick	Treatment	502.4	14.5	2	Ash Wt.	
Wick	Number of Ply	679.4	19.6		Count	5
Wax	Shape	472.5	13.6		Scale	
Wax	Diameter	455.3	13.1	43	cm.	
Wax	Melting Point	351.3	10.1		Deg. F.	150 F
Wax	Viscosity	188.5	5.4		Poise	

Bottom column headers:
Visually Attractive; Be Fragrant; Have Large Flame; Be Smokeless; Be Dripless; Long Burning; Score (sum) Σ = 3469.6; Percent Score; Percent Cost; Measurement Units; Target Units

Sell Functions: Aesthetics, Lighting
Use Functions: Convenience, Efficiency

(Handwritten annotations:)

⊙ – 9
○ – 3
△ – 1

$188.5 = (3 \times 13) + (1 \times 16) + (8 \times 3) + (3 \times 37)$

SCORE = PERCENTAGE SCORE × WEIGHT GIVEN

FIGURE 8.7 House of quality matrix for a candle.

1. *Customer Needs/Wants.* Using the various collection methods mentioned previously, the team collects data about what the customers want. A hierarchy of customer needs is constructed from this data using the affinity diagram and the tree diagram described earlier.

2. *Importance of Customer Needs.* Customer needs are rated by the team for importance using a rating scale of 1 to 5, where 5 represents very important and 1 signifies unimportant. Many teams also use a 1 to 10 scale. The basic question being answered is: How important is this feature to the user? Rating the customer needs creates considerable constructive dialogue among team members. Many viewpoints are expressed and many notes are taken. Most of the literature on QFD and the HOQ states that a consensus must be reached among team members. This is great but could consume considerable time. Sometimes a majority decision with a minority report is a good compromise. When rating importance, the team must establish whose point of view the ratings represent; which market, segment, user, or purchaser.

3. *Competitive Analysis.* The company's current candle product is rated against each customer need, using a 1 to 5 or a 1 to 10 scale. The basic question being answered here is: How well does this manufacturer satisfy the listed feature(s)? That is, how well does our current candle meet the customer's needs. In Figure 8.7 this is represented in the column labeled "us today." Next, the competitor's candle products are rated using the same scale (columns labeled "OM1" and "OM2").

4. *Future Goal.* Using the same rating scale as above, the team again rates the company candle as to where they desire to be in the future with respect to each customer need (example column "us in future"). The question addressed when doing so is, How well do we want to satisfy customer needs if we offer the candle we are now studying? In the example, visual attractiveness, with a score of 9, is currently being met with our product and is also in a strong position with other manufacturer 1(OM1) and other manufacturer 2 (OM2) products. There is no need to change our future candle on this customer need. The need to be dripless also satisfies the customer need but is in a weaker position compared to OM2. Therefore, we would like to be a 9 in the future. There is no need to be a 10 because the team believed that competitor OM2 exceeded the customer need. That is, improving our candle to a 10 on this customer need would probably result in overdesign. However, for the need be long burning, our current product does not satisfy the customer need. Our candle burns too fast and the competitor's candles burn considerably longer. We should make a significant improvement on this customer need. The team decided to be a 9 on this feature so we can be as good as our competitors. They also believed that a 10 would be unnecessary. Likewise, the fragrance of our candle is less than the competitors, so a score of 8 was selected for the future.

5. *Improvement Ratio.* An improvement ratio is computed by dividing the future goal rating (us future) by the current rating (us today). From our example, fragrance has an improvement ratio of 2 and long burning has a 3. These are significant changes from the current candle design.

6. *Sales Point.* The column labeled sales point gives marketing a chance to rate whether or not they would get leverage out of any improvements. That is, would the

improvements influence a sale? Sales point is rated as 1.5 indicating a strong or significant sales point, 1.2 denoting a moderate sales point or 1.0 indicating status quo or no additional sales impact. The basic question being addressed here is: Given the importance of this feature to the customer, and considering the magnitude of the improvement ratio, if we in fact make a change in this feature, can you, marketing get some leverage from such a change?

7. *Score.* After completing the quantifications in steps 2 through 6, a customer score is computed for each individual customer need by multiplying customer importance, improvement ratio and sales point. The products of these three numbers provide a hierarchy of customer needs based on the team's assessment of the three variables. These raw scores are finally normalized to a percentage (column labeled "% Score") which will later be used as weighting factors. Although the HOQ is not finished at this point, the information and communication that has resulted so far is more than many companies normally have had. Even stopping here would be time well spent.

8. *Product Features/Engineering Characteristics.* Using team brainstorming and expertise, a list of product features/engineering characteristics is generated. These, too, are arranged into a hierarchy using the affinity diagram and tree diagram. At this point, it is advisable to develop a glossary of terms for each feature. Doing so will make it easier for the team to communicate. Many teams also develop a glossary for customer needs as well.

9. *Customer Needs–Product Features Relationship.* Product features are entered from the tree diagram as columns adjacent to the customer needs. This creates a relationship matrix (Figures 8.6 and 8.7). For each cell of the relationship matrix, the team estimates whether or not there is a relationship between the column and the row. The question addressed is: Will this product feature/engineering characteristic have an affect on satisfying the customer need? The amount of relationship is represented by the following symbols: A double circle indicates a strong relationship and implies a numerical score of 9; a single circle represents a moderate relationship and uses a score of 3; a triangle signifies a low or possible relationship and is given a score of 1; a blank cell signifies no relationship at all and implies a score of zero. Symbols can be substituted with numbers depending on the preference of the team.

10. *Product Feature/Engineering Characteristic (PFEC) Relationship Score.* In each cell containing a numerical relationship, the relationship score (1, 3, or 9) is multiplied by the corresponding percent score for the corresponding customer need. In our candle example, the cell representing visual attractiveness and shape contains a product of $16.2 \times 9 = 145.8$. Likewise, the other relationship under the column shape is multiplied by the other corresponding percent score. A final feature (column) score is calculated by summing all the products in the column. The column totals represent a rank order of PFEC's weighted by customer needs. They indicate how much influence the PFEC's have on meeting customer needs.

11. *PFEC Correlations.* We are now ready to complete the final part of the HOQ. It is known as the roof because of its triangular shape atop the customer need–product feature relationship matrix. (See Figures 8.6 and 8.7). The roof matrix allows the team

to identify and quantify the impact, if any, that a change to one PFEC may have on other PFECs. Each cell in one row of the roof matrix represents the intersection of one PFEC with every other PFEC represented in the column headings.

As with the relationship matrix in the body of the house, symbols are entered into those cells where a correlation between two PFECs has been identified. Again, a double circle, represents a strong correlation, a single circle signifies a moderate correlation, and a triangle denotes a possible correlation. The question being asked when addressing each cell in the roof matrix is, Will a change in the specifications of one PFEC affect the specifications of the other PFEC in the pair? This is equivalent to a cross-impact analysis of PFECs to determine any potential problems that may arise if significant changes are made to any one PFEC. Correlations may also be negative as well. Consequently, different symbols may be used accordingly.

Any cell identified with a high correlation is a strong signal to the team, and especially to engineering, that significant communication and coordination are a must if any changes are contemplated in a targeted PFEC. Sometimes an identified change impairs so many others that it is advisable to leave it alone. In our candle, any change in the treatment of the wick will have little affect on any of the other PFECs except for a slight interaction with the number of ply. This is ideal because it is a high scoring PFEC, has little interaction with other PFECs and relates to four customer needs. It is also a low cost PFEC (discussed later) so any small amount of money spent to change the treatment may benefit four customer needs with little adverse interactions. However, a change in the diameter of the wax can affect six other PFECs. So even though it is a high scoring PFEC that can influence four customer needs caution should be exercised before any changes are made. Finally, number of ply can significantly affect four customer needs and slightly affect a fifth. It also has a strong interaction with diameter and some interaction with treatment. It too is a likely candidate for change.

The 11 steps complete the construction of the HOQ. However, the literature abounds with many variations. Each company modifies the HOQ process to fit its individual needs. For example, in our candle example, we have added additional information below the PFEC column percent scores. Either real cost or percent cost is added. In our example, cost was aggregated by component.

The current units of measure of each PFEC may be listed. That is, melting point is measured in degrees Fahrenheit, diameter is measured in centimeters, and so forth. However, how does one measure treatment or type of scent? These PFECs may have to be reworded so that a measure can be used. Remember, earlier we mentioned that PFECs should be measurable and benchmarkable. Target values for these measures have also been added. Target measures may also be compared to competitor product and so forth.

Using the HOQ

Now that the HOQ is built, what do you do with it? The HOQ is a structured communications device. Obviously it is design oriented and serves as a valuable resource for designers. However, engineers may use it as a way to summarize and convert data into information. Marketing benefits from it because it represents the voice of the customer. Upper management, strategic planners, and marketing or technical intelligence can use it to surface strategic opportunities. Again, the HOQ serves as a vehicle for dialogue to strengthen vertical and horizontal communications. Issues are addressed that may never have been surfaced before. The HOQ, through customer needs and competitive analysis, helps to identify the critical technical components that require change. The critical technical issues will then be driven through the other matrices to identify the critical parts, manufacturing operations and quality control measures to produce the product to fulfill both customer needs, and producer needs within a shorter development cycle time.

Other "Houses" On the Block of QFD

The same procedures used to build the HOQ may be used to construct other HOQ-type matrices to relate other interconnected variables. The HOQ data can be used to connect other matrices to determine the right parts, the right manufacturing process, the assembly operations, the right quality variables to measure along with the right statistical quality control techniques for measurement. Ultimately, the interconnected matrices lead to the right production plan to build the product.

To do this, the critical columns (PFECs) of the HOQ become the rows of a second matrix that has product/part characteristics as columns. Note that not every column is transferred to the second matrix. Only the issues critical to the success of the product are brought forward. The same procedures used in the HOQ are used to qualify and quantify this new matrix. The PFECs (rows) are used to generate relevant parts characteristics (columns). The purposes of this matrix are to select the right design concepts by determining the critical parts and characteristics. The deliverable emanating from this matrix provides critical data for parts characteristics.

Functional product requirements are often derived by using FAST diagrams. FAST diagrams are expanded far enough to produce function characteristics. Product parts can then be assigned to the function blocks of the FAST diagram. These data can then be entered in this second matrix along with the other design requirements to isolate critical parts and part characteristics.

This process continues to a third matrix where the columns (critical parts and their characteristics) of the second matrix become the rows of a third matrix. The columns generated for the third matrix contain the key process operations that influence the part characteristics. The objective for using this third matrix is to select the best manufacturing process by determining the critical process operations and parameters. Function analysis and FAST diagrams, again, are excellent tools to determine the manufacturing process steps needing attention.

Finally, in the fourth house and last phase of the QFD process, the columns (key process operations) of the third house become the rows of the fourth matrix. The columns of the fourth matrix will contain production operations and controls, such as statistical process control procedures, operator training, mistake proofing, preventive maintenance procedures, and education and training procedures.

The purposes of this last matrix are to determine critical production control and maintenance requirements and the necessary education and training to assure that critical requirements will always be met. The goal is to manufacture a consistent product which, in turn, necessitates minimum production variation. The better a team constructs and completes the three prior houses the easier it will be to construct the last matrix.

What To Do With All The Real Estate

Because of its breadth, QFD is more than just a quality tool. It is a method for planning that uses many tools to structure interrelationships in order to make decisions about the future. It engenders team work and leadership through focused dialogue. It both compliments and integrates VM. QFD by no means replaces VM, COVE, and COPD although some will argue about its similarity[24,25] and yes, some VM proponents do fear a takeover.

Engineers and designers are being pushed to think about who their customers are, and what they need the product to do. This is also a special opportunity for the value engineer to become involved. By becoming part of the QFD team, they no longer have to sell VM on their own. They can introduce the VM techniques as a natural part of the quality process. By coming under the cloak of quality, VM practitioners can take advantage of the momentum and attention of in-house quality programs. Because QFD starts at the beginning of the product life cycle, there is a greater opportunity for early involvement and a greater opportunity for significant cost prevention.

Although QFD can be a lengthy process, its characteristics appear to be less foreign to participants compared to VM, which appears to use uncomfortable terms, such as basic and secondary functions, as well as difficult and frustrating tools such as FAST diagrams. Value Management is much more than a cost reduction tool, but many VM practitioners propose it mainly for cost reduction. VM should be referred to as a value improvement process that has cost reduction as one of its many assets. QFD, from the beginning of the HOQ, is more evident as a value improvement process that transcends more than cost reduction and quality. The result of a QFD project is a more value-added product for both customer and producer. In this respect, byproducts of QFD are reduced lead time, improved quality, reduced costs, and increased market share. These, in turn, affect the bottom line and increase shareholder equity.

Because it is a value improvement method, we have included it in this book. It will provide the basis for other hybrid value improvement models, Technology Roadmapping and "Customer Oriented Product Concepting," developed by us which are discussed in Chapters 9 and 10. We also include it to draw attention to the fact that the practitioner of the VM process should integrate or adopt other leading-edge value improvement processes to better meet the increasing demand for value-added products and the increasing pressures to develop better more value-added manufacturing and production processes. We have purposely not gone into more detail on the QFD process because there are many existing references that cover this subject in detail.

But We're Already Doing It!

Quite often we hear people from many areas say that they are already QFD. They probably already work with data that looks like the type that is used in QFD. This would make pieces of the QFD process look familiar. Generally they are working with isolated sets of QFD type data. However, they are not structuring the data to reveal and study relationships between the data sets or individual items. It is these relationships that are important in addition to content. The users of the data have no process to make connections between the interfaces. QFD forces one to deal with these relationships through the structure and the discipline of the process.

SUMMARY

In this chapter we have traced the evolution of the application of VE from the use value part-piece manufacturing emphasis to the user oriented total product orientation known as COVE. We note that practitioner's myopic focus on cost reduction has kept VE in a guarded channel in a new age of customer oriented value improvement. For some time there has been a need for customer focused value engineering type methodology. That void is being filled by QFD which was imported from Japan, and is spreading rapidly throughout the United States. VE and QFD compliment each other and those who integrate them both in process and application will find many rewards in a new world of applications. Those who view QFD and similar methodology as a threat will continue in the rut of cost reduction.

REFERENCES

1. Bryant, J., "Customer-Oriented Value Engineering-COVE," *Value World*, pp. 7–12, April/May/June 1986.
2. Fowler, T. C., "40-Hour Value Analysis vs. Modern Value Analysis," in *Proceedings, Miles Value Foundation Conference*, April 1988, Vol. 1, pp. 201–204.
3. Snodgrass, T., and Kasi, M., *Function Analysis: Stepping Stones to Good Value*, University of Wisconsin, Madison, 1986.
4. Cook, T. F., "Welcome to Value Analysis and Value Engineering," *Proceedings, Society of American Value Engineers* 19, pp. 75–82, 1984.
5. Cook, T. F., "Determine Value Mismatch By Measuring User/Customer Attitude," *Proceedings, Society of American Value Engineers* 21, 145–156, 1986.
6. Wasserman, M., "Sales Market Value," *Proceedings, Society of American Value Engineers* 12, 105–114, 1977.
7. Snodgrass, T. J., and Fowler, T. C., "Customer Oriented FAST Diagramming," *Proceedings, SAVE Regional Conference*, Detroit, 1972, pp. 9.1–9.10.
8. Fowler, T. C., "The User Oriented FAST Diagram," *Proceedings, Society of American Value Engineers* 18, 89–94, 1983.
9. Jarboe, R. R., and Ferguson, J. E., *Function Analysis, A Value Engineering Handbook*, University of Wisconsin-Extension, Madison, 1977.

10. Snodgrass, T. J., and McKently, C., "An Advanced Degree in Value Engineering A Major Career Opportunity," *Proceedings, Society of American Value Engineers* **23**, 97–104, 1988.

11. Shillito, M. L., "Function Morphology," *Proceedings, Society of American Value Engineers* **20**, 119–125, 1985.

12. Shillito, M. L., "Competitive Analysis Using Value Measurement Techniques," *Value World*, pp. 21–23, April/May/June 1988.

13. DeMarle, D. J., "The Use of Value Engineering Methodology in Forecasting Future Technology," *Proceedings, Society of American Value Engineers* **20**, 94–98, 1985.

14. King, R., *Better Designs in Half the Time, Implementing QFD Quality Function Deployment in America*, GOAL/QPC, Methuen, Mass., 1987.

15. Schubert, M. A., "Quality Function Deployment," *Proceedings, Society of American Value Engineers* **24**, 93–98, 1989.

16. Sullivan, L. P., "Quality Function Deployment," *Quality Progress*, pp. 39–50, 1986.

17. Kogure, M., and Akao, Y., "Quality Function Deployment and CWQC In Japan," *Quality Progress*, pp. 25–29, October 1983.

18. Hauser, J. R., and Clausing, D., "The House of Quality," *Harvard Business Review*, pp. 43–53, May–June 1988.

19. Akao, Y. (Ed.), *Quality Function Deployment*, Productivity Press, Cambridge, Mass., 1990.

20. Bossert, J. L., *Quality Function Deployment: A Practitioner's Approach*, ASQC Quality Press/Marcel Dekker, New York, 1991.

21. Eureka, W. E., and Ryan, N. E., *The Customer Driven Company: Managerial Perspectives on QFD*, ASI Press, Dearborn, Mich., 1988.

22. Adams, R. M., and Gavoor, M. D., "Quality Function Deployment Its Promise and Reality," in *Proceedings, 5th GOAL/QPC Conference*, November 1989, GOAL/QPC, Methuen, Mass.

23. Mizuno, S. (Ed.), *Management for Quality Improvement*, Productivity Press, Cambridge, Mass., 1988.

24. Snodgrass, T. J., "Quality Function Deployment Versus Value Information Techniques Using Analytical Language," *Proceedings, Society of American Value Engineers* **24**, 99–100, 1989.

25. Park, R. J., "Value Engineering in the Modern World of Taguchi, QFD et al." *Proceedings, Society of American Value Engineers* **24**, 81–85, 1989.

9

THE TECHNOLOGY ROAD MAP

M. Larry Shillito

INTRODUCTION

This chapter describes a function matrix technique that allows you to describe and compare product concepts by a generic study of function. The morphological model is useful for designing and costing new products, especially those in the concept-ualization stage. The process is especially designed to be used best with processes, services, systems, capital, and construction projects that are internal to a company. Such internal applications have a captive user or customer. The matrix provides a common reference base, allows large amounts of information to be stored in concise form, and is an excellent comparison and concepting method. In this respect, it is a powerful communications device to view all possible options and document decision rationale. We have called the process *technology road mapping* (TR).

BACKGROUND

Value engineering has long been used to improve the value of products. The definition of product as used here is again the one developed by Arthur Mudge,[1]"any-thing which is the result of someone's effort." Therefore, *product* is not limited to hardware, but includes services, processes, systems, procedures, and so on. What has made VE a unique process and distinguishes it from traditional cost reduction is the study of function. That is, the VE process concentrates on a detailed examination of utility rather than a simplistic examination of components. There are numerous FA techniques all of which employ a semantic clarification of function. The basic method requires functions to be described with only two words, a verb and a noun. A well-known and powerful structural technique of FA is FAST diagramming,[2] discussed in Chapter 7. This technique allows two-word function definitions to be ordered in a hierarchy based on cause and consequence. It is used extensively with considerable success. The various techniques and variations of FA such as FAST diagramming work well with existing products and some products that are in the breadboard or drawing board stage. However, the techniques of FA, and in particular FAST diagramming, are not as easily applied to products that are in the conceptualization stage, where few, if any, drawings or breadboards exist. What is needed is a modified future-oriented FA technique that allows you to design new products from mental or conceptual constructs. Such a technique, which we have termed a *technology road map* (TR), will be described. Traditional FA is oriented more to the present and answers the question, How *do* we perform the basic function of the product? The TR, on the other hand, is more future oriented and answers the question, How *might* we perform the basic function of the product?[3]

CONCEPT

The TR will be used to dissect issues into manageable pieces so that the individual parts can be investigated thoroughly by

1. Forcing people to concentrate on one task at a time rather than jumping ahead to a quick conclusion.
2. Discouraging people from disregarding pieces, because they don't fit easily into a preconceived format.

The process allows you to synthesize potential solutions later when there is more information about individual pieces and when we have a clearer understanding of interrelationships. The intent of the process is to stimulate creative identification of alternatives.

BASIC MODEL

The basic TR arrays the operational functions necessary in accomplishing the basic function of some process or transformation against the various methods of performing those functions. Figure 9.1 is a schematic of the basic structure. By connecting various function methods as shown to form a technology path, you can

1. Describe any current system, method, process, or service.
2. Describe or create new (nonexisting) concepts.

The matrix then serves as a repository of methods to permit you to engage in a collective mapping process to build and describe new products and/or system con-

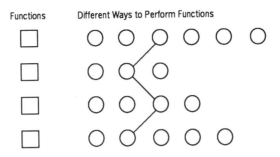

FIGURE 9.1 Basic function morphology matrix.

cepts based on functions. The final TR model is completed by augmenting the basic structure with (1) a documented set of constraints (musts and wants) representing the user of the system and (2) a documented set of decision criteria representing the manufacturer or designer. See Figure 9.2.

The addition of the two sets of constraints to the basic structure provides the filters to guide you in making the final selection of technologies that best matches customer and manufacturer needs. Sometimes defining and documenting these two sets of constraints proves more valuable than the final selection of the technology path! Too often design teams do an inadequate job defining decision criteria and design constraints. Even if there is agreement on the labels of the criteria, there usually is considerable disagreement about how important they are. Considerable anxiety is evoked from study team members when the TR process consultant has the team document and evaluate decision criteria for relative importance. Teams generally are looking for quick solutions and don't want to take the time to document their decision-making framework.

Product:		Basic Function:								
		Market:		Segment:		Country:				
Operational Functions	User		Manufacturer							
	Musts	Wants	Criteria	Wt.			Technologies			

FIGURE 9.2 Technology road map—final model.

BUILDING THE MATRIX

The process begins by the team developing answers to the following questions, which will be discussed in detail in Chapter 13, in the section Initializing the Process.

1. What is the purpose for doing the TR process?
2. Who is the company decision maker that will approve the team recommendations?
3. What is the scope of the TR project?
4. Who is the customer?
5. What is the time horizon for product introduction?
6. What is the completion date for the TR study?
7. Who are the study team members?
8. What are the assumptions upon which the TR project is based?

Having documented answers to the above, the following steps are performed using an interdiscipline team in an off-site workshop. Figure 9.2 and/or Figure 9.3 may be used for reference. Figure 9.3 is an actual application of the TR process to a manufacturing process.

1. Begin constructing the matrix. Determine the final output coming from the product being studied, for example, a color picture, packaged orange juice, an enzyme, or dyed resin. What is the deliverable?

2. Determine the overall basic function of the product. The basic function is the prime reason for existence of the product. It can also be thought of in terms of the task that the user of the product is trying to accomplish. For example, the basic function of a copier is "reproduce image," the function of the fuser assembly within the copier is "attach toner," the function of an overhead projector is "project image." All functions, including the basic function, must be stated in two words, a verb and a noun.

3. Using two-word function definitions, determine all of the operational functions that must be performed in order to accomplish the basic function and allow the user to accomplish the intended task. Functions are listed vertically in sequential order of performance (where practical). This can be considered equivalent to a function flow diagram (see Figure 9.3). Be careful not to define functions in a microlevel of detail; keep them at a higher level of aggregation, which many times is at a subassembly level. If functions are defined at too fine a detail, the project can be dissected into oblivion and collapse from information overload.

4. Based on items 3 and 4, document the customer musts and wants.[4] This usually comes from existing internal data and specialists, literature, and surveys. Additional external or internal surveys, interviews, focus groups, and so on may have to be conducted to complete the list. Gaps in data will readily surface during this phase of building the matrix. Sometimes the customer is a member of the TR team.

A *must* is a criterion or constraint that, if not satisfied, the product or operational function will not perform to the user specifications if at all. A must is dichotomous, it is either satisfied or it isn't, there is no partial fulfillment. Musts usually specify some minimum level of performance or fulfillment below which the user will not accept the product or its performance. For example, musts for a copy machine might be productivity at least 1000 copies/hr, no vinyl transfer of toner, and must accept 8½ × 11 *and* 11 × 14 paper. On the other hand, a want is a desired additional feature, selling point, or perk; it is a "nice to have" criterion, but the product will perform user needs satisfactorily without it, and the product will sell. Wants are not dichotomous but possess a range of fulfillment. They sometimes are stated as goals. Examples might be copier productivity equal to 1500 to 2000 copies/hr and waste between 1% and 3%. It is interesting to note that many times current wants become future musts, especially if there are going to be future generations or model changes in the product. Documenting musts and wants is not an easy process and is usually mentally exhausting. Most likely some team members were never before forced to write them down. Too often they are taken for granted. These musts and wants will be used later as one of two measures to select function technologies to build a concept path across all functions.

5. Once all operational functions, along with their respective musts and wants, are identified and documented, the team then lists all of the possible ways (proven or unproven, developed or undeveloped) of performing them. This is equivalent to a minibrainstorm session for each function. Only one function is expanded at a time. During the technology expansion step, the team is initially asked for a *core dump* of all existing, obvious, known, or practical ways of accomplishing the function including the current method, if one exists. After the purge, the rules for creativity are used to create other possible ways to accomplish the function. After each function has been creatively expanded, the ideas are edited to eliminate redundancy and nonsensical or silly ideas. There is a question of when to brainstorm technologies. Should one brainstorm before or after establishing the customer musts and wants? Some people believe that brainstorming after establishing the musts and wants inhibits participants' creativity. They may be restricted in their thoughts by passing judgment on ideas imposed by the musts and wants. Brainstorming technology before discussing musts and wants may produce more and possibly better ideas. The process consultant should consider the timing of the brainstorming based on observation of the team members and team atmosphere.

6. The list of technologies must now be reduced to those few that will be considered candidates for final selection. The first step in the screening process is to judge the technologies against their respective set of musts. That is, any technology that will not satisfy 100% of the musts will not be chosen as a candidate for further consideration. Those technologies that are not chosen for further consideration are not, however, eliminated from the matrix document itself. The document is to be a composite of all possibilities, including those that were not chosen for further consideration. Retaining all ideas lends credibility to the process and shows that all possibilities were at least considered in selecting the final concept(s). Another reason

		TOY MOLDING DEPT.:	

PRODUCT: Dyed Resin   BASIC FUNCTION: Deliver Resin

FUNCTION	MUSTS	WANTS	CRITERIA
A. PREPARE DYE	1. Not change dye hue 2. Not intro. contaminants 3. Meet dye specs. 4. Have good flow characteristics 5. Be safe 6. Meet dye conc. specs. (TBD)	1. Clean method for preparing 2. Easier handling in Dept. 'X' 3. Use existing equipment	Weights 1. Install. Cost 2.4 2. Operation Cost 4.0 3. Dev. Time, $ 4.8 4. Compatibility 16.1 5. Time to Complete 0.9 6. Precision 24.2 7. Reliability 21.8 8. QWL 9.7 9. Simplicity of Oper. 12.1 10. Simplicity of Design 4.0
B. PROVIDE RESIN	1. Have good precision 2. Not intro. contaminants 3. Be reliable 4. Handle all machine flow characteristics 5. Good flow chars. 6. Not segregate particle size 7. Not affect machine room clean. 8. Compatible with direct re-use	1. Easily maintainable 2. Locate inside bldg. 3. Minimize housekeeping 4. Use existing equipment	Weights 1. Install. Cost 2.4 2. Operation Cost 4.0 3. Dev. Time, $ 4.8 4. Compatibility 16.1 5. Time to Complete 0.9 6. Precision 24.2 7. Reliability 21.8 8. QWL 9.7 9. Simplicity of Oper. 12.1 10. Simplicity of Design 4.0
C. PROVIDE DYE	1. Have good precision 2. Not intro. contaminants 3. Be reliable 4. Handle all machine flow characteristics 5. Good flow chars. 6. Not seg. particle size 7. Not affect machine room clean. 8. Comp. w/ direct re-use 9. Be safe 10. Be able to supply continuous product	1. Easily maintainable 2. Locate inside bldg. 3. Minimize housekeeping 4. Use existing equipment 5. Be able to analyze mixture thru process 6. Be able to change from one dye mix to another quickly 7. Minimum waste 8. Ease in introducing future products	Weights 1. Install. Cost 2.4 2. Operation Cost 4.0 3. Dev. Time, $ 4.8 4. Compatibility 16.1 5. Time to Complete 0.9 6. Precision 24.2 7. Reliability 21.8 8. QWL 9.7 9. Simplicity of Oper. 12.1 10. Simplicity of Design 4.0
D.1 INTRODUCE/ MIX DYE BATCH	1. Have good precision 2. Not intro. contaminants 3. Be reliable 4. Handle all machine flow characteristics 5. Good flow chars. 6. Not seg. particle size 7. Not affect machine room clean. 8. Comp. w/ direct re-use 9. Be safe 10. Be able to supply continuous product	1. Easily maintainable 2. Locate inside bldg. 3. Minimize housekeeping 4. Use existing equipment 5. Be able to analyze mixture thru process 6. Be able to change from one dye mix to another quickly 7. Minimum waste 8. Ease in introducing future products	Weights 1. Install. Cost 2.4 2. Operation Cost 4.0 3. Dev. Time, $ 4.8 4. Compatibility 16.1 5. Time to Complete 0.9 6. Precision 24.2 7. Reliability 21.8 8. QWL 9.7 9. Simplicity of Oper. 12.1 10. Simplicity of Design 4.0
D.2 INTRODUCE/ MIX DYE CONTINUOUS	1. Have good precision 2. Not intro. contaminants 3. Be reliable 4. Handle all machine flow characteristics 5. Good flow chars. 6. Not seg. particle size 7. Not affect machine room clean. 8. Comp. w/ direct re-use 9. Be safe 10. Be able to supply continuous product	1. Easily maintainable 2. Locate inside bldg. 3. Minimize housekeeping 4. Use existing equipment 5. Be able to analyze mixture thru process 6. Be able to change from one dye mix to another quickly 7. Minimum waste 8. Ease in introducing future products	Weights 1. Install. Cost 2.4 2. Operation Cost 4.0 3. Dev. Time, $ 4.8 4. Compatibility 16.1 5. Time to Complete 0.9 6. Precision 24.2 7. Reliability 21.8 8. QWL 9.7 9. Simplicity of Oper. 12.1 10. Simplicity of Design 4.0

FIGURE 9.3 Technology road map for a plastic molding process.

194

RESIN DELIVERY SYSTEM VALUE ENGINEERING

TECHNOLOGIES

PURE DYE SOLID	DYE & CHEMICALS	CO-POLYMERIZE	CONCENTRATES	TODAY PROCESS
3			5	4
3.5			2	3
4			3	5
5			5	5
4			3.5	5
5 (448)			3 (355)	5 (428)
5			4	5
3			3	2
4			3	3
4			3	3

PRE-WEIGH	METER BY WT	VOLUMETRIC METER	INCREASE PART SIZE OF RESIN	INTRO DYE IN MIXER	PROVIDE IN MOLTEN STATE	TODAY	DIR FEED BELOW HOPP WHITE POWDER
3	1	3				5	5
3.5	3	3				3	3
3	3	3				5	5
5	5	4				5	5
5 (396)	3 (373)	3 (338)				5 (334)	5 (492)
4	4	3				2	5
4	4	4				3	5
4	3	3				4	5
3	3	3				3	5
4	3	3				3	5

MANUAL (BATCH) WEIGH; POWDER	VOLUMETRIC METER	LOSS IN WEIGHT; METER	MOLTEN STATE; CONC	WAX SLICER
5	3	2	1	
3	2	2	1	
5	4	4	2	
5	5	5	3	
5 (463)	3 (381)	3	2 (299)	
3	4	4 (377)	4	
2	4	4	3	
5	4	4	3	
5	4	3	2	

MECHANICAL MIX	BATCH FLUIDIZING	SPRAY COAT	VAPORIZE COAT	USE BONDING AGENT	BATCH ELECTROSTATICS	MECH INTRO HIGH-TECH BUCKET BRIGADE	TODAY
2	1	2					5
3	1	2					3
4	3	2					5
5	4	3					5
3 (341)	2 (336)	2 (327)					5 (331)
4	4	5					3
3	3	3					3
2	2	2					2
2	2	3					2
4	3	3					4

BELOW HOPPER; NO MIX	ABOVE HOPPER; NO MIX	ABOVE HOPPER; STATIC MIX	ABOVE HOPPER; ZIG-ZAG MIX	CONTINUOUS FLUIDIZED BLEND	SPRAY COAT	IMPROVE SURFACE BONDING	CONVEY DYE MIXTURE	TODAY
3.5	5	4	2	1				5
5	5	5	3	2				3
3	4	4	4	2				5
4	5	5	4	3				5
5	5	4	3	2				2
3 (406)	5 (449)	4 (458)	4 (381)	5 (312)				2 (241)
5	3	5	4	2				n/a
5	4	4	3	3				2
4	5	5	4	3				2
4	5	5	4	3				4

for retaining all ideas is that some of the ideas that will not currently satisfy the musts may, with future technology breakthrough and Research and Development, satisfy the musts in the near future, at which time they should definitely be reconsidered for application. This composite of all ideas, especially if updated and kept current, serves as an excellent reference document when developing future similar products. Too often ideas are not documented and kept in a common source. People are often shifted around so that when it is time to build another product a team must start all over again and reinvent the wheel.

7. At this point a considerable number of technology candidates still remain. It is now necessary to choose, for each remaining function, the one or two technologies that will be given final consideration for inclusion in the final concept path. To do this, it is necessary to define and quantify a set of decision criteria. This set of decision criteria represents the manufacturer's point of view. Typical decision criteria are development cost, operational cost, installation cost, time to develop, quality, reliability, maintainability, quality of work life, simplicity of operation, simplicity of design, waste level, durability, repeatability, availability, technology leadership, serviceability, accessibility, and compatibility. Having identified the criteria, the team then quantifies the relative importance of each. Please note that all of the criteria are important, but some are obviously more important than others. It is important to know and understand this quantitative relationship, because the numbers will be used later as weighting factors to be multiplied against technology scores.

The method used to quantify importance is to allocate 100 points to the team and ask the team as a group to distribute the points across all of the criteria. The team is asked to do this verbally as a group and not silently by individual ballot. The interaction required to do it as a group is well worth the time, information transfer, and team member education that ensues. Some teams use numerical flash cards for voting and then discuss the distribution of votes and revote until there is convergence around some agreeable group number. An electronic voting apparatus, such as the Q-System 100 device[5] also works very well.

As discussed elsewhere, the 100-points allocation method is a more accurate method to determine the importance of each criterion relative to all the others, because the team must consider the trade-offs and interrelationships when distributing the points. The team members have only a finite amount of importance to allocate so what is awarded to one criterion must be taken away from another and so on.

This method is really not the same as allocating each criterion a weight on a scale of, for example, 100 and then normalizing to a percentage because the team members do not have to consider any trade-offs between criteria. It is important also that the decision criteria be as mutually exclusive as possible in order to reduce confounding and interaction. For example, if two criteria were operating costs and waste level, waste level most likely is a subset of operating cost. If both criteria were used, there would be bias introduced in favor of operating cost. It may be counted twice. Criteria should all be at the same level to reduce redundancy through the elimination of subsets. There are cases where teams, after lengthy discussion, have selected to retain both criteria and acknowledge the overlap between the two.

8. The technologies that survived the musts requirement are now quantified as to how well they satisfy each decision criterion. Rating scales of 1 to 5 or 1 to 10 with category descriptors can be used for this purpose. Another method is to build a scoring model for each individual criterion.

Such a scoring model might look like Figure 9.4. The criteria are listed, and a rating scale range is selected (in this example, 1 to 5). For each criterion, a set of customized rating descriptors is established at selected numerical anchor points (in the example, 1, 3, and 5). The purpose of the scoring model is to use verbal descriptors for each criterion so that the criteria ratings are calibrated to the same level. That is, referring to Figure 9.4, a score of 5 for installation cost should be equivalent to a score of 5 for precision, quality of worklife, and so on. Scoring models are covered in Chapter 4.

The scoring model is used to score all of the remaining technologies that survived the first cut of satisfying all musts. The procedure is to score each technology one criterion at a time. That is, all remaining technologies are scored for; in this example, installation cost using the scoring model. Then all remaining technologies are scored for operating costs. Experience shows that scoring the technologies one criterion (row) at a time is much easier, because it provides a mental reference base and makes comparison easier compared to evaluating all the criteria at one time for one in-dividual function. The former method makes it easier to see relationships across technologies. The reader is referred to Chapter 4 for details about rating scales and scoring models.

9. A weighted technology score is now computed for each technology by multi-plying each decision criterion weighting factor times the corresponding technology score and summing the products. The sum of products is the final weighted composite

Criterion	Rating Scale				
	1	2	3	4	5
1. Installation Cost	Significant		Some		Little/None
2. Operating Cost	> Today's Cost		Today's Cost		< Today's Cost
3. Development Effort	Unknown Technology		Effort Needed		Little/None
4. Compatibility	Low		Medium		High
5. Time to Completion	3–5 Years		1–2 Years		< 1 Year
6. Precision	Low		Medium (PCL = 1.3)		High
7. Reliability	Low		Medium Today		High
8. Quality of Worklife	Low	←Today→	Medium		High
9. Simplicity of Operation	Low	←Today→	Medium		High

FIGURE 9.4 Scoring model for technology ratings.

score. It will be used to assist selecting one technology from each function row to build a technology path.

10. The objective of the entire process is to define one or several technology paths through the matrix. Each different path has its own set of characteristics and represents a different concept. Completing the matrix at step 9 provides a morphological matrix arraying all operational functions needed to accomplish the basic function of the product against all of the identified and evaluated methodologies/technologies of performing them. At this point the matrix provides a very valuable and comprehensive reference document for creating new concepts. By connecting a technology for every function and creating a technology path through all of the functions, you can create a considerable number of paths, each of which describes a new concept.

At first glance, you might think there are an almost limitless number of paths. However, reality takes over, and the number of practical implementable paths is limited. What happens is that there are many illogical or impossible combinations of technologies that restricts the number of paths. The matrix, however, provides the basis for maximizing the number of logical alternatives.

USING THE MATRIX

During the process of concepting (connecting technologies through the rows), many types of paths can be constructed. For instance, separate paths can be developed by combining technologies in each row that have the least total cost, highest quality, best reliability, greatest probability of success, lowest operating cost, and best precision. At this point, it is necessary to reflect back to the purpose of the exercise. Is this a cost-driven project? Is it a quality/reliability-driven project? Revisiting the purpose can give direction as to what type of path to build. The matrix puts the team in a position to customize concepts for whatever combination of criteria are desired.

The most frequent paths constructed are least cost, highest quality, and highest overall score. To provide a reference base as well as to validate the matrix for completeness, a path should be constructed representing the current product, if one exists. The matrix should be comprehensive enough to allow you to draw a path for current product as well as other configurations including those of other manufacturers. In this respect, the TR matrix serves as an excellent comparative document for benchmarking.

Connecting the least costly technology in each row to build a least-cost path may produce a path that is impossible or impractical. As stated earlier, there is not an unlimited number of path concepts, because electromechanically, chemically, and physically, there are numerous illogical or impossible combinations of technologies. However, if this concept could be implemented, it would be the lowest cost product that could be built. This cost, then, becomes the target cost. Likewise, connecting the highest-scoring technologies for quality and/or reliability provides yet another reference base.

The first path always to be developed is the *curiosity path*. This path connects the highest overall scoring technology in each row. Although it may not be a contender, it does provide a great reference for discussion. Sometimes surprises surface. A competitive analysis may also be performed by drawing the path(s) representing the competitors' products. Strong and weak points quickly become visible and are easily discussed and also provide the basis for an opportunity analysis. Many teams document their option paths by constructing a separate minimatrix arranging functions against selected technologies.

Each column of this minimatrix represents a separate path (option). The actual names of the selected technologies are written in the minimatrix cells. The columns of the minimatrix show the boundary of path options. By scanning the function rows of the minimatrix one can easily locate technologies common across path options. Discovering such common technologies can prove very valuable. Figure 9.5 illustrates a minimatrix, a strategy matrix, of path options. Note for this example that for the function "provide resin" that the technology "direct feed below hopper" is common to all options that have so far surfaced. Figure 9.6 shows yet another variation, a strategy map, for showing the interconnectedness of technologies and paths. Both devices show the same thing; some people may prefer a matrix, others may prefer a map.

Value graphs of all technology options may be constructed for each function. The plots are made by plotting the combined weighted cost scores versus the remaining sum of the combined weighted scores for the other criteria. The simplest way to do this is to subtract the sum of all weighted cost scores from the total cell score for each function and plot the two figures. For simplicity the remaining total after subtracting cost is called *importance*. Value graphs allow designers to view the relative value of the various technology options and are described in Chapter 3. Optimum value zones can be drawn on these graphs to help screen options and work nicely for presentations to management. Figure 9.7 is a value graph of different technology options.

Concept paths may also be time-phased by developing the path(s) for today and the path(s) for tomorrow. In many companies, schedules are sacred cows not to be touched. So, the preferred concept path may not be achievable within the scheduling

Function	Technology Path Options			
	Least Cost	Best Overall Score	Best Precision & Reliability	Best Quality of Worklife
Prepare Dye	Today's Process	Pure Dye Solid	Pure Dye Solid	Pure Dye Solid
Provide Resin	Direct Feed Below Hopper	Direct Feed Below Hopper	Direct Feed Below Hopper	Direct Feed Below Hopper
Provide Dye	Manual Weigh	Manual Weigh	Manual Weigh	Valometric
Introduce & Mix Dye (Batch)	Today's Process	Mechanical Mix	Spray Coat	Batch Fluid

FIGURE 9.5 Strategy matrix (minimatrix of paths).

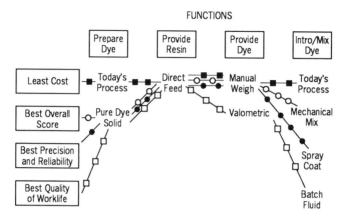

FIGURE 9.6 Strategy map.

time frame. Consequently the preferred path becomes the future generation of the product and a less optimum path is selected to meet schedule and get the product out on time.

In Chapter 8, we described an intercorrelation roof on top of the House of Quality. A similar intercorrelation can be used with the function rows of the TR matrix. Such a function roof can highlight any sensitive function intercorrelations. See Figure 9.8. These intercorrelations may be helpful when considering various technology path options.

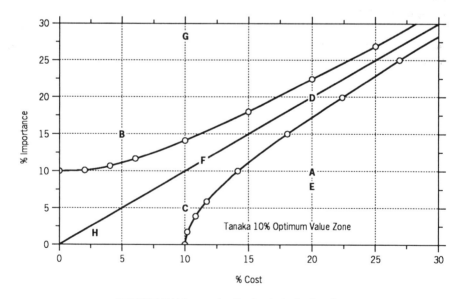

FIGURE 9.7 Value graph of technologies by function.

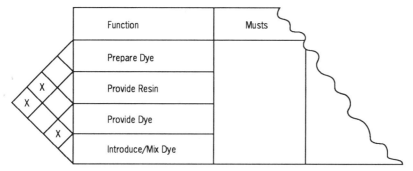

FIGURE 9.8 Technology road map with function intercorrelations.

The matrix may also be used for value planning. Often current wants become future musts. In fact, it has been our observation that wants become musts in about 2 years. Current criteria weighting factors may also change with time. This is often the case regarding current and future regulations. New criteria may evolve over time. The matrix then can be evaluated on a best-guess basis around some future date. Trends can be introduced and quantified, and the result on technology scores can be observed.

Ease of manipulation of the matrix is desirable, and such a convenience lends itself nicely to building the matrix on an electronic spreadsheet using a personal computer (PC), especially a lap-model PC. The matrix can be built on-line. Once on-line, the matrix can easily be manipulated in a simulation "what-if" mode. Decision makers can even interact with the model and change numbers and watch the resulting changes and impacts.

The matrix also makes decision makers feel comfortable that all possible options were considered in deriving the concepts and recommendations. Every technology idea, regardless whether or not it survived the criteria or selection, is kept in the matrix for documentation and reference.

LAYERED INTEGRATED APPLICATIONS

The TR process has been discussed so far as an independent process. To view it only this way would be a short-sighted interpretation and application. It can also be used in concert with other methodologies like decision analysis (referred to as criteria analysis in Chapter 4) and traditional Milesian cost-reduction-oriented VE (see Figure 9.9.)

For example, each function and/or each chosen technology in a function row can be expanded in more detail. A sublevel FA or a FAST diagram can be developed for each individual function or technology at the next level of detail. The team would then proceed to use the traditional VE model or job plan. Essentially each function or chosen technology can be considered a separate new mini-VE project, whose primary purpose is more detailed cost reduction; that is, TR is used to derive the concept and VE is used to reduce the cost. Primary resources, such as Miles,[6] Mudge,[7] and the Department of Defence[8] cover the traditional VE model very well. Chapter 12 also details this process.

Decision analysis can also be applied prior to the TR process (Figure 9.9), that is, if several alternatives are already available. Criteria analysis can be applied to narrow down the choices at a very macro level of detail. The selected alternative can then be

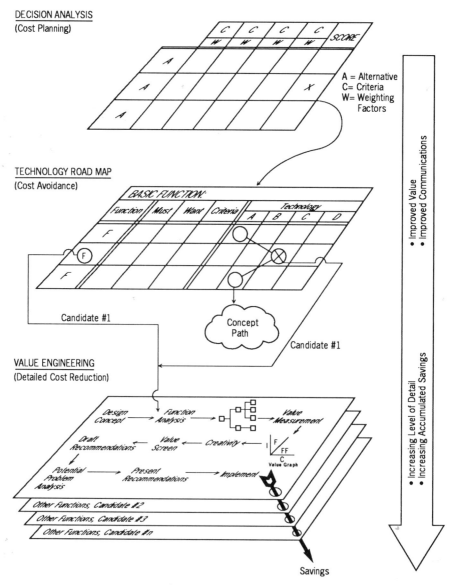

FIGURE 9.9 Value management applications hiearchy.

expanded in more detail in a TR matrix to develop the best concept that will serve the best alternative. The concept, in turn, would be subjected to a traditional VE process to remove more cost.

We strongly emphasize creative applications using combinations of various methodologies or parts of methodologies. Some of our greatest successes have been with integrated applications of whole or parts of various methodologies. We are providing detailed descriptions of some useful basic methodologies such as QFD, TR, and, in the next chapter, customer-oriented product concepting. It is highly likely that you will not apply these technologies exactly as they are presented in this book. You must be creative and flexible in their use and adapt them for best fit with your particular application.

RELATED PROCESSES

Bartlett[9] has devised a similar TR process which he calls "Functional Value Deployment and Synthesis" (FVDS). Using function analysis he dissects a product/system into a logic tree that displays the hierarchy of function relationships. The function hierarchy further links component parts with the overall task function of the product/system. The logic tree provides the basis for developing a function cost comparison table which he uses for competitive analysis of other manufacturer's products. This table can also be expanded by adding performance data and features (similar to musts and wants). If used for competitive analysis it becomes similar to a House of Quality matrix discussed in Chapter 8. He discusses several variations of FDVS but they all branch from the basic logic tree which itself is quite similar to a TR.

SUMMARY

The TR process evolved out of a necessity to describe complex systems in manageable pieces. It has been used to scope out projects and define problems. More recently, though, it has been used to design, create, and cost new product design alternatives. It has been useful in this capacity, because once the basic function flow has been established, all products can be described by the matrix. That is, a product, be it a CD player, a copier, or a computer program, can be described by functions. How a product performs the functions, the methods used, or the order in which the functions are performed does not matter, because the basic functions still must be performed. Functions are generic to a product whereas methods may vary widely. Once functions are identified, any product can be compared relative to another. This is particularly useful for cost estimating and cost comparison.

Advantages of the matrix are the following.

1. You can paint a picture of the product under study and provide a common reference base, which, in turn, promotes consistency.

2. There is a progressive buildup so you can concentrate on small manageable pieces one at a time and then integrate them.

3. It allows you to tie the flow of functions into one path to test out various combinations.

4. It can be used in a creative sense to locate areas needing breakthrough; look for bottlenecks, and check for reliability, compatibility, and costs.

5. It is an excellent communications device. Everyone has a common reference point. It provides visibility, and a dialogue can be opened up between decision makers.

6. Large amounts of information can be compiled in compact form in one document, which allows for detailed comparison and quick access to concepts.

7. Because it is generic and is based on function, the matrix provides a valuable reference document for designers and researchers when designing or developing similar products. The matrix provides a seedbed of ideas and serves as a technology library if continually updated. This library document, in turn, can reduce "reinventing the wheel."

8. The process of constructing the matrix itself is very educational and quickly gets a large team of people communicating and working together.

9. It is comprehensive and enhances completeness of current decisions, which can easily be reconstructed and explained.

10. When put on an electronic spreadsheet computer program, the matrix may be used in a "what-if" mode for simulation.

CONCLUSION

Technology road mapping has been used with considerable success on a wide variety of applications. It has proven to be very effective in the conceptualization stage of revolutionary products as well as in the redesign of existing products. In this respect, it has also been an effective communications device providing a vehicle for dialogue and a focal point for technical discussion and creativity. Finally, it has proven to be a valuable tool for launching an assault on complexity.

REFERENCES

1. Mudge, A. E., *Value Engineering, A Systematic Approach*, p. 16, McGraw-Hill, New York, 1971.

2. DeMarle, D. J., and Shillito, M. L., "Value Engineering," in *Handbook of Industrial Engineering* (G. Salvendy, Ed.), Wiley, New York, 1982, pp. 7.3.1–7.3.20.

3. Shillito, M. L., "Functional Morphology," *Proceedings, Society of American Value Engineers* **20**, 119–125, 1985.

4. Kepner, C. H., and Tregoe, B. B., *The Rational Manager*, McGraw-Hill, New York, 1965.

5. Q-System Model 100, electronic voting apparatus, Reactive Systems Inc., Teaneck, N.J.

6. Miles, L. D., *Techniques of Value Analysis and Engineering*, McGraw-Hill, New York, 1961.

7. Mudge, *Value Engineering: A Systematic Approach*.

8. U.S. Department of Defense, *Principles and Applications of Value Engineering*, Vol. 1, U.S. Gov. Printing Office, Washington, D.C., 1969.

9. Bartlett, R. L., "Functional Value Deployment and Synthesis," *Proceedings, Society of American Value Engineers* **24**, 9–25, 1989.

10

CUSTOMER-ORIENTED PRODUCT CONCEPTING

M. Larry Shillito

BACKGROUND

Producing a product "on the first try" that has high end-user value should be a goal of any manufacturer. Such a desire is universal but, unfortunately, not often achieved. Product design is too often executive or producer driven as opposed to customer driven. Such an approach usually leads to massive redesign and/or postintroduction problem solving. Often lacking is sufficient homework in the conceptualization or preliminary design stage in the front end of the product life cycle. This lack of homework merely increases product cost and delivery time.

DEFINITION

A method for assisting design concepting is customer-oriented product concepting (COPC), which is an interdisciplinary team process to derive recommendations for product design in a way that

1. End-user musts and wants are identified and quantified for importance.
2. Company and other manufacturers product features are evaluated for current and future performance (competitive and performance analysis).
3. Product design features are correlated with marketing and business plans.
4. Appropriate technology or methodology is used to provide necessary product functions for accomplishing the user's task.
5. Appropriate manufacturing methods are utilized to build product.

These, in turn, are integrated in a matrix structure to assist one in concept selection so that:

1. Products satisfy user needs in performance, price, and delivery.
2. Products satisfy company needs and plans.

A customer purchases a product in order to accomplish some task within certain performance requirements. The product provides utility to the user; utility, in turn, is provided by product features and functions. The manufacturer, therefore, produces functions by designing the proper features and performance into a product that the customer is willing to purchase. Value is determined by how well the product allows the customer to perform his task for the price paid. In this respect, value is determined by the user, not the manufacturer. Selling price is directly related to user-perceived product utility and performance and is not related to manufacturing costs.[1] COPC is a method used to design a product with value added to both the user and the producer at a cost and price that is acceptable to both. The COPC process is especially useful for concepting consumer-type products, services, and systems. The process was designed especially for revolutionary products that were never made before.

THE PROCESS

The process about to be described is an integrated model that combines (1) applied behavioral science and stake building philosophy to initialize the process and (2) a matrix incorporating value measurement, function definition, competitive analysis, a function-technology morphology, and elements of QFD. The team deliverable resulting from using the COPC process is a series of recommendations (technology paths) representing different product concepts that have been screened by customer and manufacturing criteria. Here again, as with TR discussed in Chapter 9, the process begins by the team addressing study purpose, decision maker, scope, study completion date, correct study team members, and assumptions.

The team consists of members with diverse interdiscipline backgrounds, such as marketing, sales, customer service, design engineering, manufacturing, and quality assurance. Acceptance and credibility of results can be affected by who or who is not on the team. Unfortunately, power and politics are real issues to contend with. It is recommended to use a core team of knowledgeable people throughout the project. Many problems can be encountered if the makeup of the group frequently changes. For example, considerable time can be lost just getting new members up to speed in both content and process. Also, ratings and evaluations will become much less consistent. The reader is again referred to Chapter 13 for detailed discussion on these issues.

The COPC process is used to design one product. It is not used to develop a product strategy for a family of products. If the product serves many different customers, a separate COPC should be performed for each individual user. It is also based on the assumptions that (1) the company or business unit strategy is clear and (2) that the country, market, segment, buyer, and user have been clearly defined and documented by the business unit (a big assumption indeed!). If the above strategy and marketing documentation is not available or doesn't exist, the COPC team must develop the data themselves. If this is the case and the team has developed this data, it must be presented to upper management and the business unit for verification before progressing further in the process. Sufficient time must be allocated in the COPC process for verifying (or developing) country, market, segment, buyer, and user, because markets can be complex. Too many times answers to these items are taken for granted. This merely increases the chances of the COPC team designing the wrong product. Caution must be exercised on these topics because marketing and sales can be offended or embarrassed, especially if they really don't know the answers! A structured process like COPC surfaces many gaps in knowledge and information across many areas from distribution to manufacturing to marketing. This can, of course, be threatening to those who speak in mysteries with a high fog index and whose career has been founded on deception and turf protection. Consequently, it is advantageous if the process facilitator is experienced in the behavioral science techniques discussed in Chapter 13, such as interpersonal skills, team building, and leading groups and workshops.

CONSTRUCTING THE COPC MATRIX

The matrix is composed of three parts: (1) product functions and components, (2) the marketing/customer evaluation, and (3) the manufacturing/technology evaluation. An example matrix using a videocassette recorder (VCR) as the product to be designed will be used (see Figure 10.1).

The following background information is offered to inform you more about the example product. Our product competes with all the myriad other VCRs on the market. Our model is used to record television signals as well as playback the programs. It is also widely used to playback other prerecorded videotapes, such as those rented or purchased at videotape rental stores. The following hypothetical marketing data are defined.

Country: U.S. only

Market: Video

Segment: Home entertainment

User: Family

Decision makers: Parents

Chief buying influence: Primary income producer

For the following explanation of developing the COPC matrix, please refer to Figure 10.1.

CUSTOMER ORIENTED
PRODUCT CONCEPTING

PRODUCT: VCR
TASK: Play Video on TV
DELIVERABLE: Electronic (TV) Image
BASIC FUNCTION: Display Electronic Image

COUNTRY: U.S. Only
MARKET: Video
SEGMENT: Home Entertainment
USER: Family
DECISION MAKER: Primary Income Producer

MARKET/CUSTOMER

Operational Functions	Features	Importance	Current Us	OM1	OM2	OM3	Desired Us	Improve Ratio	Sales Point	Score	Percent Score
Load Tape	Ease to Load	10	5	8	7	10	10	2.00	1.5	30.00	58.7
	Ease to Orient	10	9	9	9	8	9	1.00	1.2	12.00	23.5
	Instructions	8	7	8	8	8	8	1.14	1.0	9.12	17.8
										51.12	100.00
Transport Tape											
Record Image											
Playback Image											
Rewind Tape											
Remove Tape											

Header spans: CUSTOMER REQUIREMENTS (Features, Importance); COMPETITIVE ANALYSIS (Current Us, OM1, OM2, OM3, Desired Us, Improve Ratio); PLANNING (Sales Point, Score, Percent Score)

MANUFACTURING/DESIGN

Mfg Criteria	Percent Weight	Tech 1 Mfg	Tech 1 Cust	Tech 2 Mfg	Tech 2 Cust	Tech 3 Mfg	Tech 3 Cust
Cost	15	5	7	5	10	5	10
Dev Time	10	3	9	5	9	5	10
Durability	15	5	5	5	6	5	9
Time to Complete	5	1		5		4	
Quality	20	5		4		4	
Reliability	20	4		4		4	
Maintainability	15	4		4		4	
	100	425	711	445	905	440	982

Header spans: MANUFACTURING CRITERIA (Mfg Criteria, Percent Weight); TECHNOLOGIES (HOW TO) (Tech 1, Tech 2, Tech 3, each with Mfg and Cust)

FIGURE 10.1 Customer oriented product concept matrix (VCR example).

CUSTOMER-ORIENTED PRODUCT CONCEPTING PROCESS

Part 1: Product Functions and Components

Step 1: Define Basic Function

Purpose: To establish the reason for existence of the product.

To begin building the matrix, the interdisciplinary team first defines the task that the user wants to accomplish by purchasing the product. To define the task or deliverable, the team members envision themselves as the user and ask why they are purchasing the product. For this example, the customer task is "play video tapes on the TV set." The team then defines the deliverable produced by the product, in our example, "an electronic (TV) image." Based on the user's task and product deliverable, define the overall basic function of the product. The basic function is the prime reason for existence of the product. It is what makes the product work or sell. It is what allows the customer to accomplish the task. As in traditional VE, the basic function is described in two words: one verb and one noun. Personal analogies are helpful in defining functions. That is, you actually pretend that you are the product itself and ask, I am a _____, what do I do? Why do I exist? The answer must be stated in the verb–noun format. In our example, the basic function of the VCR is "display (electronic) image." Some other examples of basic functions are wire, transmit signal; screwdriver, transmit torque; wall panel, partition space.

Step 2: Define Operational Functions

Purpose: To establish the operational functions that must be performed to accomplish the basic function defined in Step 1.

The operational functions that must be performed to allow the customer to accomplish the task and obtain the deliverable are defined. The basic question to be answered is, What are all of the necessary operational functions that must be performed in order to accomplish the basic function (and allow the user to accomplish the task)? Operational functions are also defined in the verb–noun format. They are listed where practical in sequential order of operation, the way the product is produced and used today. This is equivalent to a linear time-sequenced function flow. This linear format is dissimilar to a traditional FAST diagram, where functions are ordered by cause and consequence and not by time sequence.

Operational functions are defined at the highest (macro) level of aggregation. Their number should be approximately 10 to 20. Personal analogies are again helpful in operational function definition. The questions asked are, I am a _____. My basic function is_____. How do I_____? In order to permit my customer to accomplish their task, I must *(function), (function), (function), ..., (function)*. Operational functions are listed as the product performs or operates today. This function flow may not be valid for future designs that may use different technologies from today. Worry about that later. This is why you are building the matrix. In our example, the operational functions are load tape, transport tape, record image, playback image, rewind tape, and remove tape. Note that all functions have been listed. We included the function "record image" to account for the customer task of recording something. This function would not be used by the customer if the VCR is used only in a playback mode. However, we include it because the customer will also be recording TV shows. The basic function, however, is still display (electronic) image. A secondary function happens to be "record image" and is one way of acquiring an image to display. The other way to acquire an image is to rent or purchase a prerecorded videotape. The operational functions are listed vertically along the left side of the COPC matrix as in Figure 10.1.

Listing operational functions works best with processes, machines, procedures, and services. For products like packaging, consumables, durable goods, and so on, it is easier to list major components first and then convert the components to functions at a later time. For example, a 35-mm film package might be listed as case, box, plastic can, can lid, case graphics, box graphics, can label, can label graphics, carton physical characteristics, or instructions. Underneath the component name would be listed its function. Sometimes two separate columns are used in the matrix, one for components, the other for functions. Listing the function is especially important for the creativity phase, where the team brainstorms to function and not components. Determining the basic and operational functions is not so simple a task for the team. Function definition is a new language to which participants must become accustomed. In addition to being the foundation of VE, the function definition provides the structure for completing the matrix. It acclimates the team members and provides focus to concentrate on what the product must or must not do.

Part 2: Marketing/Customer Evaluation

Step 3: List Customer Features by Function

Purpose: To list for each function all of the features and wants that the customer may encounter through the use of the product.

Once all functions have been defined and sequenced, Part 2 of the COPC matrix is started by listing for each individual function all of the features or customer requirements pertaining to that function. In our example, some, but not all, of the features of the function "load tape" might be "ease to load," "ease to orient cassette," "clear definitions," and "load in low light." Direct quotes from customers are great.

Customer features can be obtained from marketing intelligence, research data, focus groups, surveys, interviews, and internal company sources. The ideal source would be to invite and pay customers to be ad hoc members of the team! Typical categories of features are ease or difficulty of use, convenience, dependability, availability, appearance, information/instructions, maintenance, and service. Tom Snodgrass of the University of Wisconsin, at Madison, and president of Value Standards, Inc., has developed four categories of functions that have been used with considerable success in his *customer-oriented product design* method.[2] They are assure dependability, assure convenience, satisfy user, and attract user. These categories are also helpful as stimulators for thinking about features.

It is best to record the features in the customer's own words when possible. If customer quotes are used, it is also recommended that a glossary of terms be established that includes precise translations of the features into terms that are more meaningful to the team and technical community. For example, the customer might say they "want a sharp image." This is important to know, but the technical community needs to know how sharp is sharp. The descriptor would be recorded both on the COPC matrix and in the glossary of terms along with a descriptive translation. The feature as translated in the glossary might read, The image on the TV screen should have clear definition around the edges. Perception of sharpness may be influenced by contrast." The glossary is especially useful if there is turnover in team members or if there are ad-hoc members. It helps get members up to speed.

The question of what to do with things like government regulations/standards and other regulatory requirements arises quite often. These standards are more than customer wants. They are givens and absolutes. They cannot be overlooked. To always keep them visible one can add them as a grouping in a separate row at the bottom of the list of customer functions and needs. The functional equivalent can be labeled as satisfy regulations. There is no compromise with regulatory requirements. They are not brainstormed or tested, but they need to be resolved. Technologies, in addition to scoring them against customer needs and manufacturing criteria, must be checked to see if they fulfill 100% of the regulatory requirements. This is equivalent to the TF process (Chapter 9) where the technologies were checked for compliance with customer *musts*.

Step 4: Quantify Customer Features for Importance

Purpose: To determine how important each feature is to the customer.

After listing all of the features for all functions, they are quantified for perceived importance to the customer. One method that can be used is to establish a simple rating scale, say, 1 to 10. If a scale is used, descriptors should be written by the team for at least three or four anchor points in the scale (e.g., 1, 5, 10). The descriptors should, as best as possible, be stated in the words of the customer. Such a rating scale might be as follows:

Rating	Descriptor
10	I must have this feature. I expect it and would definitely switch brands and pay more to get it.
8	I would like to have it; all else being equal, I would probably change brands or pay more to get it.
5	It would be nice to have; it would make me happy, but I wouldn't go out of my way to get it. I might switch brands to get it but would not pay more to get it.
1	I am apathetic about this feature; it really doesn't influence my buying decision.

Descriptors written by the team members are used in order to promote consistency across raters. For example, without scale descriptors, what one person rates an 8, another person may rate a 10, and so on. Developing the descriptors encourages a dialogue among team members that results not only in more consistent ratings, but a sharing of viewpoints that might not have surfaced. In turn, this dialogue leads to better understanding. Caution must be observed, because there is a tendency when establishing rating descriptors to state them more in marketing terms like "must have this feature in the product to have a superior competitive advantage." This is a producer-oriented phrase. The actual user of the product could probably care less about whether or not the company's product has a competitive advantage to the producer. The user is concerned about their own competitive advantage. Rating scale descriptors must be tailor-made by the team for each project/product. There is *no* universal rating scale! Rating and scaling are discussed in Chapter 4.

The actual rating should be discussed verbally as a group. Numerical flash cards or an electronic voting system can be used to get discussion started. The full-scale range of rating numbers (1,2,3,4, . . . ,10) is used for rating. It is advisable to have a team recorder/secretary record pertinent comments resulting from team discussions that lead to the final team rating. For example, it is not uncommon to have a bimodel or even a rectangular distribution of ratings. When this happens, the facilitator asks those who had high ratings to share their viewpoints leading to their ratings. Likewise, those who had low scores share their views. After the ratings discussion, the team is usually asked to vote again. The discussions and feedback generally lead to a convergence of votes in the second round. There is considerable information generated during these rating discussions that can be valuable later on. Such notes document the rationale behind a rating. This is especially important months after the team sessions when memories start to fade.

What happens if the study team wants to list more than one customer for determining importance of customer needs? Will all customers have the same needs? Will each customer have their own measure of importance for each need? Our experience shows that the labels on the customer needs generally do not change across customers. What does change across customers is the importance of those needs. One way to capture all of this is to have an importance column for each customer. A separate importance rating would be given to each. In addition, weighting factors could be assigned to each customer according to their importance, volume of business, and so on. To do this 100 points would be allocated across the customers instead of using a rating scale. The weighting factors would then be multiplied times the importance ratings. The products would be summed for each customer need row. The final sum of products would then represent an overall weighted importance across all customer columns. It is then possible to interpret the customer matrix to better understand the weighted importance scores. Caution is advised if you include a large number of customers, say more than 10. Including every possible customer can dilute the overall weighted importance score so that it can be almost meaningless.

It is obvious that marketing, customer representatives, or service and repair people should participate in rating the features for importance. The ideal state for determining customer importance would be to interview real customers or have them participate in the team exercise. This could prove difficult depending upon confidentiality and product disclosure. Having the right people participate in this exercise will also enhance the credibility of the COPC project. Another advantage resulting from customer feature identification and quantification is that gaps or missing information readily surface. If answers are missing, question marks are inserted, and the topic is addressed later. The gaps in information can be very useful input for designing surveys or focus groups to obtain the needed information. Therefore, identified missing information can be just as useful and important as known information. Information gaps that surface highlight the fact that "we didn't know we didn't know the information was missing."

Step 4a: Optional Rating Method

An optional and often-used method for rating importance is DME[3] similar to that discussed in Chapter 4. It entails locating the most important feature in the feature group being considered and giving it the highest rating, say, a 10 (assuming a 1 to 10 scale is used). All other features are rated relative to this feature. For example, a rating of 3 would be about one-third the importance of a 10. It is possible and legitimate for other features to also have a rating of 10. Regardless which rating method is used, it is important that the ratings reflect only one customer/user. One set of feature ratings can not be used for a family of products representing different users. If this is the case, separate importance ratings must be performed for each individual user. Very often the majority of the ratings are the same, but it is still important that every rating is considered. Scale descriptors should also be checked for appropriateness for rating each user or product combination. It may also be necessary to add, delete, or change feature descriptions.

Step 4b: Special Case

Experience is showing in some cases that when rating customer features for importance the team discovers that the features are not at the same level for analysis. There is a hierarchy of importance. This happens when functions or components differ in their level of indenture or when some features cluster as subsets under another feature. Consider an automobile, for example. A feature such as "withstand force" most likely would get a high rating, say, a 10, regardless of the function or component to which it pertains. However, a score of 10 for the feature pertaining to the bumper or seat-belt system is considerably different than a score of 10 for the same feature pertaining to the arm rest on the door. When this situation occurs, and it often does, the analytical hierarchy method,[4] also known as the *nested hierarchy method*, of scoring should be used to mathematically convert all feature importance ratings to the same equivalent level of indenture. This produces a treelike hierarchy whose number of levels of features may vary from two to four. Our VCR example has two levels, one level being functions, the other level being features. It may happen that the first level functions may vary in their relative importance. All of these functions are important, or the product would not work. However, some of the functions, relative to one another, are more important than others. When this happens, the features relating to the more important functions also take on a more important significance than those features belonging to the lesser important functions. So, even though some features may have the same second level rating, they may not in actuality be of equal final importance. The final importance rating is dependently factored by the relative importance of the function to which it pertains.

The following procedure (a two-level tree) is used to rate and stratify two levels of functions and features (see Table 10.1).

1. Rate the highest level functions (or components) for relative importance using a rating scale established by the team, as discussed earlier. The optional DME method may also be used.
2. Normalize the function ratings so that the normalized ratings sum to 100. To do this, sum up the individual rating scores from Step 1 and divide this sum total into each individual function rating. The result is original rating scores converted to a percentage that sums to unity.
3. Repeat Step 1 by rating the features (second level) for each function.
4. Normalize the second level features within each function such that the normalized second level ratings sum to 100 for each function.
5. To obtain the final feature importance score, multiply the normalized feature scores (level 2) within that function by the corresponding normalized function

TABLE 10.1 Nested Hierarchy of Feature Importance Ratings (Two-Level Tree)

Function (Level 1)		Feature (Level 2)		Weighted Importance Score				
Score	%	Score	%	%L1		%L2		× 100
		10	50	0.32	×	0.50	=	16.0
8	32	5	25	0.32	×	0.25	=	8.0
		5	25	0.32	×	0.25	=	8.0
		Sum 20	Sum 100					
		8	80	0.40	×	0.80	=	32.0
10	40							
		2	20	0.40	×	0.20	=	8.0
		Sum 10	Sum 100					
2	8	Sum 8	Sum 100	0.08	×	1.00	=	8.0
		5	18	0.20	×	0.18	=	3.6
5	20	5	18	0.20	×	0.18	=	3.6
		10	36	0.20	×	0.36	=	7.2
		8	29	0.20	×	0.29	=	5.8
Sum 25	Sum 100	Sum 20	Sum 100					

score (level 1). The resulting product is the weighted feature score for each feature. By weighting the individual normalized feature score by the normalized function score, we have now brought the final feature scores to the same level of comparison. Table 10.1 illustrates the two-level nested hierarchy computation. The final scores are multiplied by 100 to make the final numbers more convenient. The final weighted feature importance scores derived in Step 4b are posted to the COPC matrix instead of the original nonfactored scores.

Sometimes situations require a three- or four-level tree. This usually happens when some features are subsets of other features, which, in turn, are nested under functions with different importance. A three-level tree involves the following steps (see Table 10.2):

1. Rate and normalize level 1 functions.
2. Rate and normalize level 2 features.
3. Rate and normalize level 3 features.

TABLE 10.2 Nested Hierarchy of Feature Importance Ratings (Three-Level Tree)

Function (Level 1)		Feature (Level 2)		Feature (Level 3)		Weighted Importance Score
Score	%	Score	%	Score	%	%L1 × %L2 × %L3 × 100
				10	.40	0.32 × 0.50 × 0.40 = 6.4
		10	50	10	.40	0.32 × 0.50 × 0.40 = 6.4
				5	.20	0.32 × 0.50 × 0.20 = 3.2
				Sum 25	100	
				9	60	0.32 × 0.25 × 0.60 = 4.8
8	32	5	25	6	40	0.32 × 0.25 × 0.40 = 3.2
				Sum 15	100	
				10	59	0.32 × 0.25 × 0.59 = 4.7
		5	25	2	12	0.32 × 0.25 × 0.12 = 1.0
				5	29	0.32 × 0.25 × 0.29 = 2.3
		Sum 20	Sum 100	Sum 17	100	
				9	60	0.40 × 0.80 × 0.60 = 19.2
		8	80			
				6	40	0.40 × 0.80 × 0.40 = 12.8
10	40			Sum 15	100	
		2	20	10	100	0.40 × 0.20 × 1.00 = 8.0
		Sum 10	Sum 100	Sum 10	Sum 100	
2	8	(...)		(...)		(...)
5	20	(...)		(...)		(...)
Sum 25	Sum 100					

4. Compute the final feature importance score by multiplying the normalized function score (level 1) times the normalized feature score (level 2) times the normalized feature score (level 3). The final products of the three normalized scores is the final feature importance score calibrated to the equivalent level for analysis. Table 10.2 depicts a three-level nested hierarchy tree. Likewise, as with the two-level tree, the final factored feature importance scores are posted to the COPC matrix in place of the original nonfactored scores. In both cases, the hierarchy trees and calculations are retained for reference in case questions arise concerning their derivation.

Step 5: Conduct Competitive Analysis

Purpose

1. To quantify how well current company product satisfies each feature (current state).
2. To quantify how well competitor's products satisfy each feature (current state).
3. To quantify how well company product should satisfy features in the future (future state).

After developing and quantifying all features and functions for customer importance, a competitive analysis is conducted regarding the performance of company and other manufacturer's current product against each feature. The importance of the feature to the customer must not enter into this comparison. How important a feature is has nothing to do with how the products really perform.

To begin the analysis, a simple rating scale is used wherein the basic question being asked is, How well does this current product satisfy this feature? A rating scale is established to estimate how well products satisfy each feature. A typical scale is 1 to 10. Descriptors are written by the team for at least three numerical anchor points. A satisfaction rating might be as follows:

Rating	Descriptor
10	Fully meets customer need. In some cases may even exceed expectations. Gee whiz!
5	Satisfactory; not exceptional; not a problem or concern. Okay.
1	Unsatisfying; causes aggravation; customer may switch brands to avoid/eliminate this problem. Major problems.

As before, the scale and descriptors must be developed by the team so they have a scale that is meaningful and comfortable to them. The above scale is for example only.

All major competitors are listed. Sometimes there are too many competitors to test them all. Hence, only the most important are selected. Using the full rating scale range (e.g., 1,2,3,4, . . . ,10) the current company product and the competitor's product are scored for performance or fulfillment of each feature. If possible and where practical, real products should be available for comparison and evaluation. Scoring can be started with numerical flash cards or an electronic voting apparatus. It is important that team scores be discussed especially if there is wide variation in scores. Scoring products according to feature provides a competition benchmark for current company product. This benchmark will be used to rescore where the company (team) desires to be, feature by feature, at some date in the future. The same rating scale is used. The future time horizon is established by the team and heavily biased by marketing. This will usually be the product introduction date. Future-state rating should reflect the company's current standing with competitors product and, in this case, the perceived importance of the feature to the customer. The basic question being asked when rating company future product performance is, Given our current product performance in relation to the competitor's, and, based on the importance of this feature to the customer, where do we desire to be in the future with respect to this feature?

When doing the competitive analysis, and product current and future performance, only one feature is rated at a time. That is, company current product is rated for feature performance. Then, immediately, the other competitor's products are rated for the same feature while thoughts and discussion are still fresh in team member's minds. After rating company and competitor's products for that particular feature, the team proceeds to rate company future desired performance.

Doing all of these ratings consecutively by row helps maintain the mind set and reference base for the feature. It is less confusing than rating one entire column (company) at a time across all features. When scoring where company product is desired to be in the future, don't fall into the trap of thinking that there must be an improvement for every feature. The performance of many features will remain the same. Be realistic! There is a tendency to overachieve or overdesign. It is possible sometimes to have a performance rating greater than 10. This may set a new standard, or it may remain overkill. Here, as with feature importance ratings, recording of comments from team dialogue necessary to score the products is very valuable.

Rating desired company future performance is generally done by using the current company products as the reference. Sometimes this may not be so simple. For example, the current company product may be a manually operated product where as all competitor's products are automatic operation. In this case, what is the reference base? What some teams have done in this situation is to do two comparisons to produce two desired future scores. The first comparison is with the company's current manual product. The second comparison is based on the average across all competitors scores as the reference base. Both desired future scores can be useful. Reference with the current manually operated product gives an indication how much the company has to change based on current design and current manufacturing methods. The other score indicates how much the company has to change the product based on the average competitor's product. Both scores can be used to compute two improvement ratios, discussed next in Step 6.

Step 6: Calculate Improvement Ratio for Features

Purpose: To compute a ratio signifying how much change is desired for current company product for each feature.

In order to determine the amount of improvement needed for each feature, an improvement factor is computed. This can be done two ways. First, an improvement ratio can be calculated for each feature by dividing the desired future performance rating by the current performance rating. The resulting ratio is the amount of improvement needed for each feature. The ratio highlights those features needing attention and improvement in relation to customer need and competitive standing. There has been some controversy about computing a ratio. The argument has been, for example, that an improvement ratio of 9 divided by 3 is much more important than an improvement ratio of 3 divided by 1. The ratio does not really reflect where along the continuum of change and competitive position the ratio lies.

An alternative to the ratio is to use the arithmetic difference between desired future state and the current state. Using this method with the same above numerical example creates an improvement rating of 9 − 3 = 6 and 3 − 1 = 2, respectively. The feeling has been that the arithmetic difference maintains the focus of improvement and, in addition, places emphasis on the more important changes.

To avoid situations where the arithmetic difference might produce a zero, a 1 can be added to all resulting arithmetic differences. Thus, 1 + (future state − current state) = change factor.

Step 7: Quantify Market/Sales Point (Leverage)

Purpose: To quantify how much advantage marketing may have if changes are made for particular features.

Marketing and sales team members are asked for additional input called *market leverage* or *sales point.* Additional ad-hoc marketing personnel may have to attend the team meetings to give additional input, or the marketing team member may canvass pertinent personnel outside of team meetings and bring the data to the team at a later date.

The basic question being asked of marketing for each feature is, Given the importance of this feature to the customer, and considering the amount of improvement (improvement ratio) needed, if we in fact make a change in this feature, can marketing take advantage of it? The questions are answered with numbers from a rating scale. The following sales point rating scale has been used.

Rating	Descriptor
1.5	Significant market leverage if change is made; most likely would mention it in advertising.
1.2	Some market leverage; technical reps would have advantage on customer premises.
1.0	Status quo; no significant leverage; no negative impacts.

The ratings 1.5, 1.2, and 1.0 have been researched, studied, and used thoroughly over time. They seem to be empirical. These ratings have been carefully developed by the Japanese in their efforts to develop and apply the QFD process[5] discussed in Chapter 8. They provide sufficient information and, at the same time, lessen the impact of this rating on prior ratings of customer importance and improvement ratio.

Again, marketing must participate in this evaluation. If the team does this rating without marketing input, the COPC project will lose credibility. There have been cases where the sales point quantification step has not been used. Sometimes the customer importance rating, improvement ratio, or the competitive analysis is sufficient. To use or not to use sales point is a judgment call by the team on a per-case basis.

Step 8: Compute Feature Score

Purpose: To combine customer importance, improvement ratio, and market leverage into one figure to represent the merit of each feature for each function.

The feature score is a product of three numbers:

$$\text{(feature importance to customer)} \times \text{(improvement ratio)} \times \text{(sales point)}$$
$$= \text{feature score}$$

After raw feature scores are computed for each feature, they are normalized to a percentage by dividing the total of all raw scores across all features for one function into each individual feature score. Note that normalizing is done separately for each individual function and is not based on the grand total of all scores across all functions.

The product of these three ratings, the feature score, draws attention to those customer features needing the most attention in designing a product based on customer needs. The purpose for normalizing is that these percentages will be used as a weighting factor for multiplying technology scores later in Part 3 of the matrix.

At this point in the matrix, we now have quantified customer features and functions. We have addressed the question, What are we trying to do for our customer? This is very useful information, but to stop here would be an incomplete analysis. What we have at this point is an identification of those features that will play an important role in designing the product from the customer's perspective. We now need input from manufacturing.

Part 3: Manufacturing/Technology Evaluation

Step 9: Develop Technologies (Methodologies)

Purpose: To generate for each individual function a list of ways, methods, or technologies that might be used to accomplish or perform the operational functions.

It is now time to configure product concepts. To do this, we array against each operational function all of the known or possible ways of performing the function. This entails listing all of the technologies, methods, and procedures that might be used to construct a new product concept. One function is expanded at a time. Brainstorming or other creativity techniques are used to generate a list of all possible options that might be used. Each function will, in fact, require its own individual minicreativity session. The usual rules of creativity apply: go for quantity, no judging of ideas, hitchhiking is welcomed, and so on. The team is first encouraged to have a core dump of all of the obvious off-the-shelf ideas for achieving the functions. After this purge, the team then develops more blue sky ideas. Function-technology expansion is continued until the team feels they have saturated or exhausted the function-technology possibilities. The creative expansion process is done verbally as a group, so all team members benefit from interaction. It also encourages hitchhiking on other's ideas. Sketching ideas on chart pads also helps. Some teams have even added sketches on the COPC matrix.

It is during this technology expansion that the team finally gets into the nuts and bolts technology of building the product concept. It is now time for designers, manufacturing engineers, researchers, and so on to contribute their content knowledge to build a product concept. It has been suggested that the technology brainstorming be performed immediately after generating the operational functions in Step 2. The feeling is that doing the importance ratings, competitive analysis, improvement ratio, and sales point before the creativity session may inhibit the individual's creativity. The correct procedure is the one that the facilitator and team feel most comfortable with. Again, the COPC facilitator must have facilitating and interpersonal skills to handle these situations.

After this creativity step has been finished for all operational functions, the team will have created a function-technology morphological matrix.[6] This is a useful matrix. It can be considered a technology library catalogued by functions. It will provide the landmarks to chart a concept path to future design. The mechanisms for charting the path will be discussed next.

Step 10: Establish Manufacturing-Type Decision Criteria
Purpose: To document and quantify a set of manufacturing-type criteria to use for selecting technologies.
After expanding all the functions for technologies, it is now necessary to choose one technology for each function to create a technology path through the matrix. Each different combination of technologies creates a different path and thus creates a different concept.

How does one choose which technology to use for which function? It will eventually be done by scoring and weighting the technologies against two sets of criteria: one for the marketing/customer features, which we have already developed, quantified, and normalized, and one for manufacturing and design, which must now be developed.

A set of manufacturing/design decision criteria will be established for each function. Typical criteria are development cost, operating cost, installation cost, development time, time to completion, quality, reliability, maintainability, ease to automate, and quality-of-work life (QWL). After identifying and defining the criteria, they are weighted by the team for relative importance. This is done by allocating the team 100 points to distribute across the criteria such that the total sums to 100. The team does this by verbal interaction as opposed to individual distribution. The interaction necessary to reach group agreement is very information rich and worth the time. Sometimes the discussion can become very heated with strong arguments. This sum-to-unity method is more representative of the team's decision criteria than scoring each criterion individually on a scale (e.g., 1 to 100, 1 to 10) and then normalizing them to a percentage by dividing each individual criterion score by the sum of all scores. With the 100-point sum-to-unity method, the team is forced to deal with trade-offs across criteria; what they add to one criterion they must take away from another! This trade-off gives a more realistic picture of relative importance. All the criteria are important, but we want to know which are more important than others.

It is important that decision criteria be as mutually exclusive as possible in order to reduce confounding and interaction. For example, the criterion "waste" can be a subset of the criterion "operating cost," and using them both can give double accounting to operating cost.

These weighted criteria will be used as a set of weighting factors to multiply times technology scores to produce an overall weighted technology feature score that can be used for choosing appropriate technology. Next, how do we score the technologies?

Step 11: Score Technologies

Purpose: To quantify how well each technology satisfies (1) the customer features and (2) the manufacturing decision criteria.

Thus far in the process, two sets of criteria and their weighting factors have been established. They are customer features and manufacturing criteria. The next step is to rate each technology against both sets of criteria. A rating scale is established to allow team members to quantify how well the technology satisfies the manufacturing criterion and the customer features. The basic question asked is, How well does this technology satisfy the criterion? As seen before, a typical rating scale is 1 to 10. Descriptors are also written for several anchor points. A typical scale might be

Rating	Descriptor
10	Satisfies feature/criterion in all respects. ideal.
5	Satisfies feature or criterion. Okay.
1	Satisfies feature/criterion little or not at all.

The team decides which to rate first, customer features or manufacturing criteria. It is easier if the technologies are all rated against one set of criteria or features at a time. After scoring all technologies for customer features, the scoring process would then be repeated for the manufacturing criteria or vice versa.

This is less confusing than scoring the technologies for both customer features and manufacturing criteria at the same time. Also, when rating features and criteria, it is best to do so one row at a time. That is, rate all technologies for function 1 and feature

1, then rate them all for function 1 and feature 2 and so on. The same procedure would be used for rating against manufacturing criteria; function 1 and criterion 1, function 1 and criterion 2 and so on. Scoring one row at a time is much easier, because it provides a mental reference base that makes comparison easier. The same rating scale and descriptors are usually used to rate both sets of features and criteria, although separate scales could be established for each.

Step 12: Compute Technology Scores.

Purpose: To calculate (1) a customer feature score and (2) a manufacturing criteria score for each technology to provide a basis for comparing the attributes and merits of technology options for each function.

At this point, each technology cell in the matrix has two sets of ratings, one for customer features and one for manufacturing criteria. These two sets of ratings will be multiplied by their respective weighting factors. Their products will be summed to produce two separate total weighted scores. That is, the customer feature score is obtained by summing the products of the individual normalized customer feature scores, developed in Step 8, and the technology feature score, developed in Step 11. Likewise, a manufacturing criteria score is calculated by summing the products of the individual manufacturing criteria weighting factors, developed in Step 10, and the technology manufacturing score, developed in Step 11. This process is repeated for each technology for each function row.

At this point there is a natural tendency to combine these two sums into one overall technology cell total score. The two summed scores for each technology should be kept separate. Combining them consolidates too much information. Important signals are easily missed. With one overall score, you are inclined to choose the technology with the highest combined score, which many times is not the best choice. Both scores are needed to search for trade-offs and to balance the right combination of technologies for customer and company.

Step 13: Create Technology Paths

Purpose: To choose one technology in each function row to create a technology path representing the best combination of options for both customer and manufacturer.

The matrix has now been completed both qualitatively and quantitatively. The team is in a position to start choosing technologies to form a technology path across functions. The path is really a concept and not necessarily a detailed design. The path will be based on the best combination of function technology scores. Strong and weak points of each technology become obvious. The scores allow the team to discuss and choose the best combination of technologies that best satisfy both customer and manufacturer needs. Trade-offs will have to be made. Each different combination of technologies represents a different concept. At first glance, you might expect that there are an infinite number of possible paths through the various combinations and permutations of technology cells. In reality, this is usually not the case, because there are combinations of technologies across function rows that are illogical or impossible for both customer or manufacturer criteria.

For example, sometimes it is electromechanically or physicalchemically impossible to connect certain technology combinations. In any case, there are still many options available, and picking the best combination of technology scores will help surface options and identify the most likely candidates. To begin, several paths are created that have different objectives. For example, a least-cost path could be constructed by connecting the lowest-cost technology in each row regardless of whether they are logical or optimum combinations. This path becomes the least costly concept that could ever be achieved and still meet customer and manufacture needs. It is highly likely that this path may be technologically impossible to achieve, but, if it were possible, it would be the lowest-cost alternative. The cost of this path then becomes the target cost. The same principle can be used to construct other paths for best quality, reliability, maintenance, QWL, and so on.

Selection of technologies may not be easy. Constructing the various paths begins to bound the various options. Some technologies will consistently surface for all paths. This does ease the selection process. As discussed in Chapter 9 on TR, a mini-strategy matrix (Figure 9.5) or a strategy map (Figure 9.6) can be used to condense the path options to a more simpler document for presentation or communication.

With some products, selection of technology is confusing. What happens is there is an interdependency among different function technologies. That is, whichever technology is selected for function x will constrain what technologies can be selected from function y and z, and so on. Some teams select the technology from the most important function and let this selection determine the technology selections for the remaining functions. To do this, they assigned importance weighting factors to all functions using the 100 point sum to unity rule. They then let the higher weighted functions guide their selection of technologies.

Step 14: Select the Best Technology Path

Purpose: To select the path(s) that will become the final recommendation(s) for product design.

The team narrows down the options. After presentation and recommendations, the concepts are eventually turned over to design and manufacturing who will start prototyping, testing, and designing. It is not uncommon that two paths are developed for final recommendation, one for "today" and one for "tomorrow." The today path is based on the best combination of technologies that fit into current time constraints to get the product out on schedule. The tomorrow path is clearly the best choice, but still requires some research or development that will not permit meeting current schedules. It is this path that sets future direction for marketing and manufacturing. It is also the concepts that will bring added value to customer and company and that will allow best fit with the company strategic plan. All of the other least-cost and quality paths will still be retained for reference for presentation to management. Retaining these paths in an appendix will document and support current decisions. They can be used to show management that all possible options were considered in deriving the current recommended concepts.

Okay, You've Built the Matrix, Now What?

At this point, several technology paths have been developed. However, there may be gaps in information. For example, there may still be questions about some of the importance ratings or some parts of the competitive analysis. These kinds of questions frequently arise. So, it will be necessary to verify some of the COPC matrix information. Many times verification will require customer data collection of some sort.

The COPC process helps define areas of needed information and directs the surveys and data gathering into areas of greatest payback. In this respect, the data and signals developed in the COPC process provide the input to design better questions for focus groups and customer surveys. We recall a classic case where a focus group questionnaire was developed independent of the COPC project. At the last minute, a representative from market intelligence decided to review the questionnaire with the COPC team. Much to everyone's shock and amazement, more than 50% of the focus group questions were inappropriate and way off target. Little would have been gained if the original set of questions would have been used. After this embarrassing discovery, the market intelligence representative participated on the COPC team almost to project completion. The result was an almost completely revised set of focus group questions.

In some COPC applications, prototype products have been built and used at focus groups. Focus groups may involve videotaping customer reactions from behind a two-way mirror. By observing a focus session, a COPC team can obtain first-hand information on customer reactions to the prototypes and/or to a prepared set of questions. The data from the interviews, questionnaires, and focus groups are summarized. The summary information is mapped back into the original COPC matrix. Importance and/or competitive analysis ratings are revised and updated. New features or criteria are added as necessary. The technology scores are recalibrated to reflect the data updates. The original technology paths are checked for validity. New technology paths may also surface. The matrix is fine-tuned to the point where a decision can be made on the technology path(s) to recommend.

The matrix should be continually updated and considered a living document. If the matrix is properly updated, the COPC process is self-correcting. Feedback loops often require the team to go back and update prior data, which, in time, provide some missing links and course corrections.

IS THERE LIFE AFTER COMPLETING A COPC MATRIX?

After a technology path has been chosen the following methodologies, and many others, can be used to refine it and bring the concept closer to design.

1. Failure Mode Effects Analysis (FMEA)
2. Cost estimating
3. Value Graph of functions

4. Decision and Risk Analysis (DRA)

5. Focus groups, market surveys, conjoint analysis

Used as a potential problem analysis, FMEA can be applied to the technologies of certain critical functions to determine the liklihood of failure, failure mode, cause of failure, frequency and seriousness of the failure and recommended measures for improvement. Both the technology and functions from the COPC matrix help define all the interfaces in a system. Once determined, the interfaces provide the foundation for conducting a FMEA. Generally a separate small team is established for each interface. Critical areas from each team are integrated into an action plan for project management.

The chosen COPC path should also be subjected to a more rigorous cost analysis. This cost estimating will subject the concept path technologies to more scrutiny and surface any need to select another concept path.

A value graph can be constructed by plotting the percent importance of the functions versus their percent cost. The function importance can be estimated using many of the value measurement techniques discussed in Chapter 4. The importance measures are normalized to a percentage. Function cost may be derived from the cost of the technologies chosen from one of the preferred paths. The cost analysis mentioned above can also provide this information. Costs, too, are normalized to a percentage. These function value components are plotted on a value graph.

An addition can be added to the right side of the COPC matrix. That is, product technical requirements can be developed and added as another matrix. A relationship matrix can then be developed for technologies and product technical requirements. The relationship matrix is described in detail in Chapter 8 for the House of Quality. This relationship would be based on how well the selected path technologies satisfy the product technical requirements.

The technology path will also provide enough information for the manufacturing people to begin a Design For Assembly (DFA) and/or Design For Manufacturability (DFM) analysis.

If any of these activities result in a negative outcome, a new path may have to be developed or certain technologies may have to be changed for particular functions. However, in choosing a new path or new technology most of the homework has already been done. You merely go back to the matrix and search for the next most appropriate technology scores.

HOW DOES COPC RELATE TO THE BUSINESS CASE?

In Chapter 13, we discuss the need for a design team to have access to the company business plan when developing products. Because COPC is a design process whose output is a product concept it should certainly connect with the business plan. The recommended product concept should be checked with the business to be sure both are congruent with each other. Such a comparison can work both ways: Does the concept fit the plan and does the plan fit the concept? Use of the COPC process can provide the dialogue for challenging the focus of the business plan.

As with TR, the technology path(s) information can be used as input to a decision and risk analysis (DRA) process for building a business case. Some of the technologies surfaced in the COPC process become candidates for Research and Development projects. These projects are generally approved based on their fit to a business case. DRA and portfolio analysis are quite often the processes used to derive the business case. The COPC process integrates very nicely with this type methodology. That is, the use of the COPC process results in a concept/design and the application of DRA can help one develop a business case based on the COPC design both favorably and unfavorably.

GENERAL OBSERVATIONS OF THE PROCESS

The process of building the matrix is fairly straightforward. The behavioral and scoping activities to initiate the process are more complex. The perceived simplicity of the matrix can also be a detriment, because many teams believe they can run the process and build the matrix themselves. The entire process from launch to finish is best done with a neutral third-party facilitator to lead both team building and process. Otherwise, the teams can easily get mired in turf protection, drown itself in details, get lost in the woods, and quickly rush to solve unclear problems and shortcut the process. The process facilitator provides the COPC model and keeps the team on track. The facilitator's function is to keep the team well focused and make decisions on how long to dwell on a particular problem. This person does not get involved in content discussion and ratings but does ask devil's advocate questions while, at the same time, keeping the pace moving.

Meeting length and frequency are important. As with the other processes discussed in this book, the initial launch meeting is best scheduled for a full day. Regularly scheduled follow-up meetings should be 4 hr and held at least once per week (twice a week is better) until the COPC project is finished. Because of the complexity and detail in designing products, infrequent meetings less than 4 hr do not work as well as longer duration meetings. Team members spend too much time getting back to speed to where they previously left off in the process.

Using the right number and blend of core team members is important for developing a well-rounded knowledge base and for minimizing the effects of bias and political influence. Keeping the same people on the core team throughout is very important. It takes time to develop team cohesiveness and a team benchmark/framework for making judgments. Excessive changes in core team membership fractures the team bond and the mental reference base used in decision making.

The credibility of the COPC process and output is highly dependent on the consistency of team decisions, ratings, and applications of the process. Group consensus is the preferred method of decision making. However, where reaching consensus may take considerable time, the process facilitator should guide the team to proceed with a majority decision and write a minority report. Good meeting notes should be compiled so that the reasons behind group decisions can be retraced if changes have to be made at a later date. Many facilitators maintain an ongoing list (referred to as a "parking lot" in Chapter 13) for recording ideas, questions, and action items that surface prematurely out of sequence in the COPC process. This is very helpful and helps bring focus on these issues later at a more appropriate time. The parking lot list of items is always kept visible at all team meetings either on wall charts or typed notes.

The COPC matrix is to be considered a living document that is continually updated with both qualitative and quantitative information. Product variables, customer needs, environments, and technology change with time. The matrix should be updated accordingly. Building the matrix on an electronic spreadsheet greatly simplifies the updating process. We have even constructed matrices on-line during the team meetings using a lap-top or tabletop PC. An electronic screen coupled to the PC to project the computer image on a room-size projection screen has also proven very useful.

With the matrix on a spreadsheet, the team now has the opportunity to more easily interact with other team members and decision makers in a simulation mode. Numbers can be changed and scores recomputed instantly in a "what if" capacity. This is very appealing especially to decision makers, and such interaction serves as stake building to foster ownership in the process, which in turn increases chances for acceptance of recommendations.

An interesting variation in using the electronic spreadsheet is to use it with customers/users. The entire COPC matrix would be developed by the team and put on the spreadsheet. Customers would then be asked to interact with the matrix. They can verify the customer needs by entering their numbers for their importance of the needs. They also would add additional needs that have not been included in the team-generated matrix. After direct customer input, the program would be run to recalculate technology scores. The on-line customer interface is particularly useful for those kinds of products that must be tailor-made for specific customer needs. In this case, the technologies in the matrix may be various interchangeable modules that can be integrated to form various customized configurations. The customer's importance ratings would highlight those modules best suited for the customer application.

In this respect, the glossary of terms is very useful. It provides a consistent language in the midst of change and amendments. The scoring and rating processes require a good understanding of the item(s) under discussion. Experience shows that considerable time can be lost due to confusion when terms are not specifically defined. Good recorded definitions not only speed up the rating processes but promote better consistency as well. Communication is improved through a better shared understanding of information. Believe it or not, information can sometimes be a barrier to initiating a COPC process. We refer to this as *information paralysis.* This is the condition where teams think they know nothing about customers and product performance parameters. They believe conducting customer surveys and collecting data have to be done before beginning a COPC project. The belief is usually unfounded, because it is highly unlikely that intelligent people involved in the sale of products know nothing about its use and performance. The COPC process draws out the collective knowledge of the team and brings all team members to a higher level of knowledge.

ATTRIBUTES OF THE PROCESS

The process of initializing the matrix activity, constructing the matrix, and choosing a concept path provides a vehicle for dialogue that has many advantages.

1. It is probably the first time marketing, research, manufacturing, and other individuals involved have a mutually developed common-source reference base displaying all of the parameters necessary in making design decisions.
2. Both vertical and horizontal company communications are enhanced.
3. Because the matrix was constructed by an interdiscipline team, it lends credibility to the process and the resulting recommendations.
4. Because it is a common-source reference base, it permits one to see interrelationships and highlights information gaps. This, in turn, provides valuable input for designing surveys and focus groups to obtain the right information.
5. Because it is a structured process, it becomes an assault on complexity and forces participants to deal with complex issues one piece at a time.
6. When kept up-to-date, and because it is based on functions, the matrix will provide a technology reference base for designing future similar products.
7. The matrix provides the basis for designing a product that meets both customer and company needs.
8. The process highlights features important to the customer.
9. The competitive analysis provides the structure to pinpoint strengths and weaknesses for both company and competitors' product features and functions.
10. The process highlights areas needing improvement and attention.
11. Because it is function based, the process of expanding the technologies allows for greater use of creativity.
12. The two sets of criteria and the technology ratings allow you to choose the best combination of technologies to satisfy both customer and company.
13. Quantifying the matrix also promotes communication because numbers are easily debated and they highlight areas of disagreement.

CONCLUSIONS

The COPC process has proven to be successful in a wide variety of applications. The reason for many of the successes is that they were also designed and planned politically and had top-down management support and funding. The process as designed and initially presented to management and sponsors had the following attributes that contributed to successful application:

1. There was a perceived advantage to using the process.
2. It was compatible with the existing company/department structure and operating philosophy.
3. The process is based on the existing language and vernacular of the organizations. There were no new threatening terms or fads.
4. The process does not have a high price in terms of anxiety, emotion, and comfort level. It appeared "safe."
5. There is little to lose if the process is terminated before closure.

Finally, the COPC process as well as technology roadmapping, Quality Function Deployment, and other value management processes, are vehicles for dialogue. To be able to communicate people and teams need a structure and a language. The value management processes like COPC provide the structure through the matrices and the format procedure used to construct them. A language is provided by the numbers that are derived and put into the matrices by the team members. Team members can debate different choices for numerical inputs. Considerable dialogue is required to merely derive the rating scales used to select numerical inputs to the matrix. The combination of the COPC matrix and the numerical language of value measurement provide the basis for enhanced communications.

REFERENCES

1. Cook, T.F., "Determine Value Mismatch by Measuring User/Customer Attitudes," *Proceedings, Society of American Value Engineers* **21**, 145–156, 1986.
2. Snodgrass, T.J., and Kasi, M., *Function Analysis: the Stepping Stones to Good Value*, University of Wisconsin, Madison, 1986.
3. Meyer, D.M., "Direct Magnitude Estimation, a Method for Quantifying The Value Index," *Proceedings, Society of American Value Engineers* **6**, 293–298, May 1971.
4. Saaty, T.L., *Decision Making For Leaders*, RWS Publications, Pittsburgh, Pa., 1988.
5. King, R., *Better Designs in Half the Time, Implementing QFD Quality Function Deployment in America*, GOAL/QPC, Methuen, Mass., 1987.
6. Shillito, M.L., "Function Morphology," *Proceedings, Society of American Value Engineers* **20**, 119–125, 1985.

IV

THE MANAGEMENT
OF VALUE

This section describes several methods for managing value. Valuism, an evolving worldwide system for value improvement is described. Value analysis, value engineering, and value management methodologies, which are central to valuism, are described in detail. Value analysis, value engineering, and value management all rely on interdisciplinary teams to analyze functions and improve the value of products and services. The organizational and behavioral aspects of this teamwork are treated in detail. Value planning, a powerful new methodology based on function and technology planning, is described and illustrated. We conclude with a forecast of the future of valuism and value management.

11

VALUISM

David J. De Marle

VALUISM

As the last decade of the twentieth century unfolds, the signs are unmistakable, communism is in decline, capitalism is nearing bankruptcy, and a widely used but unnamed philosophy is growing. Let's call it Valuism. Valuism is a collection of management principles and methodologies used by industrial organizations to produce value. Centered in the entrepreneurship of business, it originated during the Industrial Revolution and has evolved into an efficient and effective management process used by corporations around the world. Based upon studies of market needs, it focuses the resources of corporations upon the creation and production of products and services to fill human needs. Utilizing a large number of modern techniques, such as value measurement, cost–benefit analysis, and project management, it marshals capital, labor, and creativity on the production and constant improvement of value. In the factories, financial institutions, markets, and universities of the world, valuism couples science and technology with entrepreneurship in the creation of value.

Hidden in the internal working of business, valuism is more pervasive and powerful than capitalism (which fostered its growth), communism or socialism, which are unable to compete with its performance. Today, valuism lies at the heart of competitive enterprise. It governs the creation of new goods and services and rules the economies and markets of the entire world. Valuism ended the Cold War, led to the fall of communism in Eastern Europe and Russia, and is inherent in the drive toward a world wide market economy. The quest for value is universal, companies and nations that embrace valuism grow, while those that do not die.

The goal of valuism is the efficient production, distribution, and use of high-quality, low-cost goods and services to satisfy human wants and needs. Driven by the

233

consumer's demand for value and the producer's need to produce goods competitively, valuism is a dominant factor in our economy. Valuism consists of a set of tools that corporations use to measure, create, and manage value in a rational manner. It is based upon the philosophy that products and services *must* meet the needs of customers. Valuism includes a growing number of value-oriented business techniques and practices that allow organizations to improve their internal effectiveness and efficiency. Valuism's methodologies are centered around a growing number of professional disciplines and societies including industrial engineering, quality assurance and control, and value engineering.

VALUISM AND VALUE: THE FOREST AND THE TREES

"Wait a minute!" you say. "I've never heard of valuism. Why haven't I heard of it before?" Perhaps because valuism has been hidden in a forest of many trees. Each tree was so valuable that we failed to see the forest itself. Through the ages people have worked for, connived for, cheated for, and died for things they value. We have been so absorbed in cutting the trees in this forest that we have failed to recognize the forest in which we live. Preoccupied with goods and services, we have failed to recognize valuism as the forest around us.

It has been said that the last thing a fish will be aware of is the water it swims in. Only when a fish is dragged from the water does it sense its necessity. Like fish pulled from an unseen ocean, we have only recently begun to understand the vitality of valuism. Pulled from profitability to loss by unforeseen competition, we devise value based techniques that allow us to survive and grow. Yet these techniques have not been seen in aggregate. Often those who invent new methodologies create parochial practices to obscure similarities among methodologies. This in turn obscures valuism. Valuism is a new concept. It maps the world in a new way. I hope that awareness of valuism will grow and stir further thought and discussion. Study of its elements will lead to the integration of many business techniques into a comprehensive methodology that will enable both industry and government to prosper. Chapter 15 offers some thoughts on how this integration might occur.

THE GROWTH OF VALUISM

Chapter 2 described the evolution of a number of different products and services. Valuism has also evolved and a brief history of valuism is in order.

The Evolution of Value Measurement

In the late 1700s and 1800s, the English philosophers Jeremy Bentham, James Mill, and John Stewart Mill developed a utilitarian theory of behavior based on hedonism that associates the rightness of an act with its consequences. According to Bentham's ethical philosophy, human behavior—although hedonistic—should be conditioned to approve or disapprove of actions based on their tendency to increase or decrease the pleasure of all parties concerned.[1]

Bentham proposed that the utility of actions should be measured to determine which actions would produce the greatest good for the most people. Bentham suggested the development of a *felicific calculus*, a mathematical system that he claimed, would allow economists to make such measurements and comparisons. He suggested that government could promote the well being of its citizens by funding projects that would satisfy the needs of the average citizen, and he tried to develop a method to measure the value of government actions in economic terms. Following Bentham's suggestions, attempts were made to measure the marginal utility of goods and services.[2] In 1881 the great English economist Francais Y. Edgeworth published his famous monograph, *Mathematical Phychics*, in which he used Bentham's utility theory to describe market behavior.[3] Bentham's *felicific calculus* and Edgeworth's mathematical phychics were early attempts to introduce value measurement methods to management. Cost–benefit analysis, frequently used today to help determine the consequences of decision making, is one descendent of these early methods.[4] The value measurement techniques described in Chapter 4 are other derivatives of Bentham's philosophy.

The Growth and Development of Methods to Improve Industrial Efficiency

With the industrial revolution, business became the center for the development of a large number of methods designed to improve efficiency. Perhaps the first person to report the development of these methods in industry was Charles W. Babbage, a professor of mathematics at Cambridge University. In 1832, Babbage published his book, *On the Economy of Machinery and Manufacturers* in which he described how manufacturing plants in England had saved time and improved productivity by subdividing tasks into elements that could be constantly repeated. Babbage was the first person to study the time a factory worker required to learn a particular task (today's learning curve). His studies of the effect of educating workers about the amount of waste were a century ahead of the first publication on quality control (QC).[5] Babbage was a genius well ahead of his time. Intrigued with the application of mathematical analysis to industrial problems, he is perhaps best known for his attempt to build an "analytical calculating machine." His mechanical device is now recognized as a major historical forerunner for the computer.

Industrial Engineering

At the turn of the century, most factories were still very labor intensive and methods were needed to reduce the labor content of production. This need led Taylor and the Gilbreths to develop time and motion study to improve the efficiency of the factory. Taylor referred to his methodology as *scientific management*.[6] Frank B. Gilbreth and his wife Dr. Lillian M. Gilbreth extended Taylor's time study methodology to motion analysis.[7] The Gilbreths subdivided motions into *therbligs* (Gilbreth spelled backward), which could be used to analyze motion in a more scientific manner. Soon *efficiency experts* were timing the motions involved in human labor, and management was using worker incentive systems based upon piecework to pay their employees. Taylor and the Gilbreths had launched a management practice based on time and motion study that was to dominate industrial practice for over 60 years.

Other creative individuals added new innovations. Alan G. Mogenson developed a *work simplification* process for training workers to analyze and challenge their own work. Morgenson's method of training has been copied by many other consultants who have devised proprietary management techniques for improving value. The *Scanlon plan* is an example of this. The Scanlon plan is a franchised incentive method used by several major corporations. Under a Scanlon plan, groups composed of employees and supervisors meet regularly to discuss suggestions to improve productivity and lower costs. The workers receive a bonus equal to 75% of the savings with the company getting the extra 25%.

As time went on, many people were trained in methods study. To accommodate the growth in *scientific management* the original Taylor Society was followed by the Society for the Advancement of Management and eventually by the American Institute of Industrial Engineers (AIIE), which was founded in 1948 in Columbus, Ohio. AIIE defines industrial engineering as a field of engineering

> concerned with the design, improvement and installation of integrated systems of people, materials, equipment and energy. It draws upon specialized knowledge and skill in the mathematical, physical and social sciences together with the principles and methods of engineering analysis and design to specify, predict and evaluate the results to be obtained from such systems.[8]

The usefulness of time and motion study and the incentive system began to decline as mass production and automation replaced piecework. In the 1950s, industrial engineers began to mathematically model the performance of manufacturing systems composed of people and machines. The development of the computer and the advent of linear programming allowed companies to do *operations research* on complex manufacturing systems. Inventory control systems were studied to allow manufactures to reduce the cost of in plant inventories. This new focus of value thinking increased the productivity of the plant and reduced factory costs further.

Market research methods based on statistics allowed laboratories to design products and services to better satisfy customer needs. Statistical sampling theory was applied to QC and product design and development. These are a few of the many management techniques that were developed to help companies improve the value of the products they create and sell.

Quality Control

Work at Bell Telephone in the 1920s[9] on the application of statistical sampling to manufacturing process control led to a process of statistical quality control (SQC). W. Edward Deming[10] and Joseph M. Juran[11] while working at Western Electric successfully applied SQC to manufacturing. Both left Western Electric and began to consult on QC. They were early members of the American Society for Quality Control and consulted extensively on SQC. Deming was invited to Japan in 1950 to consult on SQC by the Union of Japanese Scientists and Engineers. This society spread Deming's 14-point QC method throughout Japan, where it blossomed. Subsequently Juran and other QC gurus were invited to consult in Japan, and this led to the transformation of Japan into a nation known for the excellent quality of its goods and services. In the 1950s, the cost of manufacturing was much lower in Japan than in the United States or Europe. By improving the performance of its products and selling them at low prices, Japan was able to offer products of superior value to consumers around the world. Japan's success rejuvenated interest in QC and led to worldwide acceptance of QC.

Value Analysis/Value Engineering

Value analysis and value engineering are very effective value improvement techniques that had their origin in studies of product changes that occurred during World War II. Larry Miles, then director of purchasing at General Electric Company, developed VA/VE, a function-based methodology for improving value in 1947. VA/VE can be used to reduce cost while simultaneously improving functional performance. Its use was taken up by the military who introduced value engineering change proposals (VECP) incentives into Department of Defense contracts. VECPs allow contractors to receive a portion of the VA/VE dollar savings. VECP incentives, discussed in Chapter 12, caused VA/VE to spread throughout the defense industry. Today VA/VE is used around the world and is a key ingredient in valuism. Chapter 12 describes the VA/VE process in detail and shows how it evolved into a worldwide process of value management (VM).

Today, VM provides an explicit value focus upon which the other methodologies play. It does this by

1. allowing a manager to quickly assess the value of any management methodology to specific organizational needs.
2. helping a manager form and staff a value team to meet these needs.
3. providing a value team with an efficient process for improving value.

We will describe these methodologies at length in Chapters 12 and 13.

ORGANIZATION AND STRUCTURE OF VALUISM

Figure 11.1 is a diagram that shows some of the important components and linkages within valuism. Valuism links people together in a network of different professional organizations. The diagram is based upon Eric Berne's[12] cellular model of an organization. To the left of the diagram is a circle labeled A, which represents an industrial corporation or business establishment. The circle represents Berne's membership boundary, and all of the employees of the company are contained within this circle. Professional employees of the company who have expertise in a value-centered

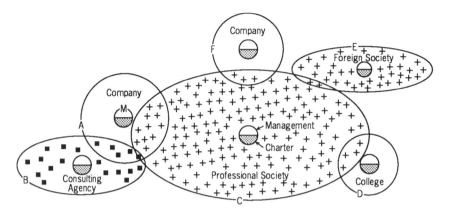

FIGURE 11.1 Structural elements of valuism.

management service such as industrial engineering, QC, or VE are shown by the + marks within circle A. These value methodologists act as consultants within the company in their area of expertise.

A second circle labeled M is located at the center of this circle and is used to designate the organization's top management structure. This circle is bisected by a line drawn through its center to show that management consists of two separate components. Following Bern's organizational model, the individuals who manage the organization are designated by the upper semicircle, whereas the company's charter and its associated managerial rules and procedures are designated by the lower shaded structure in which the people in the organization are controlled by a management nucleus composed of management people and a set of management practices. Leadership is vested equally in top management and in these practices which provide checks and balances. Berne believes that every human organization has a cellular structure. A second company is shown as circle F at the top of Figure 11.1. This company has different employees and management than company A but also contains a number of value methodologists like those in company A. These individuals are again designated by +'s. They are experts in a specific methodology that forms the basis for membership in a professional society drawn as the large ellipse C at the center of Figure 11.1. This professional organization is composed of members who are employees of a large number of different companies like companies A and F. In addition to people employed by industry, other members are employees of various colleges like that shown as circle D. Other members come from small consulting agencies and businesses such as those shown by ellipse B. Finally some members are also members of other professional societies like that shown by ellipse E, used to designate a foreign professional society closely affiliated with the domestic professional society C.

Note that each of these different organizations have a top management nucleus composed of officers and charters that work in concert to run each organization. Each organization's management is responsible for the conduct of the organization and the administration of rules and regulations that are defined in the organization charter. These rules define membership and professional practice within the organization.

Professional societies and management consulting businesses furnish effective management practices to corporations and are a key element in valuism. Today there are a large number of consulting businesses and professional societies that foster the growth and dissemination of value-oriented business methods throughout business. Some of the larger professional societies, such as the American Institute of Industrial Engineers, are closely linked to academia where they help define and control the academic requirements for training in their specific professions. In our global society, many American professional societies are also linked to overseas organizations, which foster the growth of a profession in other nations. Thus, the Society of American Value Engineers (SAVE) interfaces with SJVE, the Society of Japanese Value Engineers, and with other value societies in 28 different nations. Valuism is thus a massive movement with an international constituency. It contains many of the intellectual leaders who shape and manage free enterprise in our world.

VALUISM IS A MULTINATIONAL INDUSTRIAL MOVEMENT

I want to emphasize that valuism developed as a methodology for managing companies in free enterprise systems. Competition in a free market provides an incentive for industry to develop and use tools that allow it to create products and services that are both effective and efficient. Valuism is not a political movement and has not grown in governments or agencies that lack competition. Although government agencies provide valuable services, they seldom do so efficiently or effectively. Lacking competitive pressures, government has grown increasingly large and bureaucratic. It lacks the incentive to control its own growth and efficiency. Accordingly as government grows, it often becomes predatory, depending on increases in taxes and deficits to pay for its growth.

In contrast industries have learned to downsize their operations by divesting themselves of antiquated products and services. They have learned to do this while increasing their manufacturing and marketing capacity and introducing new and more valuable products and services. Until recently government has not done anything similar. Now as large and inefficient government agencies in Communist countries fail, their leaders are privatizing government agencies and developing free market economies. In nations that lack the underpinning of efficient business organizations, governments are forced to privatize to prevent famine. Russia and Eastern Europe are selling government assets to private industries. These nations are embracing valuism. In Western democracies the size and expense of government continues to grow, supported only by the robust nature of business and the wages and profits upon which government preys. Yet in America government, deficits grow as entitlement programs expand and government-insured banks fail.

Linkages

Figure 11.1 shows how business organizations and professional societies and management consulting agencies are linked together on a global scale. A variety of entrepreneurial consultants provide business with tools for improving the value of their products and services. New management tools are constantly being developed by people to fill business needs. Consulting organizations are formed around these new procedures that are often incorporated into the practice of corporations and professional societies that embrace them. In aggregate, this entrepreneurial system allows business to continuously improve value. This diffuse and heterogeneous network constitutes an ad-hoc system of value management.

From decade to decade valuism is dominated by specific methodologies that industry most needs. In the early part of the 20th century it was dominated by efforts to lower labor costs and increase the efficiency of manufacturing. Industry concentrated on mass production processes and time and motion studies governed piecework and associated incentive systems. In contrast, the later half of the twentieth century has been dominated by efforts to improve the effectiveness of industry. Today valuism is dominated by quality, design, and scheduling methodologies.

Corporations

Hundreds of thousands of large companies and millions of small companies are part of valuism. The *Wall Street Journal* annually publishes a list of the 100 largest publicly held corporations in the world.[13] Listed below in Table 11.1 are 15 multinational companies that had the largest fiscal profit in 1989. All figures are in billions of U.S. dollars.

It is interesting to compare the annual sales and profits of these multinational companies with the gross national products (GNP) and budgets of different nations

TABLE 11.1 The World's Most Profitable Corporations (1989)

Company (Country)	Profit ($\times 10^9$)	Sales ($\times 10^9$)
Royal Dutch Shell (Netherlands)	$6.539	$85.412
IBM (USA)	5.260	62.700
General Motors (USA)	4.220	110.000
General Electric (USA)	3.940	54.600
Ford Motor (USA)	3.840	82.900
Daimler-Benz (West Germany)	3.802	45.170
Exxon (USA)	2.980	95.200
Philip Morris (USA)	2.950	44.800
British Petroleum (U.K.)	2.807	47.800
AT&T (USA)	2.700	36.100
Fiat (Italy)	2.609	41.057
British Telecom (U.K.)	2.523	17.858
Dow Chemical (USA)	2.480	17.600
Dupont (USA)	2.480	35.500
Toyota (Japan)	2.405	55.759

in the world. Multinational companies are often wealthier than many populous nations. Listed in Table 11.2 are the budgets and GNP of 15 different countries of the world.[14] Population figures are in millions and budgets and GNP figures are in billions of U.S. dollars.

Large multinational corporations that use valuism to produce products that satisfy human needs have grown so powerful that their annual profits and sales are larger than the budgets and GNPs of many nations. Yet in the United States, the richest nation in the world with a 1990 GNP of $4.8 trillion and a budget of $1.2 trillion, the absence of valuism in the federal (and state) governments, is producing huge deficits that are bankrupting America.

Consulting Agencies and Professional Societies

Often professional societies originate from the entrepreneurship of single individuals who originate new and useful management methods. These entrepreneurs form small consulting companies that sell the entrepreneurs' methodology to business clients. As other consultants begin to emulate the success of the original guru, the methodology is often incorporated into a professional society. Originally, the society is run for the benefit of a small group of consultants who use it to attract clients and restrict practice to members certified by the society. As professional societies grow, they assimilate peripheral technology and broaden their range of services. As the technology grows,

TABLE 11.2 Population, Budget, and Gross National Product of
Various Nations (1987–8)

Country	Population ($\times 10^4$)	Budget ($\times 10^9$)	Gross National Product ($\times 10^9$)
Argentina	32.3	$9.5	$74.3
Bangladesh	117.9	3.3	18.1
Bolivia	6.7	2.8	4.6
Brazil	153.8	40.1	313.0
China	1130.1	66.1	350.0
Colombia	32.6	4.7	39.0
Egypt	54.1	12.2	25.6
Ecuador	10.5	2.6	9.4
Ethiopia	51.4	2.0	5.7
India	850.0	56.0	246.0
Indonesia	191.3	21.1	75.0
Madagascar	11.8	0.48	2.1
Mozambique	14.7	0.427	4.7
Nigeria	118.9	4.8	78.0
Pakistan	113.2	9.1	39.0

large companies form departments composed of in-house consultants to spread the practice throughout the companies. Other companies become aware of what their competitors are doing and copy them. In some cases this bandwagon effect sweeps across an entire nation. The resulting movement allows national corporations to increase their productivity and competitiveness.

The current international quality movement is a good example. As mentioned earlier, work at Bell Telephone on applying statistical sampling to manufacturing process control led to a methodology of statistical process control.[15] Quality control was promoted by W. Edwards Deming[16] and Joseph M. Juran,[17] both of whom worked as statisticians at Western Electric, the manufacturing arm of the old Bell Telephone System. Deming left Western Electric to form his own QC consulting business. Juran was employed for 17 years at Western Electric before he formed the Juran Institute Inc. to consult on QC. They were early members of the American Society for Quality Control, which was formed prior to World War II to promote QC. Deming and Juran introduced QC to Japan in the 1950s where it blossomed. The result was the revolutionary transformation of Japan from a nation known for shoddy products into a nation known for excellent quality. By improving the performance of its products and selling them at low prices Japan became a center of valuism and an economic giant.

Today there are thousands of management consulting agencies and hundreds of professional societies in the world. Only those that provide management-oriented services that improve value should be included as components in valuism. A partial list of societies that are directly involved in providing value-oriented management services to industry is:

Institute of Industrial Engineers (IIE)[18]

American Production and Inventory Control Society (APICS)[19]

American Society for Mechanical Engineers (ASME)[20]

American Society for Quality Control (ASQC)[21]

Human Factors Society[22]

International Material Management Society (IMMS)[23]

Operations Research Society of America[24]

Society of American Value Engineers (SAVE)[25]

Society of Manufacturing Engineers (SME)[26]

I recognize that this short list excludes many important scientific societies, such as the American Chemical Society and the American Association for the Advancement of Science. These societies certainly provide important professional services to industry. However, they are usually not directly involved in supporting management in its task of planning, directing, and controlling the creation of value by industry or government. Accordingly, I exclude them from consideration here. Only professions, consulting organizations, and methodologies that directly effect the performance cost ratio of corporations and their products and services are included in the definition of valuism.

Menu of Services

Through the years a large number of specialized services have been developed to assist industry. Today the number and variety of these services can confuse managers who feel the need for management assistance. Table 11.3 shows some of the management services that are available today. Each of these methods can help improve the value of a company and its products or services. Value increases if performance is improved or if cost is reduced. Some methods improve value by improving the performance of a company's products and/or services. Other methods improve value by reducing the cost of a company's products or services. Other methods focus on improving the value of the company itself. Still others focus on the future, and plan ways to improve the value of new products and services.

Table 11.3 classifies a number of different methodologies based upon their area of applicability. Although experts in the individual methodologies may object to this classification and point to examples where their methodology has been effective in areas I have not designated, I believe the classification is generally valid and can help managers select methodologies based upon their specific needs. This can be difficult to do, because the proprietary nature of these methodologies often means that an "expert" consultant skilled in a specific methodology is frequently unskilled in other methodologies that could be used to improve value. Value measurement techniques, described in Chapter 4 can be used by a manager to select appropriate methods from this list.

An astute manager and good friend of mine does not think kindly of many of the management consultants that offer these proprietary services. He likens them to the patent medicine men who at the turn of the century sold a wide variety of elixirs to the unwary from town to town throughout the United States. With stirring words and heartfelt passion, these hucksters told one and all of the multitude of benefits that would follow the consumption of their special patent medicines. An ounce of this potion would cure rheumatism or arthritis, while weekly consumption of another potion would cure any affairs of the heart and would surely improve your muscle tone and general well being.

Most of these patent medicines disappeared and are now forgotten. The peddlers carts and horse drawn wagons are nostalgic things of the past—stage props in *The Music Man*. Yet in my friend's view, the peddlers have not disappeared. He calls them management consultants, who substituting air travel for the buckboards of yesteryear, ply the boardrooms of the United States with briefcases and lap-top computers. In his gloomy view, most of their management elixirs will fail. He had a bad experience with some of these elixirs and thinks that everyone else will rue the day they use a management consultant. He points out that without the Food and Drug Administration to insist on truth in advertising, there is little control over today's management consultant who is here today and gone tomorrow.

The Effects of Valuism

I do not share his pessimistic view. I see each of these management tools as providing useful services. I feel that the record is good and that valuism has created an explosion

TABLE 11.3 Management Methodologies and Their Primary Uses

Abbreviation	Method	Performance Improvement of		Cost Reduction of		Planning
		Product	Organization	Product	Organization	New Value
CR	Cost reduction			1	1	1
DFA	Design for assembly			1	1	
DRA	Decision and risk analysis					1
DTC	Design to cost			1		
FA	Function analysis	1	1	1		1
HK	Hoshin Kanri planning					1
IL	Integrative learning		1			
JIT	Just in time		1		1	
KT	Kepner Tregoe		1			
LP	Linear programming		1			1
MBO	Management by objectives		1			1
MS	Methods study			1		
NA	Needs analysis	1		1		1
OD	Organization development		1			
PA	Portfolio analysis	1				1
PM	Performance management		1			
PM	Performance measurement			1	1	1
PM	Project management		1			
PM	Participative management		1			
QA	Quality assurance	1	1			

QC	Quality control	1			
QFD	Quality function deployment	1			
SD	System dynamic modeling	1	1	1	1
SE	Simultaneous engineering	1	1	1	1
SP	Strategic planning	1	1	1	1
SP	Scanlon plan	1	1		
SPC	Statistical process control	1			
ST	Statistical tolerancing	1	1		
ST	Sensitivity training	1	1		
STS	Sociotechnical systems	1	1		
TF	Technology forecasting	1	1		1
TM	Taguchi method	1	1		
TQC	Total quality control	1	1	1	
TQM	Total quality management	1	1	1	
VA	Value analysis	1	1	1	
VE	Value engineering	1	1	1	
VM	Value measurement	1	1	1	1
VM	Value management	1	1	1	1
VP	Value planning	1	1	1	1
WS	Work simplification	1	1		
ZD	Zero defects	1			

in the number and types of products and services available in the market. If citizens of the 19th century could come back to witness the world of 1990, they would find an unbelievable assortment of valuable new products. We not only fly, we fly from continent to continent in minutes. We travel in space. Wires and cables crisscross land and sea, carrying telephone and facsimile messages around the world. Radio and television enter our homes and transmit video signals from distant planets. Radar deciphers the topography of Venus. Computers, which not long ago replaced slide rules and adding machines, have transfigured business practice. A plethora of products made in Europe, Japan, and America compete in markets around the world. Oil transported by supertankers fuels our cars, while atomic power fashions electricity and atomic weapons. These and similar earth-shaking developments profoundly effect the way we live and work—and new and bigger developments lie ahead.

VALUISM AND INTERNATIONAL ECONOMIC COMPETITION

At the present time Japan and the United States are locked in an undeclared economic war that is enriching Japan and draining America. For several decades, billions of dollars have poured from the United States to Japan annually in what economists call a *trade deficit*. The trade deficit, calculated as a net gain or loss in a balance of payments, is the difference between the monetary value of exports and imports between two countries. Table 11.4, taken from data in the *World Almanac*[27] shows the serious nature of the trade deficit. The numbers show annual trade figures in billions of U.S. dollars. Twenty-five years ago the United States had a billion dollar trade surplus, today it is running annual deficits of over $37 billion. The trade value index was calculated by dividing U.S. exports to Japan by U.S. imports. It represents the aggregate annual comparative value of each nation's goods.

Japan's creation of high-value goods has produced a severe outflow of money from the United States to Japan, which has weakened the U.S. economy and con-

TABLE 11.4 U.S. and Japan Balance of Payments

Year	U.S export to Japan ($\times 10^9$)	U.S imports from Japan ($\times 10^9$)	Balance of Payments ($\times 10^9$)	Value Index
1966	$6.5	$4.9	+$1.6	1.32
1976	10.1	14.6	−4.5	0.69
1980	20.8	30.7	−9.9	0.68
1983	21.2	40.8	−19.6	0.52
1985	22.6	68.8	−46.2	0.33
1986	26.9	81.9	−55.0	0.33
1988	37.7	93.2	−55.5	0.40

tributed to the present U.S. recession. In contrast, money flowing into Japan from the United States, Europe, and Asia has fattened Japan's corporations, filled their banks, and fed their growing economy. Today Japan's banks are so cash rich that the seven largest banks in the world are Japanese! Listed in Table 11.5 are the assets of the 10 largest Japanese and U.S. banks. The data show assets on hand in December 1989 and are in billions of U.S. dollars.[27]

Japanese industrial corporations and insurance companies have also accumulated massive amounts of capital that they are investing in their economic war with the United States and Europe. Despite a 40% crash in the Japanese stock market in 1989, almost all Japanese corporations still have huge dollar cash reserves. Mitsubishi had 15.2 billion; Toyota, 11.8 billion; Hanwa Steel, 17.2 billion; Sumito, 11.3 billion. Even larger dollar reserves are held by Japan's insurance companies. Together Japanese industrial and insurance companies have a "war chest" of cash reserves of nearly a trillion dollars, which they are investing to expand Japan's international business.[28]

Wealthy Japanese banks have been taking over U.S. banks. Today Japanese banks control over a third of California's banking assets. Japanese banks and corporations are purchasing our failed savings and loan banks at bargain basement prices. Because interest rates in Japan were lower than those in the United States, Japanese banks could loan money to their new U.S. affiliates at low rates. By undercutting U.S.-owned commercial bank rates, the Japanese banks hope to obtain equity in vital segments of U.S. industry. Japanese banks have been happy to loan Japanese companies money at very low interest rates. Nissan was able to expand the production capacity of its Smyrna, Tennessee, Sentra plant by borrowing $1.5 billion from Japanese banks at a rate of 3.375%. At the time U.S. interest rates were over 7%. This loan allowed Nissan to circumvent car import quotas that the United States imposed to help control the trade imbalance. These quotas restrict the number of automobiles

TABLE 11.5 Comparison of U.S. and Japanese Bank Deposits (1989)

Japanese Banks	Deposits ($\times 10^9$)	U.S. Banks	Deposits ($\times 10^9$)
Dai-ichi Kangyo,	$314.8	Citybank	$111.5
Sumitomo	288.2	Bank of America	73.3
Mitsubishi	278.8	Chase Manhattan	61.7
Fuji	278.6	Manufacturers Hanover	42.7
Sanwa	275.9	Morgan Guaranty	40.9
Industrial Bank	213.3	Security Pacific	40.6
Mitsubishi Trust	192.3	Wells Fargo	36.4
Tokai	179.0	Bank of New York	34.4
Norinchukin	176.3	Chemical Bank	32.3
Sumitomo Trust	171.6	Bankers Trust	27.9

Japan can export to the United States but allow Japanese companies to build plants in the United States to build Japanese (and American) cars. To escape import restrictions, seven Japanese auto manufacturers using part of their war chest have built enough new U.S. factories to allow them to build 2 million autos annually. Major U.S. car manufacturers have entered into contracts with these companies and now sell some of these cars under their own brand names. In 1990, the Honda Accord was the best selling automobile in America and cars manufactured in Japanese owned factories captured a record 32% of the U.S. new car market. Other Japanese products, such as TV sets, video cameras, and camcorders have become so dominant that the United States no longer produces them. Japan produces over 85% of all the new copiers sold in the United States and over 50% of all the semiconductors made in the world.

Attempts to stem this trade imbalance have been marginally effective. In August 1988 Congress passed the Omnibus trade bill, which threatened to impose severe trade restrictions on Japanese goods unless Japan opened its markets to more U.S. products. Japan agreed to open more of its market to U.S. goods and the international banking community supported the dollar at rates that favored it versus the yen. As a consequence of these actions, the trade deficit dropped to $44 billion in 1989 and $37 billion in 1990. Japan's government also announced that it would take steps to increase its domestic spending. In 1991 Japan will allocate about 28% of its entire GNP to upgrading and renewing its current industrial and municipal base. Much of this money will be spent on capital construction projects.

Industry and government cooperate in the development of valuism in Japan. These efforts are continuing. Long overpopulated, Japan plans to distribute its new population growth to new municipalities under construction throughout Japan. After identifying a number of new technologies that it considers vital to it's future growth, Japan began funding major research programs aimed at developing its scientific and technical competency. Working together, industry and government developed a plan that calls for the construction of *science cities* devoted to these technologies. The idea was to "seed" new technologies by establishing research centers at rural locations within Japan. As world-class research projects grew at these centers, Japanese industry would begin to locate factories near the science centers and "science cities" would grow as each technology developed. Employment opportunities would draw people from overcrowded areas to the new cities and these new cities would act as incubators for Japan's twenty first century economy. From these "seeds" Japan would dominate areas such as biotechnology, oceanography, fusion and space.

Fifty years after its defeat in World War II, landlords have replaced warlords in Japan and valuism has turned Japan into an economic powerhouse. By developing high-value products and selling them at reasonable prices, Japan has had astonishing success. Today Japan is the world's second largest economy and is the world's leading lender. Forty years ago Japan's GNP was only 5% of the United States's. Today despite sustained growth in the United States, Japan's GNP is 40% of America's. Japan's investment in valuism has strengthened Japan and Japanese products continue to offer outstanding value.

SUMMARY

This chapter describes valuism, a worldwide management process that brings technology, people, and resources together and focuses capital, labor, and creativity on the creation and production of new products and services. It examined the structure of valuism and described how valuism links the management practices of corporations together in a heterogeneous and ad-hoc system centered in the private enterprise system. It examined the role of managers, value methodologists, business practices, and professional societies in valuism and pointed to the lack of valuism in government. In the next chapter, I will describe VM methodologies that lie at the center of the valuism movement.

REFERENCES AND ENDNOTES

1. Utilitarianism recognized that hedonism's attempt to equate behavior with only pleasure and pain was incomplete. One person's pleasure could cause another person pain. Many beneficial activities, such as acquiring an education or acting with compassion and virtue toward others, were not explained by hedonism.

2. For a time, utility theory was taken very literally, and economists believed that there were definite units of utility that could be ascribed to the consumption of specific quantities of goods.

3. Unfortunately Edgeworth and Bentham were ahead of their time. Hedonism and utilitarianism came under intense fire from religious groups who objected to a theory that saw humans governed by animal instincts. Statistical analysis and the computer were not available to help in the development of Bentham's *felicific calculus* or Edgeworth's mathematical phychics. Human behavior was too complex and human needs were too varied to fit the mold of the pleasure and pain continuum both believed in. Economics and psychology were embryonic sciences awaiting the development of people like Freud and Pareto. Freud had yet to propose his theory of the subconscious, and the Italian economist, Vilfredo Pareto, had yet to investigate the maldistribution in society of wealth and goods. In time, attempts to measure and explain the utility of goods or services were abandoned. Today, however, market research and computer simulations allow us to measure value and utility in new and exciting ways.

4 Utilitarianism led to the description of the law of diminishing marginal utility. It is now well known that as the quantity of money increases, its utility increases less rapidly. Utilitarianism attempted to determine the utility of various goods and services where the utility of most goods and services follow a law of diminishing marginal utility.

5. Shewart, W., *Economic Control of the Quality of Manufactured Product*, Bell Telephone Laboratories, 1931.

6. Taylor, F. W., *Shop Management*, Harper & Row, New York, 1947.

7. Gilbreht, F. B., *Motion Study*, Van Nostrand, 1911.

8. Salvendy, G., *Handbook of Industrial Engineering*, pp. 1.1.1, Wiley, New York, 1982.

9. Shewart, *Economic Control of the Quality of Manufactured Product.*

10. Walton, M., *The Deming Management Method*, Perigee.

11. Main, J., "Under the Spell of the Quality Gurus," *Fortune*, pp. 30–34, August 18, 1986.

12. Berne, E., *Structure and Dynamics of Organizations and Groups*, Grove Press, New York, 1966.

13. *"World Business Report," Wall Street Journal*, p. R 28, September 21, 1990.

14. Data are for 1987–1988 and were taken from the section Nations of the World, pp. 684–771, *The 1991 World Almanac and Book of Facts*, Scripts Howard Co., New York.

15. Shewart, *Economic Control of the Quality of Manufactured Product.*

16. Walton, *The Deming Management Method.*

17. Main, "Under the Spell of the Quality Gurus."

18. Institute of Industrial Engineers (IIE), 25 Technology Park/Atlanta, Norcross, GA 30037.

19. American Production and Inventory Control Society (APICS), Watergate Building, Suite 504, 2600 Virginia Ave., NW, Washington, DC 20036.

20. American Society for Mechanical Engineers (ASME), Management Division, 345 East 47th Street, New York, NY 10017.

21. American Society for Quality Control (ASQC), 161 W. Wisconsin Avenue, Milwaukee, WI 53203.

22. Human Factors Society, P.O. Box 1369, Santa Monica, CA 90406.

23. International Material Management Society (IMMS), 310 Bardaville Drive, Lansing, MI 48906.

24. Operations Research Society of America, 428 E. Preston Street, Baltimore, MD 21202.

25. Society of American Value Engineers (SAVE), 60 Revere Drive, Suite 500, Northbrook, IL 60062.

26. Society of Manufacturing Engineers (SME), P.O. Box 930, Dearborn, MI 48128.

27. *The 1991 World Almanac and Book of Facts*, Scripts Howard Co., New York.

28. Robertson, P., *The New Millennium*, Word Publishing, 1990, Dallas, TX.

12

VALUE MANAGEMENT METHODOLOGIES

David J. De Marle

THE VALUE DISCIPLINES

A number of disciplines have been developed that are specifically aimed at reducing the cost and improving the value of products and services. The oldest and most common of these is cost reduction. Conventional cost-reduction exercises require products, services, and organizations to reduce the cost of their components while maintaining quality and performance. Inspection of value equation 1.7 shows that if performance is held constant and price (or cost) is reduced, value improves. Cost reduction is a special, although limited, form of value improvement.

A much more effective value improvement technique had its origin in studies of product changes that resulted from material shortages during World War II. The challenge of substituting materials without sacrificing quality and performance caught the attention of Larry Miles and purchasing people at General Electric Company (GE). Miles, then director of purchasing at GE's Schenectady, New York, plant, was asked to study several hundred product changes that GE had been forced to make due to material shortages encountered during World War II. Miles, the inventor of VA and VE was surprised to find that these changes had produced substantial product improvement and cost reduction. Necessity had been the mother of invention. GE engineers had successfully substituted new materials and manufacturing technology in older designs. The resultant products often had fewer parts and higher quality and cost less to manufacture than their predecessors. Miles found that many of these changes required the participation and cooperation of different organizations within GE. Many of these changes might not have been authorized by these departments were it not necessary to replace materials in short supply. A typical example cited by Miles was the substitution of aluminum and plastic for wire-formed stainless-steel

251

shelving in refrigerators. This change required many changes within GE, was well received by consumers, and saved GE millions of dollars.

Value Analysis

Based upon the results of this investigation, Miles developed a formal methodology in which teams of people reexamined the design of products manufactured by GE. Miles urged these teams to determine why products were made the way they were. He developed a basic questioning logic that people could use to determine the function of components in a design. He urged teams to consider if new materials or manufacturing processes could substitute for older materials and methods. This type of substitution was termed *function* substitution, one in which lower-cost materials provide the same function as those originally used in a product or manufacturing process. Miles developed function analysis into a methodology that teams could use to improve the value of products and services. Using team-oriented creative techniques, changes were made in products which lowered cost without reducing performance. This new methodology was called Value Analysis (VA). Although most of the techniques used by Miles were not new, the philosophy of a functional approach was unique. From the experiences of GE and other organizations, several fundamental concepts evolved that are at the heart of the value disciplines. These basic concepts are:

1. Use interdisciplinary teams to effect change.
2. Develop change through the study of function.
3. Use a basic questioning logic.
4. Use creativity to spur invention.
5. Conduct studies in an organized manner (Miles's job plan).

Over the years, the number of value analysis studies has expanded, as has their areas of application. Today, value analysis studies are a recognized methodology for improving the value of products, *products* being defined in the broadest sense. Value analysis was defined by Miles as "an organized creative approach which has for its purpose the efficient identification of unnecessary cost; i.e., cost which provides neither quality, nor use, nor life, nor appearance, nor customer features."[1] Value analysis traditionally refers to the analysis of an existing product.

Value Engineering

Historically, VE grew out of VA. Value engineering refers to a process that a design team trained in value analysis uses to design a new product. Although the terms VA and VE are often interchangeable, they are different and should not be confused. Miles was asked by the U.S. Navy to see if VA could be developed into an engineering method that could be used to produce new designs. He applied the principle of VA to first-time designs successfully and organized a 40-hr workshop course on

VA/VE that has remained a standard in the field. These workshops have been applied very successfully to many different types of products and services since Miles first developed them. Value Engineering is used extensively in the Department of Defense, and both VA and VE spread from the Defense industry to a number of different commercial and governmental establishments.

Value Planning

As VA and VE grew, they spread overseas. In recent years, Japan has instituted a nationwide training effort designed to spread value thinking and the value disciplines throughout Japan. It is estimated that over a million workers in Japan have been trained in VA/VE. As VA/VE spread throughout Japan it became the site of the creation of another of the value disciplines, value planning. Value planning is a process where management and the technical community jointly plan for the invention, manufacture, and marketing of products and services of ever-improving value over a 5 to 20 year horizon. Value planning uses learning curves to specify functional cost and performance and price ratios within which future products and services will develop. Target costs are developed on a functional basis and given to VE teams to implement as part of a competitive corporate strategy. Chapter 14 describes the value planning process in detail.

Value Management

Value management is another of the value disciplines. It originally was developed in the U.S. auto industry and consists of upper management support and involvement in VA and VE studies. In VM, upper management selects projects where value studies should be conducted. Management also selects a team of people to participate in the study and reviews the proposals of the team that result. Experience with many of the early VA and VE studies pointed out the importance of management support for the studies. In Japan VP is an integral part of VM.

EVOLUTION OF THE VALUE IMPROVEMENT METHODOLOGIES

Through the years, value studies have evolved from a technique-oriented process to a philosophy supported by a battery of unique methodologies that allow value thinking to be applied to the improvement of any item of value. Alex Cunningham, executive vice president of General Motors, underscored this when he gave the keynote address at the 1985 meeting of SAVE. Addressing the audience at this meeting, he said,

> What you do has many names: value management, value engineering, etc. That's because the process has expanded well beyond its original scope. It started out as a technique, a specific way to approach engineering problems. Then it became a discipline, a set of principles that could be applied to a wide variety of business tasks today.

Today, it's both of these things and more. It is a philosophy which I'll call value thinking. You can take an essential kernel of thought, an essential kernel of truth, and transplant it almost anywhere, and it will grow, making virtually every organized effort more effective.[2]

In his speech, Cunningham reported that from 1982 to 1985, General Motors employed VA/VE in all 23 of its North American divisions. In those years, over 10,000 people participated in VA/VE studies, which improved the performance of GM cars and produced over $340 million of actual savings. He estimated the total potential savings from this VA/VE effort at $823 million. The evolution and characteristics of different types of value studies are pictured in Figure 12.1

Growth of VA/VE in Japan

Today, value thinking spearheads the VM movement around the world. In Japan, over 71% of the 698 Japanese companies listed on the Tokyo, Nagoya, and Osaka Stock

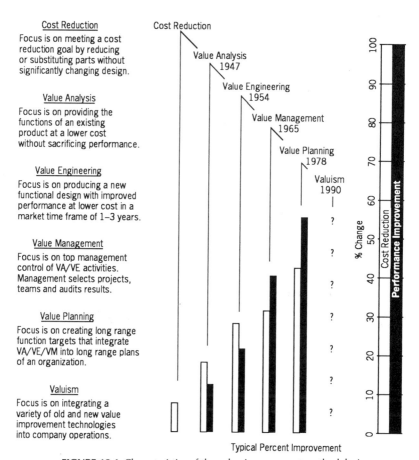

FIGURE 12.1 Characteristics of the value improvement methodologies.

Exchanges had instituted formal VA/VE programs by 1984.[3] A study conducted from 1982 to 1984 among these companies indicated the following implementation by industry group.

Electric	97.9%
Transportation	91.3%
Precision equipment	90.0%
Machinery	84.5%
Other industries	71.4%
Chemical	50.0%
Construction	39.0%
Food	37. 5%

In addition to their internal VA/VE programs, 47% of these companies augmented their programs with outside vendor VA/VE programs.

The Japanese have used VA/VE extensively, and most of the products produced in Japan today are the products of formal value studies. Japanese Victor Corporation (JVC) in Japan described how VA/VE was used to design a camcorder in record time. VA/VE teams redesigned the GX-59 camera and produced the Gz–Se camera. This new camera was 68% smaller, 54% lighter, had 31% fewer parts, and cost 18% less to manufacture.[4] JVC trains its people extensively in VA/VE. All of its engineers have participated in a half-day orientation and a 30-hr course devoted to basic VE techniques. After this training, 500 people participated in workshops totaling 100 hr, and 100 people received 30 hr of training as VE project leaders. The company is endeavoring to foster a comprehensive VE climate throughout its organization by offering thorough corporate-wide training in VE with a special emphasis on product planning and design (value planning).

Another example of the spread of VE in Japan comes from Matsushita. Mitsigi Kanaya, vice president of the Matsushita Electrical Works, described the development of value studies in his company:

> Historically, following Industrial Engineering and Quality Control, in 1961 a study of VE was initiated, and subsequently from 1971, a full scale introduction of development was started within our company. Specifically with regard to VE, a company-wide effort was organized, and 41 full-time personnel were trained. Because of this, the number of employees able to use VE increased to 5000. The latest accumulated cost reduction amounted to $27,000,000 [in 1979].[5]

Fuji Photo Corporation has a company-wide value program that includes VM, VP, VE and VA. This program includes extensive training in value planning, VE, and Value Analysis. Courses last from 1 to 8 months and are offered at a training center that Fuji uses to spread the use of VA/VE throughout the company. Like many other Japanese companies, this value program extends throughout the entire company and its associated suppliers. An annual VE conference is held within the company, and a special column in the company's newspaper describes VE activities through the company. At Fuji the industrial engineering, QC, and VE functions are integrated together in a special department that includes all of Fuji's management technologies.[6]

In 1983, Japan instituted the Lawrence Miles award to recognize outstanding value accomplishments. When introducing this award, Ishiro Ueno, president of the Society of Japanese Value Engineers (SJVE) said:

> The propagation of the value disciplines in Japan is one of the major reasons for receiving the medal of Honor from the emperor. The extent of the importance of the value disciplines to Japan is clearly indicated by this award.

Today in Japan, VE is thriving. It was originally incorporated into the Japanese quality program as a means of cost control. However, in many companies, it has grown into a major business practice that has augmented the QC effort. The Japanese and others have recognized its broad potential and see it as a way to improve quality and value.

Application of VA/VE in Different Areas

Value engineering was originally developed and applied to hardware products and components. Today, the term *product* is defined as "anything which is the result of someone's effort."[7] A product, therefore, can be a process, procedure, or service; an objective or organization; as well as an item, a device, component, subsystem, and so on.

In recent years, value studies have been used to examine organizations as part of management downsizing operations. If an organization is under analysis, organization charts are converted to function diagrams. The importance, cost and value of function in the organization are measured and low value units are modified or are eliminated.

Many items can be the subject of value analysis studies. An educational program can be broken up into a set of courses and course sequences. Individual courses can be broken down into the topics that make up the courses.

The use of a value study is limited only by the imagination. There are many examples of value improvement applications in construction.[8] administration,[9,10] training,[11,12] nonhardware,[13] management,[14] information systems and procedures,[15] venture analysis,[16] technology forecasting,[17] marketing,[18] and resource allocation.[19,20] Obviously, value studies can apply to far more than products and procedures.

This book contains examples of real-life applications of VA/VE. You are encouraged to examine the diversity of applications reported here. The subjects range from candles to icebergs!

THE VALUE PROCESS

Value improvement through the value disciplines is the result of a systematic use of recognized techniques that identify the basic performance functions of a product and ensure that they are provided at the lowest total cost. Many techniques used to accomplish value improvement are not unique to a value study and are taken from other fields. Ernest Bouey, past president of the SAVE, has said, "VE . . . holds no respect for proprietary concepts that are merely ways of thinking. VE will adopt any technique or method (or piece thereof) for use in any of its procedural phases."[21]

Although a value study may involve conventional cost-reduction analysis, it is usually much more comprehensive. Because value is a function of what a device does and how well it does it an extensive analysis is involved. This analysis concentrates on a detailed examination of function and function cost. Value is improved without sacrificing quality, reliability, or maintainability. Collateral gains are often realized in productivity, parts availability, lead time, reliability, serviceability, and quality.

Value studies employ a structured process. They use an interdisciplinary team to

1. Analyze components and the functions they perform.
2. Gather and interpret cost data.
3. Measure value in terms of functions that fulfill customer needs, goals, or objectives.
4. Develop and evaluate alternatives to improve or eliminate low value.
5. Develop ways to implement the best alternatives.

Value studies make use of a structured approach that Larry Miles referred to as a *job plan*. There are six phases in the job plan, each performing different functions and using a variety of techniques. Throughout the first five phases of the job plan, detailed answers are sought to five questions developed by Miles that must be answered before an improvement in value can be made. The five questions developed by Larry Miles are

1. What is it?
2. What does it do?
3. What does it cost?
4. What is it worth?
5. What else will do the job?

These questions are asked and answered repetitively in any value study. Initially they are part of the origination phase of a value study and are used to help select subjects for study. They are used a second time in the study of an existing (VA) or intended (VE) design to target areas for value improvement. They are used a third time to evaluate ideas and suggestions for future improvement. The questions are then asked and answered relative to improvement suggestions from the value analysis. The questions are so germane that they have been found to apply to the study of all manner of products, services, procedures, organizations, and so on. Table 12.1 shows the job plan.

TABLE 12.1 Value study job plan

Origination Phase

Value question	What is it?
Functions performed	Select project, form team(s), organize study, define mission, set market goals (performance/price), set manufacturing goals (benefit/cost)
Techniques used	Team building, value measurement, value planning

Information Phase

Value questions	What is it?
	What does it cost?
	What is it worth?
Functions performed	Analyze components, component levels, costs; determine performance needs and tolerances;
Techniques used	Explosion diagrams, flowcharts, gozinto analysis, cost analysis, cost and importance estimation

Analysis Phase

Value questions	What does it do?
	What does the function cost?
	What is the function worth?
Functions Performed	Analyze functions, dysfunctions; determine function costs; measure value; graph value; set target costs
Techniques used	Verb/noun description, function mapping, FAST diagrams, function matrices, dysfunction analysis, value measurement techniques—value graphing

Innovation Phase

Value Questions	What else will do the job?
Functions performed	Generate alternatives
Techniques used	Brainstorming, nominal group technique, Gordon technique, Synectics, checklists, catalogue technique, morphological analysis.

Evaluation Phase

Value questions	What is it?
	What does it do?
	What does it cost?
	What is it worth?
Functions performed	Screen alternatives, measure value, prioritize alternatives, determine feasibility
Techniques used	Value measurement techniques (Pareto analysis, direct magnitude estimation, paired comparison, Q-sort, cost metric analysis, nominal group technique, scoring models, T charting, etc.)

Implementation Phase

Functions performed	Summarize findings, assemble report, plan program, design audit process, present recommendations.
Techniques used	Value comparison, impact analysis, potential problem analysis, critical path analysis, report writing.

Origination Phase

The VE process begins with the origination phase. A VE study team is formed and a project is selected and defined. Performance and cost targets are set, as is a completion date for the value study. Team members are selected and trained in the VA/VE process.

Project Selection Many value studies arise in specific, well-defined areas. In these cases, it may not be necessary to follow a formal project selection process. However, the time and resources that can be allocated to a value study are often limited, and it is wise to follow a formal selection process. Value management and value planning are processes in which management regularly reviews a list of suggested projects and selects VA or VE projects that it wants done. Traditionally, the main criteria for identifying a VA project has been high dollar volume and high costs. VA/VE studies should be viewed as an integral part of the business, rather than a part of employee training.

In evaluating possible VA/VE projects, management should aim at improving the value of its products in the market. Identifying items that have poor value is the first step. Subjects selected could range from products, manufacturing processes, or marketing operations. A VA study can be expected to reduce the cost and improve the performance of such items. A value team can measure the value of such items and quickly identify areas where improvements can be made.

Personnel Selection After a project is selected, personnel need to be assigned to the study team. To be effective the team must contain a mix of people from areas in the company that can contribute effectively to its mission. A schedule of meetings needs to be developed. If people don't have time to attend these meetings, they should not be placed on the team. The success of a value study is enhanced if organizational and political aspects of the study are considered when selecting team members. Problems can be anticipated before they occur. Team-building exercises can be used to get the team acquainted with one another. Often team building can center on the team's mission and goals. Teams range in size from three to seven members and make extensive use of part-time assistance from others in the organization.[22-26]

Management endorsement of the team and its mission are critical. An initial meeting of management with the team and others in the organization who will be effected by the study is a good way to show management's support. Subsequent team meetings will generally follow the Miles job plan, and the specific subject under investigation will determine what data have to be gathered and what pace the study will follow. Chapter 13 addresses a large number of organization and behavioral factors that experience proves are important to the success of VA/VE studies.

Information Phase

In the information phase, the product and its use are examined in detail to obtain a thorough understanding of its nature. The price and performance of the product is compared to other competitive offerings in the market. Constraints that dictate design, choice of materials, or components or procedures are challenged for validity. This analysis produces a detailed breakdown of a product which shows components, their costs, and performance specifications and tolerances.

Component Information After a team is formed the team has to obtain detailed information about the item under study to answer Miles's What is it? question. The value team conducts an item by item analysis of the product under study. In a VA study of an existing product, the team obtains a sample of the item under study and disassembles it. Each component in the item is listed and categorized at an appropriate system level. Samples of analogous or competitive products are examined and compared to the product under study. In a VE study, a list of probable components or procedural steps is prepared. In either VA or VE, components should be itemized by level of indenture, listing items on an equivalent basis (i.e., assemblies, subassemblies, component parts, etc.) If a procedure is under study, a flow chart of the process is constructed. Manufacturing operations are examined and the manufacturing sequence is defined using flow charts, etc.

Component analysis is not restricted to multiple component systems. A single piece of material has components that can be specified if necessary. The molded plastic ice cube trays discussed in Chapter 1 are examples of products in which the components are not separate parts—they are different locations on the part: end tabs, side rails, nesting legs, and water compartments. Each of these locations have different functions and occupy different portions of the single tray. The size, number, and thickness of these parts directly affect the performance of the final ice cube tray. Value teams can answer the "What is it?" question at many different levels of a system and for many different types of systems.

Cost Information After a list of the components in the product or service is defined, the cost of each component is determined. Component costs are derived by determining the material and labor costs of the item under study. Costs consist of actual costs such as material and labor dollars. For an existing product, they are often obtained from a bill of lading that lists material and assembly costs. Complete data for parts cost are gathered together with labor costs and overhead costs attributed to the product. Cost of operations involved in a process can be obtained by timing the operation and multiplying these times the labor rate. If manual assembly operations are involved, ergonomic data can be obtained by examining a video tape of the assembly operation. Time and motion studies of repetitive manufacturing operations will provide statistical data from which assembly costs can be accurately estimated. Cost data is then allocated to each part in the product. It is important to check the accuracy of cost information by actual observation and analysis.

In VE studies, cost data may not exist, and cost estimates may be all that can be used. These soft costs can be estimated material and labor costs derived by cost appraisal methods. Cost estimates can be established for each item or component by comparing and rating them relative to each other. Numbers are assigned to items in proportion to their perceived relative costs. These ratings are then normalized to individual percentages by dividing each rating by the sum of all item ratings. When a group of people is used to estimate cost, the average cost is obtained for an item by averaging all of the participants' ratings.

In an organizational value study, labor and overhead costs are collected and the cost of organizational components are obtained. Wages, permanent and temporary, are often posted on an organization chart by organizational unit. These are later converted to functional costs that depict the cost of the functions that the organization performs. All costs may be posted to component diagrams, Ghant charts, or function listings and diagrams. This allocation process translates the traditional item-cost viewpoint to a systems view of cost relationships, which is often quite revealing. Function cost analysis, described in Chapter 7, is especially revealing, because it shows the cost of specific functions performed in the organization.

Performance Information During the information phase of a value study, it is important to gather data on the desired performance of the item under study. What needs does the product or service fill? Who buys the product? Why do they buy this product or service? What features are most important to sales? How do the features of competitive products or services compare? What performance trends are important? Are sales seasonal? Regional? What are the performance musts and wants? The study team needs to quantify the importance of different performance factors. In a consumer product this was previously described as the relative importance of weighing factors in a value equation (see equations 1.8 and 1.10).

Constraint and Tolerance Information A value study team needs to collect data on any marketing, manufacturing, management, or regulatory constraints placed on the item they are studying. Marketing demands range from meeting customer needs to meeting the needs of sales and retailing people. Manufacturing and product tolerances should be itemized. Constraints are imposed as tolerances. You may find that design engineers, in an effort to improve quality, have simply added another decimal to their print tolerances. Specifications thus jump to ±0.0001 from ±0.001 in. Unfortunately, these tighter tolerances can dramatically increase waste, increase inventory, and add significantly to manufacturing costs. They can lengthen production run times and necessitate expensive new equipment purchases. It is important for a value team to examine the specifications placed on products and services.

A value team needs to determine and challenge all of the constraints under which a product or service can operate. Budgetary constraints that dictate funding need to be examined as should constraints on product and environmental safety. Constraints and tolerances need to be itemized and challenged. They may be the product of internal political interests and have little to do with satisfying customer needs. The value process uses teams to optimize overall value, and helps prevent short-sighted approaches that benefit selected interests.

I recommend the use of statistical tolerancing in conjunction with value studies. Statistical tolerancing is a process for rationally setting the tolerances of components in a system. It uses simulation models[27] to determine the statistical response of a system and to determine the contribution of individual components to the system's variability.[28]

Analysis Phase

In the function analysis phase, the functions of the product and its components are documented by FA techniques. Functions have importance and cost; these are quantified by VM techniques and an ordered list of function importance and value is created. Low-value functions become candidates for value improvement.

Function Analysis Once the components in the product have been itemized, the value team determines the function of the product and its components. Function analysis answers the value question, What does it do? and is unique to value studies. Many value specialists consider it to be the technique that makes their studies most productive. The method requires that functions be described by only two words, a noun and a verb. By restricting function descriptions to two words, clear descriptions unconfounded by modifying phrases, adjectives, or adverbs are obtained. Function analysis provides the basis for quantitative value measurement.

After functions are identified, they are frequently classified as basic or secondary. The basic function is the prime reason for existence of the product. Secondary functions support the basic function(s) and allow them to occur. They generally are present as a result of a specific design. Although secondary functions may improve dependability or convenience, or may serve sensory or aesthetic functions, they are often prime candidates for elimination or improvement and innovation.

FAST diagrams are also used to classify functions. These diagrams order functions in a hierarchy based on how and why a specific function is performed. They display the causal interrelationship of all functions. Function spreadsheets are also used by value teams to classify functions. They consist of a matrix in which components are arrayed against the major functions present in a product or service. They are often used to determine function costs and are described in Chapter 7.

What Does the Function Cost? Analyzing function costs provides insight into costs that satisfy customer needs. Function costs are derived by determining the cost of the items, components, or labor necessary to provide respective functions. Once function costs have been calculated for a specific product or service, these costs are compared to function costs in other items.

Value Measurement

In value measurement, the Miles' question "What is it worth?" is answered. The value of components and features in the product or service are measured. Value indices are calculated, and value graphs, which show the relative importance and cost of components in the system, are prepared.[29] These techniques are discussed in Chapters 3 and 4.

Importance Measurement The relative importance of components and functions is measured by comparison using a number of different subjective rating techniques. Numbers are used to weight the importance of components and functions, and the subjectively derived numbers are normalized to a percentage. The relative importance of an item in a system is calculated in terms of its contribution to the performance of the total system. Chapters 3 and 4 describe the value measurement techniques used to rate the importance of components.

Value Index After calculating the percentage importance and percentage cost of a component, a figure of merit, the ratio of importance to cost, is calculated for each component. The figure of merit is referred to as the *value index*.[30–32] It is a dimensionless number that expresses the relative value of items in a design. Generally, a value index greater than 1 represents good value. A value index less than 1 flags those components in a system that need attention and improvement in the innovation phase. A value index is used to select parts of a product for study and to set cost, value, or performance improvement targets. The value index is discussed in detail in Chapter 3.

Value Graph A value graph is a graphical representation of the value index obtained by plotting relative importance against relative cost.[33] By convention, cost is plotted on the x axis and importance on the y axis. If the x and y scales are uniform, then items with % I equal to % C will plot along a 45° line on the value graph. This line represents acceptable value, and items on the line have a value index of 1. The area above this line represents a region of good value, whereas that below the line represents a value improvement region where value is low. Items plotted to the bottom right of a value graph are prime candidates for improvement or elimination. Items plotted in the upper left corner of the graph offer exceptional value. Value graphs are used to portray the results of value measurement exercises.

Value of Functions The value of a function is measured in the same manner as the value of a component. The relative importance of a function is determined by comparison using subjective measurement techniques. Function value is then calculated by dividing the percentage importance of the function by its percentage cost. Alternately the value of a function is established by comparing the cost of performing the function in a design against an external standard. Function value analysis using creative comparisons of this type is covered in References 34 to 36.

Innovation Phase

In the innovation phase, creative techniques are used to generate design alternatives. Ways are sought to eliminate, replace, or improve low-value items and functions. Many different alternatives are listed to improve performance and/or reduce cost before judging the merit of these alternatives. Current ideas are collected and creative techniques such as brainstorming are used to conceptualize new ways to improve the value of a design. Eliminating or combining low value elements is the goal.

Creative Techniques A creative technique is a system, procedure, or method designed to enable an individual or group to produce a large number of original ideas.

Various creative techniques have been used successfully to promote idea fluency and originality and to help visualize new combinations of seemingly disparate objects. Their greatest value is providing alternative solutions to problems that can have a variety of acceptable solutions.

There are two precepts common to all ideation techniques. First, all criticism is eliminated from the idea-producing stage; ideas are listed without comment. This encourages maximum production of ideas and prevents premature rejection of potentially good ideas. Second, all ideas, even the apparently impractical ones, are evaluated at a later time. This is done to encourage everybody to explore even the seemingly impractical ideas and to feel free to express thoughts without fear of ridicule.

Types of Creative Techniques   There are many techniques for generating and collecting ideas. Some may be more appropriate than others for given situations. Techniques can be classified along a continuum based on the degree of freedom and structure used in the creative process. In the list below, the more structured techniques are listed first and the more free-wheeling techniques are listed last.

Analytical—logical, step-by-step approach
  Attribute listing[37]
  Checklists[38]
  Morphological Analysis[39,40]
  Nominal Group Technique[41–43]
Forced relationship—between two objects or ideas
  Catalog technique[44]
    Focused object technique[45]
  Listing technique[46]
Free association—quantity of ideas, daringness
  Brainstorming[47,48]
  Gordon technique[49]
    Synectics ©[50–52]

Choosing a Creative Technique   Experience shows that it is usually better to start with the simplest technique and progress to the more mind-stretching techniques later. The simplest is to merely collect ideas that already exist. They can be gathered via interviews, questionnaires, literature searches, and so on. Group meetings can be used to pool the members' ideas, employing a facilitator.

If no satisfactory ideas emerge by a simple gathering process, the more unique ideation techniques are tried. All are designed to circumvent negative thinking and uncover unexplored areas by stimulating creative thinking. They are designed to suppress habitual perceptual, cultural, and emotional blocks to creativity.

Evaluation Phase

The evaluation phase categorizes ideas, gathers information about their feasibility and cost, screens the ideas, and then measures the value of the best ideas. The process repeats that used to analyze the current product in the information and analysis phases of the job plan. The initial analysis is referred to as a present state analysis, repetition of this analysis on ideas from the innovation phase is referred to as a future state analysis.

This future-state analysis uses the same value measurement techniques used in the information phase. The highest-valued alternatives that emerge, now numbering only two or three, are then examined for economic and technical feasibility. Cash flow and break-even economic analysis are conducted on these alternatives, and their ability to perform the desired function satisfactorily, accurately, reliably, safely, and with ease of repair and proper environmental impact is ascertained.

A value team needs to objectively analyze the value of the ideas generated in the innovation phase. The evaluation phase accomplishes this reduction in two steps. First, a qualitative analysis of value evaluates the ideas against design objectives such as cost, ease of implementing, performance, and so on. It is very important to conduct this evaluation in an unbiased manner as excellent ideas can be lost if people quickly rate the ideas based on their current beliefs. Chapter 6 discusses several evaluation methods that keep the screening process objective. Second, a quantitative analysis using numerical value measurement techniques produces several high-value alternatives that are then studies exhaustively. It is important to test the feasibility of these ideas experimentally and to conduct these evaluations in an objective and unbiased manner.

Prescreen: Qualitative Value Analysis In prescreening exercises, the ideas are examined for redundancy, and any redundant ideas are eliminated by combining them with similar ideas. The ideas are then categorized into different design approaches and appraised against criteria constraints such as performance, cost, quality, and so on. Prescreening techniques such as Pareto voting[53] and Q-sort[54] are often used to rate the ideas in each design category. The more ideas that need to be screened, the more useful these techniques are.

Detailed Quantitative Value Analysis The alternatives judged best in each design category are subjected to more rigorous and discriminating value measurement techniques. Paired comparison, scoring models and category scaling can be used. The most promising alternatives are now extensively reviewed for cost-effectiveness, technical feasibility, and possible implementation. Experiments and tests on several simple approaches can provide objective information to speed the selection process. The assistance of knowledgeable people external to the value study team is often solicited to help judge the potential and feasibility of the best alternatives. New criteria may evolve that warrant further analysis. Techniques that are helpful in locating unforeseen problems are potential problem analysis[55] and impact analysis.[56]

When the final evaluation is complete, the remaining alternatives need to be tested thoroughly before implementation. Testing the alternatives may involve developing prototypes and should include extensive economic analysis. Mathematical modeling is often useful at this stage of a VA/VE study.

Implementation Phase

In the implementation phase, a report summarizing the value study is prepared. It contains conclusions and makes specific proposals. Action plans for implementing the recommendations are described. Project management techniques such as critical path analysis or PERT are useful here. A plan for monitoring the implementation of value change proposals helps to ensure that their benefits are achieved.

Recommendations The implementation process begins when a value team's recommendations are prepared and presented to management for approval. A report is prepared that describes the proposal(s) and lists suggested action plans for implementation. Careful consideration should be given to assure that the recommendations are clearly understood. An oral presentation is an excellent supplement to a written document. The report should contain a brief executive summary that describes each of the team's recommendations. A separate study team report containing detailed information can be prepared and used as a working document during implementation.

Special attention needs to be paid to the cost reduction and performance improvement benefits of the recommended changes. Financial analysis should include cash flow and break-even analysis. The cost of implementing the ideas should be estimated, and the financial savings resulting from implementing the ideas should be specified. It may be important to discuss the cost of maintaining the system. If so a complete life-cycle cost analysis should accompany the recommendations. A reasonable time table of financial investments and returns should be prepared. The effects of current inventories on implementation should be examined. The manufacturing and marketing advantages of the proposals should be detailed, and special attention should be paid to effects of the changes on customer acceptance of the product or service.

A major factor in obtaining acceptance and use of recommendations involves making the decision maker feel comfortable with the recommended change(s). Having the decision maker as a team member is ideal. Where this is not possible or practical, the decision maker should be periodically informed of the value team activity. Scheduling periodic reviews to solicit input is helpful and alleviates anxiety over change or potentially threatening information. Reviews allow the decision maker to contribute to the team and foster ownership in the direction and outcome of the study. Ownership increases the chances for acceptance and positive action. The behavioral aspects will be covered in Chapter 13.

Implementation Plan Implementation of each approved recommendation needs to be well planned. Although implementation may be done by the value team, it is often carried out by others. In any event, the value team needs to develop a detailed plan for implementing their approved recommendations before the team is dissolved. Project management planning and scheduling methods like PERT are very useful for implementing value recommendations[57] and help ensure that the recommendations are acted upon and implemented. Implementation is more likely when the value study team prepares a suggested plan of action and a timetable for accomplishment. A series of milestones are itemized in the implementation plan, and progress in implementing the recommendations is monitored to see that implementation takes place on time.

The phases of a value improvement study are shown in Figure 12.2 These study phases apply to VA/VE. Cost-reduction studies do not involve FA. Value management and value planning are often used before a VA or VE study is initiated and include a forecasting phase described in Chapter 14.

SUMMARY

This chapter described the origin and growth of the value disciplines. Value analysis, a process for improving the value of existing products, resulted from an analysis of manufacturing changes made by GE during World War II. Interdisciplinary teams examine the functions of existing products and recommend design changes that will improve performance and reduce cost. Value analysis led to value engineering, an effective process for designing new products and procedures. Value planning and

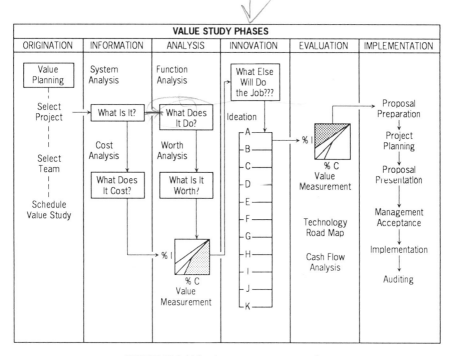

FIGURE 12.2 Value improvement process chart.

value management evolved from VA/VE in response to a need to forecast and plan products that will meet the needs of tomorrow. A structured process that VA/VE teams have used to improve the value of products and services was described. VA/VE/VP/VM are in widespread use in a number of nations around the world. In the ensuing chapters of this book, we will discuss some of the behavioral and organizational aspects of VA/VE and introduce value planning methodologies that broaden the base of VA/VE and valuism.

REFERENCES

1. Miles, L. D., *Techniques of Value Analysis and Engineering,* McGraw-Hill, New York, p. 1, 1961.
2. Cunningham, A., "Value Engineering at General Motors," *Proceedings, Society of American Value Engineers,* Vol. 20, 1985, Keynote speech.
3. Tsustomu Ohta, "Survey of the Use of VE in Japanese Companies," *Proceedings, Society of American Value Engineers,* **20,** pp. 213–221, 1985.
4. Takayuki Takubo, "Technology for Product Planning and Development," *Proceedings, Society of American Value Engineers,* Vol. 20, pp. 240–250. 1985.
5. Mitsugi Kanaya, "Value Creation as a Management Objective," *Proceedings, Society of American Value Engineers,* Vol. 14, pp. 13–17, 1979.
6. Kawamura, S., "Promoting Value Engineering at Fuji Film," Fuji Film Research and Development, No. 34, pp. 158–163, 1989. [In Japanese]
7. Mudge, A. E., *Value Engineering, A Systematic Approach,* McGraw-Hill, New York, 1971.
8. Dell'Isola, A. J., *Guide for Application of Value Engineering to the Construction Industry,* 2nd ed., McKee-Berger-Mansueto, Washington, D. C., 1972.
9. Fifield, F., "Administrative Value Analysis," *Industrial Engineering,* pp. 24–28, November 1973.
10. Fowler, T. C., and Higgins, B., "Organization Value Analysis Made Easy," in *Proceedings, SAVE Regional Convention,* Detroit, pp. 3.1–3.7. 1974.
11. DeMarle, D. J., "The Application of Subjective Value Analysis of Training," in *Proceedings, SAVE Regional Conference,* Detroit, pp. 5.1–5.6, 1971.
12. Dobles, R. W., Drost, P. M., Hazen, S. S., and Hinkleman, R. C., "If You're Not Doing It Already, Verify Your Training Objectives," *Training,* **16**(12), December 1, 1972.
13. Illman, P. E., "Value Analysis in the Non-Hardware Field: A New Approach in Design and Training," *Proceedings, Society of American Value Engineers,* **6,** May 1971.
14. Ridge, W. J., *Value Analysis For Better Management,* American Management Association, New York, 1969.
15. Valentine, R. F., *Value Analysis for Better Systems and Procedures,* Prentice-Hall, New York, 1970.
16. Rand, C., "New Venture Value Search (Making Companies Well Through VE/VA)," *Value World,* **3,** 17–23, May–June 1979.
17. DeMarle, D. J., "Use of Value Analysis in Forecasting," paper presented at James R. Bright's Technology Forecasting Workshop, Castine, Maine, Industrial Management Center, Austin, Tex.

18. Groeneveld, L., "Value Analysis and the Marketing Concept," *Industrial Engineering,* pp. 24–27, February 1971.

19. d'Ascanio, G., "Social Value Analysis," *Proceedings, Society of American Value Engineers,* **10,** 174–178, May 1975.

20. Love, S. F., "Resource Allocation by the Delphi Decision Process," *Optimum,* **6,** 39–48, 1975.

21. Bouey, E., "Value Engineering Is Unique," *Interactions,* Newsletter of the Society of American Value Engineers, **5**(8), p. 1, 1979.

22. Boothe, W., "Developing Teamwork," Golle and Holmes, Minneapolis, Minn., July 1974.

23. Lashutka, S., and Lashutka, S. C., "The Management of Team Development in Value Analysis/Engineering," *Value World,* **3**(4), 5–10, 1980.

24. Ridge, *Value Analysis For Better Management.*

25. Reigle, J., "Value Engineering: A Management Overview," *Value World,* **3**(23), 5–8, 1979.

26. Schein, E. H., *Process Consultation: Its Role in Organization Development,* Addison-Wesley, Reading, Mass., 1969.

27. Barker, T. B., "Monte Carlo Simulation by Experimental Design," in *Transactions of the 41st. Annual Quality Conference,* 1985, Rochester Section ASQC, pp. 1–10.

28. Cox, N. D., "How to Perform Statistical Tolerance Analysis," in *The ASQC Basic References in Quality Control: Statistical Techniques,* Vol. 11, American Society for Quality Control, Milwaukee, 1986.

29. DeMarle, D. J., "A Metric for Value," *Proceedings, Society of American Value Engineers,* **5,** 135–139, April 1970.

30. DeMarle, "A Metric for Value."

31. DeMarle, D. J., "Criteria Analysis of Consumer Products," *Proceedings, Society of American Value Engineers,* **6,** 267–272, May 1971.

32. DeMarle, D. J., "The Nature and Measurement of Value," in *Proceedings, 23rd Annual AIIE Conference,* May 1972, pp. 507–512.

33. DeMarle, "Use of Value Analysis In Forecasting."

34. Becker, R. F., "A Study In Value Engineering Methodology," *Performance,* **4**(4), 24–29, 1974.

35. Crouse, R. L., "Function and Worth," *Proceedings, Society of American Value Engineers,* **10,** 8–10, May 1975.

36. U.S. Department of Defense, *Principles and Applications of Value Engineering,* Vol. 1, U.S. Govt. Printing Office, Washington, D.C., 1968.

37. U.S. Department of Defense, *Principles and Applications of Value Engineering.*

38. Osborn, A., *Applied Imagination,* Charles Schribner's Sons, New York, 1960.

39. Whiting, C., *Creative Thinking,* Reinhold, New York, 1958.

40. U.S. Department of Defense, *Principles and Applications of Value Engineering.*

41. Delbecq, A. L., and Van de Ven, A. H., "A Group Process Model for Problem Identification and Program Planning," *Journal of Applied Behavioral Science,* **7**(4), 466–449, 1971.

42. Delbecq, A. L., Van de Ven, A. H., and Gustafson, D. H., *Group Techniques for Program Planning: A Guide to Nominal Group and Delphi Processes,* Scott Foresman, Glenview, Ill., 1975.

43. Huber, G. P., and Delbecq, A., "Guidelines for Combining the Judgments of Individual Members in Decision Conferences," *Academy of Management Journal,* **15**(2), 161–164, 1972.

44. Biondi, A. M. (Ed.), *Have an Affair with Your Mind,* Creative Synergetic Associates, Great Neck, N.Y., 1974.

45. Von Fange, E. K., *Professional Creativity,* Prentice-Hall, Englewood Cliffs, N.J., 1959.

46. Von Fange, *Professional Creativity.*

47. Richards, T., *Problem-Solving Through Creative Analysis,* Wiley, New York, 1974.

48. Stein, M. I., *Stimulating Creativity, Part 2: Group Procedures,* Academic Press, New York, 1975.

49. Gordon, W. J. J., *Synectics, the Development of Creative Capacity,* Harper & Row, New York, 1961.

50. "The Synectics Course in Creative Group Problem Solving," Synectics Inc., 28 Church Street, Cambridge, MA., 02138.

51. Alexander, T., "Synectics: Inventing by the Madness Method," *Fortune,* pp. 165–171, August 1965.

52. Raudesepp, E., "Forcing Ideas with Synectics, a Creative Approach to Problem Solving," *Machine Design,* pp. 134–139, October 16, 1969.

53. Shillito, M. L., "Pareto Voting," *Proceedings, Society of American Value Engineers,* **8,** 131–135, May 1973.

54. Helin, A. F., and Souder, W. E., "Experimental Test of a Q-Sort Procedure for Prioritizing R&D Projects," *IEEE Transactions on Engineering Management,* **EM-21**(4), 159–164, 1974.

55. Kepner, C. H., and Tregoe, B. B., *The Rational Manager,* McGraw-Hill, New York, 1965.

56. Shillito, M. L., "Impact Analysis," paper presented at James R. Bright's Technology Forecasting Workshop, Castine, Maine, June 1977, The Industrial Management Center, Austin, Tex.

57. American Society of Tool and Manufacturing Engineers (Greve, J. W., and Wilson, F. W., Eds.), *Value Engineering in Manufacturing,* Prentice-Hall, Englewood Cliffs, N.J., 1967.

13

VALUE MANAGEMENT: BEHAVIORAL AND ORGANIZATIONAL ASPECTS

M. Larry Shillito

BACKGROUND

Everyone today is involved one way or another in change. In fact, one constant we can always expect in the future is change. Because we have change, we will always require decision making. Decision makers very often deal with complex issues. All decision makers can always use more help in dealing with issues and generating and evaluating alternatives. Generally, there is a reluctance to deal with complexity in a structured systematic manner. According to the John Warfield model[1], which he calls the "fundamental triangle of societal problem solving", there are three basic elements to addressing an issue: the issue itself, an interdiscipline team, and an appropriate methodology (See Figure 13.1).

As in any triangular relationship, one element cannot work well without the other. They are all interdependent. The connections between the elements are just as important as the elements themselves. If a flexible simplistic methodology can be developed that can serve both the team and the issues, then there is a greater chance of connecting the team successfully with the issue. Fortunately, VM provides this connection. The more multidiscipline the VM team, the better the connection. However, the interconnections of the three elements involves people, teams, organizations, and politics and thus further compounds the relationships (Figure 13.2).

The successful application of VM projects and methodology is very often weakened, because the behavioral, people, political, and organization elements are not addressed properly. This chapter will discuss some of these issues that affect a VM project and a VM program.

An example of the many questions that should be considered in launching and conducting VM projects are; What is the scope of the study? Define what the study is or is not. Who funds the study and how much funding is required? Who approves

271

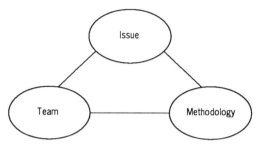

FIGURE 13.1 Basic elements of an issue.

the study? Who is the requestor? Who pays for implementation of recommendations? What are the study start and completion dates? Who should be on the study team? Who are the intended users of the study results? What is the study deliverable? What is the expected format of the deliverable? What levels of the organization should be involved? What geographic areas? What are the expected time, manpower, and cost? How do we introduce VM into the department and company? What politics are involved? Do power plays exist? Is turf threatened?

Although the VM process is a rational, systematic, and structured process, its foundation is structured around the effective use of people in teams and subsequently their interaction with the management chain in making recommendations that ultimately will lead to implementation and change. Once we add people, teams, organization structure, emotions, and power to the process, the process becomes more complex. Consider the following circumstances and roadblocks usually encountered in the VM process.

1. Value management is always performed with interdiscipline teams. Teams can waste time, be overly conservative, avoid decisions, and prematurely solve unclear problems.

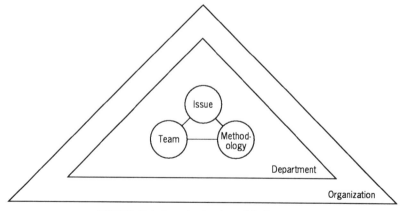

FIGURE 13.2 Organizational model of an issue.

2. Individuals involved in VM usually have other full-time jobs and are already busy.

3. Strong parochial interests are common.

4. The output of a VM study may be threatening, especially to designers, planners, and decision makers.

5. Emotional as well as rational conflicts of interest are often generated.

6. The purpose of the VM project is not always clear.

7. The final decision maker is not always obvious.

The success of the VM project is enhanced if organizational, political, and behavioral aspects of the project are addressed early in the project. Roadblocks can be anticipated and planned for before they occur. The process facilitator should emphasize to the team that they should consider the first team meeting as the first day of implementation. The following getting started topics are particularly germane to the process.

INITIALIZING THE PROCESS

When starting a VM project, the following subjects must be thoroughly addressed: purpose, alignment, scope, time horizon, completion date, market, study team members, assumptions, implementation, and project selection. Discussing and documenting these topics is one of the most important parts of the VM process. A poor job here can cause teams to be off course, lose time, and develop excellent recommendations on the wrong thing. The VM team should take as much time as needed to do a good job documenting these topics. Each topic will now be discussed separately.

1. Purpose
    A. Why are we doing this VM project?
        1. To produce a revolutionary design?
        2. To produce an evolutionary design?
        3. To redesign an existing product to
           (a) Reduce costs?
           (b) Increase productivity?
        4. Increase customer perceived value?
        5. Improve manufacturability? Assembly? Setup?
        6. Reduce cost only?
        7. Improve quality? Reliability?

Answers to these questions are not always obvious. Considerable amounts of time have been consumed by teams in generating a purpose statement. Sometimes it is necessary to check again with the requestor of the study to determine the purpose. It is better to do this up front before the team works on the wrong objective. The verbal interaction necessary to draft the purpose statement helps bring focus and quickly gets team members working together.

Two basic questions must be considered when writing a purpose statement: (1) What are we trying to do for our customer? (2) What are we trying to do for ourselves (the company)? The answers to either of these questions cannot be at the sacrifice or expense of the other. Discussion of these two questions brings focus on the purpose of what specifically the team (Project) is trying to achieve. The VM team has a task to accomplish. This task is usually to improve some product/service that meets customer and manufacturing needs and that provides some competitive advantage over some period of time. A purpose statement should be a short, broad definition of what is to be accomplished by the team and the VM project and why. It essentially creates a target or goal for the team. Sample purpose statements might be: "manufacture component x for $1.00 for delivery by 19xx; Insure that the proper tools and supplies are available when needed so cost and delivery commitments can be met; Reduce the total elapsed time required to implement a design change; "Develop product x with a revolutionary design in a way that cost is less than $k dollars and product is easy to use so that product has high appeal for market segment Y."

The time spent in defining purpose is very important. For example, we recall a VM project where the stated purpose was solely cost reduction. So much emphasis was placed on cost reduction that the team forgot or at least did a poor job of defining user/customer musts, wants, specifications, and constraints. The team did a tremendous job in reducing costs. Several weeks later during the implementation phase, the team discovered they could not implement most of their recommendations because they would jeopardize quality and reliability. It's true, they did not do a good job in the information phase. The purpose (cost reduction) consumed so much attention and mental energy that it jeopardized the rest of the study.

2. Alignment
   A. Who is the decision maker?
      1. Who is the person who will approve or disapprove team recommendations?
      2. Who pays the implementation cost?
      3. Who is it that the team will present an offer they can't refuse?

It is surprising how often teams have a difficult time answering these kinds of questions. Who the decision maker is, is not always clear. The decision maker must be identified, otherwise, the team can do an outstanding job on the VM project only to make recommendations that some decision maker never wanted. A decision maker in the context of the VM process is the person(s) to whom recommendations will be given, and who can approve or disapprove them. The decision maker may or may not pay for implementation of the recommendation. Sometimes there is a chain of decision makers. In this case, the first person in the decision chain that can say no is considered the decision maker and the person whom the team must keep informed.

The decision maker should be periodically informed of team progress and direction. In this respect, the decision maker should be used as a resource to the team.

Through the interaction of the update sessions, the decision maker builds ownership in the team project. This, in turn, builds alignment so that team and decision maker are congruent in their content, direction, and expectation. Alignment increases the chance of acceptance of team recommendations and their subsequent implementation, and reduces the chances of the team performing an academic exercise.

Experience shows that decision makers don't like surprises. Surprises generally are threatening to a decision maker and increase the chance of a veto. This can happen when team recommendations are radically different from status quo or when they make historical decisions made by the decision maker look bad. Therefore, identify the decision maker and start building ownership and alignment as soon as possible. The quickest way to get started in stake building is to review the team-generated purpose statement with the decision maker. The more decision makers are consulted or interact with the team, the more ownership they will feel in the team project. By the time the team is ready to make its recommendations, the decision maker already knows what to expect. There are no surprises. There is a feeling of ownership. It is an offer that is difficult to refuse. The likelihood of approval is very high.

3. Scope
    A. What is included in the study?
    B. What is not included in the study?
    C. What is/is not the team able to control/change?
    D. What are the boundaries within which the team operates?

It is important to bound the study so that it will be a manageable unit. The amount of time the team spends on scoping should not be jeopardized by frustration to prematurely start the VM process. The danger of improper scoping is that the team may perform an excellent job and develop an excellent design for the wrong thing! Proper scoping increases the chances of the team doing the right things right. One model used in scoping is the black box input–output transformation model (see Figure 13.3). The black box is where some transformation or action takes place. It is the area of team responsibility. The inputs to the box are outside the scope of the team to influence.

The inputs are given and are characterized by certain parameters. Coming out of the black box is some output produced by the transformation. Once this output leaves

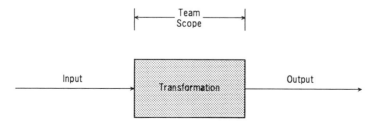

FIGURE 13.3 "Black box" transformational model.

the boundary of the black box, it is no longer in the team's sphere of control. So, the scope of the team's project is everything within the boundaries of the black box.

For more complex studies it may be necessary to conduct a preliminary evaluation of the scope itself and to modify the original, if necessary, to develop a better understanding of what is really wanted. Sometimes money or time constraints influence the scope and restrict the boundaries of the VM project. Reviewing the scope with the decision maker early in the project, is time well spent.

4. Time horizon (consumer-type products)
   A. What is the introduction or implementation year (date) for the product under study?
   B. Will there be more than one model (design)?
      1. Will the models be time-phased for introduction?
   C. Which model (design) does this current VM study consider?

Too often teams get frustrated because there are so many models to work on. Many models or design variations are dependent upon when they will be used or introduced. Input from upper management is necessary to establish model type and introduction dates. That is, management should select the product and introduction dates; the team selects the functions and features to be studied. It is obviously necessary to have proper product/model/date selected before beginning the VM process.

5. Completion date for VM process
   A. When must the VM project be completed?
      1. When will the team present recommendations to the decision maker and other management?

Establishing an end date for the VM project brings focus for completion. This, in turn, helps establish how often and for how long the team should meet.

6. Market (for consumer-type products)
   A. What country? Domestic only?
   B. What market?
   C. What segment?
   D. Who is the customer?
   E. Who is the user?
   F. Who is the chief buying influence?
   G. What product?

Answers to these questions will shape the outcome of the entire VM study. Product features and their importance will vary greatly across market segments. Segment also determines what products (both our company and other manufacturers) are considered. This can be a time-consuming and sometimes difficult task. A popular way to start the process of defining market segment is to establish a Pareto distribution of sales volumes by product line and market segment. When doing so, it is best to use unit volumes instead of sales dollars. There are various ways to correlate volumes such as percentage units sold versus market segment, percent units sold versus product line, product line volume versus market segment, and so on. Signals to watch for are things like 80% of total volume coming from one segment or 80% of total volume coming from one or two product lines. The team and project time should be spent on the vital few versus the trivial many. The team must determine whether volume is the correct indicator to establish which market segment to work on. Should the team be working on the high-volume areas to design or redesign a product to maintain market share? Should the team be working in low-volume areas to design a product to capture more market? There are many approaches to determine which product to work on. The final approach will depend upon product, market, management, and the line of business strategic plan. Sometimes market or segment is dictated by upper management.

For consumer-type products defining market segment, customer, user and, *chief buying influence*,[2] many times is not as simple as it may appear. It is important to define who these people are because they can each have their own set of needs. Tom Cook, president of Thomas Cook Associates, offers a good example:

> Consider a baby food product. The baby would be the user. The mother or father would be the purchaser or customer. However, the chief buying influence may be none of these, but would instead be the pediatrician. . . . "[2] p. 147

So, for whom are we designing the product? The resolution of customer, user and chief buying influence must be established before developing design parameters.

7. Study team members
   A. Based on the answers to the preceding six questions, are the right people on the core team?
   B. Will ad-hoc members be needed on the team at certain times in addition to the core members?

When a core team is assembled, the members will remain on this team until the project is completed. Generally the core team consists of a minimum of the chief designer, the project leader, a VM process facilitator, and the VM team leader. On numerous occasions, study teams discovered that certain key people are not represented on the core team. The missing representatives are discovered only after the teams have discussed the previous six questions. Usually the proper people are already on the core team, but additional ad-hoc members may have to participate for a limited duration at appropriate times in the future. These people will have specialized knowledge for certain areas of the project. Making them a permanent member of the core team would not be practical (Figure 13.4).

Team members must represent both the customer (user) and the producer. This would suggest the following types of disciplines: design, manufacturing, operations, marketing, customer service, user/equipment operators, manufacturing engineering,

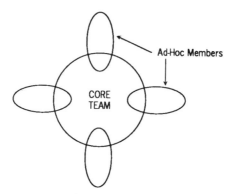

FIGURE 13.4 Core team model.

product planning, and so on. When choosing team members, you should focus on a person's expertise and not on his or her position in the organization.

A VM team normally comprises three to seven members. With more than seven members, interaction becomes complex, discussions become indistinct, and the group begins to fracture. An odd number is also helpful, because it reduces chances for split decisions. The following characteristics are very important when assembling a team:

1. It should be interdisciplinary, incorporating a good balance of background, viewpoints, and disciplines, as well as good geographic representation.
2. Members should be from equivalent levels in the organization's hierarchy in order to minimize peer pressure and politics.[3]
3. It is sometimes helpful to include a decision maker on the team, because acceptance of results may depend upon who is on the team. Caution should be taken if having the decision maker on the team. They may induce peer pressure and assume a boss role which may inhibit participation and candid discussion.
4. It is necessary that one or more members be versed in the VM process. Alternatively, a third-party facilitator or outside consultant can supply the VM methodology.
5. At least one member must be an expert on the product or subject being studied.

The team members themselves should

1. Be at least generally familiar with the product or area of study.
2. Know the sources of data for their area of expertise.
3. Have interest, motivation, and commitment to engage in the task.
4. Be able to get cooperation and assistance while representing their organizational area.
5. Have sufficient time to do the job and be engaged long enough to provide continuity to the product.

6. "Be able to create, accept, and be eager to exploit change."[4]
7. Have an open mind and be able to work and communicate with others in a team environment.

Team formation and team member selection should be directed toward creating an interdiscipline team as opposed to a multidiscipline one. Figure 13.5 illustrates the difference between the two definitions. Generally, the terms are used interchangeably. Making a distinction between the two terms helps to emphasize the importance of working together to do a high-quality job and to provide commitment and energy to make something happen. A good example of a team is that of Eric Berne's.[5]

Referring to Figure 13.6, the outer circle represents the membership boundary. This boundary encompasses the team itself, a selected group of people. The outer circle could be considered as representing a VM team. Internal to this team is another boundary or core that represents the leader. This core also consists of procedures and rules (Figure 13.7). The leader directs the team through a certain methodology or constitution. The process, or methodology, forms the framework for the leader. What should characterize a VM team is that it should not be a personal leadership role. There must be a coordinator, someone who sets up the meetings and the agenda. For the team to really be effective, the personal leadership must diminish as time goes by so that the leadership is shared by all of the team members. This is not possible unless there is a substitute for the leader, some sort of core leadership that comes from constitutions, methodologies, or rules. The VM process is such a methodology. In the case of a VM team, it is the VM techniques and methodology that fulfills the leadership role. The leader dissolves into the team so that the team members carry on through the structure of the VM process.

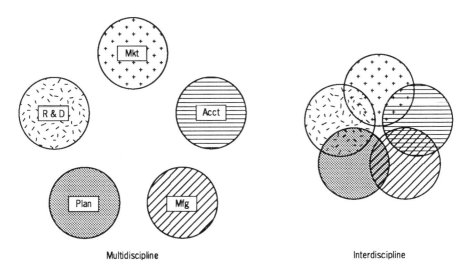

Multidiscipline   Interdiscipline

FIGURE 13.5 Team cohesiveness comparison.

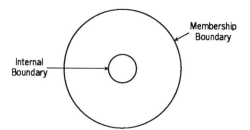

FIGURE 13.6 VM team boundaries.

The various meetings are patterned on a plan. The team devotes the first several sessions to gathering information about the problem and defining it in a better sense. The team then devotes more sessions to getting ideas how to change or improve the particular situation that is presenting the problem. It may seem trite or ritualistic, but in a system that has to function without a charismatic leader, and it must do this in order to get the team members committed, there must be some sort of goals or procedures inherent in the system. It is these goals and procedures that form the basis for carrying on. They elicit commitment from the members and provide the vehicle and momentum for overcoming obstacles. They are an operational method of putting the ideas of behavioral science to work. Other excellent references on teams are Fiorelli,[6] Dillard,[3] and Parker.[7]

8. Assumptions
   A. What are the initial assumptions?
      1. The study will proceed based on these.
      2. As assumptions change or become fact, the study content and direction will be altered to reflect the changes.

All projects begin with a basic set of assumptions. Too often these assumptions are not documented or visible. Many times they are taken for granted. Forcing documentation of assumptions in the beginning of the project enhances communications and highlights gaps in information. It can also assist as a double check on the

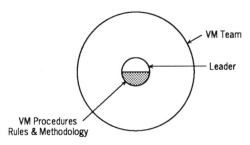

FIGURE 13.7 The methodology-team interface.

purpose statement. Assumptions are things like "the building will be occupied by the shops division"; "the building will have two floors"; "we will have in line-testing"; "we will use the current power source"; "branch offices will be located in areas x and y."

To keep assumptions visible and up-to-date, it is good practice to write them on a chart pad. This pad is then posted on the wall of every meeting. As new assumptions arise, they are added to the list. As current assumptions are confirmed, they are so noted. This on-going list is commonly referred to as a parking lot. Parking lots are also used for other subjects as well, such as premature ideas, questions and action items.

9. Company business plan

It is surprising how often teams design products without knowing the business plan or the strategic intent of the company or line of business. Too often design teams never think to ask for this information. The worst cases are those where the company business plan is not allowed to be seen by the lower ranking individuals. The team is directed to design a product and upper management will decide whether or not the team's recommendations fit the need of the company. Such noncommunication will surely be short-lived and company resources will be needlessly wasted. This is the price companies pay when individuals guard and consider information as power.

To hopefully avert such a situation, the following questions should be answered:

1. Does the team have a copy of or have access to the company business plan?
2. Do the answers to the eight questions fit or reflect the company business plan?

10. Implementation
   A. What are the necessary steps to implement recommendations resulting from the VM study?
      1. What factors could prevent implementation of recommendations?
      2. How can these opposing factors be eliminated or resolved?
   B. Consider the first day of the VM study as the first day of implementation.

Discussing these issues and their countermeasures early can increase the chances of success as well as help shape the study itself and the process. As an analogy, an artist must visualize what the final painting or sculpture will look like before she begins the work. So, too, the VM team should visualize the possible proposals and recommendations that may result from the VM study. Based on this vision of the results, the team can consider possible solutions to sources of opposition. These sources of opposition may vary with time. The method of force field analysis readily lends itself to this discussion. Basically, the team considers the approval process. It considers how opinions of management can be determined without risking a pocket veto. Other things considered in gaining commitment are; Who implements? Who budgets for implementation? Who will be affected? Who gains, who loses? Who can delay or block implementation? The answers to many of these questions begin to surface during the team discussions on alignment mentioned in item 2. Discussing the commitment issue early in the process is a step toward building implementation.

11. Project selection

Many VM studies arise out of necessity in a very specific, well-defined area. Consequently, a predefined project may obviate the need for formal project selection. However, the resources that can be allocated to a VM study are limited. Traditionally, the main criteria for a VM project have been high dollar volume and high total costs in relation to the function performed. There are other general criteria that should also be considered. The VM study should:

1. Solve a problem. The need should be real and should be supported by management.
2. Have a good probability of success and implementation.
3. Have objectives that are credible.
4. Be important to the people in the area being studied.
5. Have the commitment of the requestor and the team members.
6. Have receptivity. The sponsor or decision maker must be receptive to change.

A U.S. Department of Defense publication[8] lists additional project selection criteria. There will be other technical parameters necessary for project selection. Many lists of project selection criteria abound in the literature.[9,10] The above list is generic to all projects. This list of items should be kept in mind when selecting projects, because too often supervision or management wants to apply the VM process to the most pressing problem, which many times is also the most difficult.

If VM is being introduced for the first time, it is wise to start with an easy problem with virtually a guaranteed successful application of VM; a problem where you are sure recommended changes will be implemented. Build credibility first, then take on the tougher ones, because the first project establishes the pattern and reputation for all succeeding projects. Sacred cow products/projects should be cautiously considered for project selection if at all. What is a sacred cow? It is a product that has been around for a long time and was most likely invented years ago by the current president or other higher officer in the company. It is their pride and joy, even though it is overdue for updating and cost reduction. Usually, any time spent investigating such a product results in roadblocking and discomfort if the inventor/officer has less than an open mind. Other products that may have a propensity for VM project failure are those that have high esteem value or those that have been previously extensively investigated. Fowler[11] recommends that the VM product or project selected should be the heart of the line with maximum potential spin off to other models in the product line.

LAUNCHING AND CONDUCTING THE VM PROJECT

Third Party Facilitator

The behavioral and scoping exercise to initiate the process can be complex. The entire process from launch to finish is best done with a neutral third-party process facilitator to lead team building and process. The use of third-party facilitators also helps to keep honest people honest. Otherwise the teams may get mired in turf protection, drown itself in detail, get lost in the woods, and quickly rush to solve problems and shortcut the process. The process facilitator provides the VM working model and keeps the team on track.

From our experience, most people involved with VM projects are normally used to getting problems solved quickly. Usually in the course of their job, they are given little time and are used to quickly reaching a conclusion with little data. Consequently, the VM process, which takes more time and requires more data, will be frustrating to these types as team members. They are to be reminded and reassured that this frustration is normal and that answers and solutions will evolve quickly.[7] However, with the VM process they will have more data and ideas that will enable them to make a better, more informed decision. In fact, the hours spent on the VM project will probably be the most time the team members will have worked together in their career on a common problem and objective. They and the decision maker should feel assured that the team will have considered all possible solutions before making the final recommendations. The third-party facilitator can play an invaluable role in the dynamics of team interaction in its journey to delivering final recommendations.

Meetings

Meeting length and frequency are also important. The initial launch meeting should be scheduled for a full day. Regularly scheduled follow-up meetings should be scheduled at the launch meeting. These should be 4-hr meetings held at least once per week until the VM project is finished. Because of the complexity and detail in designing products, infrequent meetings less than 4 hr do not work well. Team members spend too much time getting back to speed. So, at the onset of the project, the team is told to expect one or two meetings per week. Each meeting will be 4 hr in length to provide continuity and will occur at regular-standing times and days (e.g., every Tuesday from 1:00 to 5:00). They can expect to spend 40 to 60 hr of team time (approximately one-third of a person-year), over the next 2 to 3 months. By now, not only the team members but you, as well, are probably wondering how all of this is going to happen. After all, this many people for this amount of time! It is very beneficial to have management endorse the VM study in a properly distributed letter to the supervisors of each of the team members. The letter should define the project, the sponsor, the scope of the project, and the priority. Most importantly, the letter should designate the team members, give them the time and priority to participate in the study, and pave the way to access needed information. Refer to Figure 13.8 for an example of such a letter.

DATE

TO: W. Johnson
         G. Smith
         W. Rowe
         R. Williams

FROM: I. M. Boss, Anywhere, Rochester, N.Y.

SUBJECT: Wigit Assembly Value Management

The Wigit Engineering Section is sponsoring a value management project to determine a least-cost alternative for a new Wigit subsystem design.

In this endeavor, we have assembled a team comprised of the following members:

1. Anthony Brown, Processing Engineering
2. Michael Rhodes, Product Development
3. Edward Jones, Industrial Engineering
4. Carl White, Customer Equipment Services
5. Mary Melone, Manufacturing Engineering
6. Charles Nelson, Reliability Engineering

Because of their knowledge and experience, I am asking if you would approve the above individuals and necessary time to participate in the project. It is expected that this may require four half-day sessions during the next four weeks. This time includes attendance at meetings as well as possible homework between meetings.

The VM process has a proven track record on other projects both internal and external to the company, and we believe the effort spent in applying it to the Wigit assembly will result in significant cost reduction and design improvement.

I appreciate the fact that your people are very busy with other assignments, and it is with this in mind that I am asking for your cooperation. This endeavor will provide timely input to new product development activity. Thank you for your cooperation.

FIGURE 13.8 Management letter announcing the VM project.

The Launch As mentioned earlier, experience shows that the initial launch meeting should be a full day. It should be kicked off by the decision maker or management person requesting the study to lend more credibility to the project and state how important it is to the company. For some team members, this may be the first time they will have had to interact with this level of management. A good gesture, and money well spent, is to have a company-paid lunch and/or an evening cocktail party to celebrate the initiation of a very important project. The location of the launch meeting is also important, and experience shows it is best to locate it off-site at a neutral site away from telephones, desks, and tendencies for team members to skip out to address external issues. If team members are easily distracted, the process facilitator, and team begin to lose control which, in turn, prolongs completion time. Team members who are present become discouraged. Local motels, hotels, or conference centers are excellent choices for this occasion.

Launch Agenda Launch agenda, pace, and atmosphere are very critical to a good start to set the feeling and engender commitment to the project. A time-tested agenda is illustrated in Figure 13.9. The decision maker introduces the project and stresses its importance and emphasizes how critical the team is to the project's success.

Another point to mention here is that the organization competes externally and not internally. The decision maker then introduces the VM facilitator and team leader. A team warm-up exercise is used to draw people in if team members do not know each other. One which we use is the "who am I?" exercise, which requires each team member to prepare answers to the following questions: Who are you? How do you like to be addressed? How do you not like to be addressed? Where do you work? What do you do? What do you like most about what you do? What do you like least about what you do? How do you feel about today? Tell us one interesting thing about yourself that might be useful to the other team members in getting to know and work with you.

The members are given 5 min to prepare the answers. Then each member presents his or her answers to the group. Only 5 min are allowed for each individual presentation. Next, the 40-min video by Joel A. Barker[12] titled *Discovering the Future: The Business of Paradigms* is shown. This is a very relaxing video about looking into the future and the effect that paradigms have on blocking our vision. A short

AGENDA		
DATE		
	WELCOME	
8:10	Kickoff	Decision maker
8:20	Who am I?	Group
8:45	Video	Group
9:20	Discuss video	Group
9:35	Teams	Shillito
9:40	Responsibilities and operating guidelines	Shillito/group
9:50	Overview of process	Shillito
10:15	Break	
10:30	Getting started	Shillito/group
11:30	Lunch	
12:30	Begin process—Customer needs	Shillito

FIGURE 13.9 VM Launch Agenda.

paradigm workshop (approximately 15 minutes) is conducted. The following types of questions are discussed by the team with process facilitator moderation:

1. What specific paradigms have an affect on our potential team output?
   (a) Positive affect?
   (b) Negative affect?
2. What are some paradigms that influence the way we think?
3. What are some paradigms that can influence our designing a product and getting it launched successfully?
4. What are some paradigms that can affect progress on this team project?

The facilitator writes the group answers and ideas on chart paper. The team may highlight the one or two most significant paradigms. One option at this point is to have the team discuss how it might minimize the affects of negative paradigms and/or maximize the affects of positive paradigms. Experience is showing that negative paradigms are the first to surface. The process facilitator needs skill at this point to keep the discussion from becoming a doomsday exercise in futility. Remind the group that negative paradigms, like problems, can be interpreted as opportunities.

During the paradigm workshop, team members quickly become interested and involved. They become aware of and sensitized to paradigms. They recognize situations where paradigms are influencing their outlook or restricting their thinking and creativity. They also are alert to other team members' paradigms. In an extreme case, surfaced paradigms may redirect an entire project, if the project purpose and/or assumptions are riddled with false paradigms. Surfacing paradigms can sometimes have stunning affects.

After the paradigm workshop and discussion, we give a short lecture about team dynamics and the stages of team development from immature and ineffective to mature and effective teams. Many times it is necessary to have a short team-building workshop as part of the VM project launch. We use a retrospective planning exercise. First we get the team to describe the ideal future state they are trying to achieve. Next we assume that this future state or goal has been achieved, and ask, how did we do it? What characterized the teams effort and their ability to pull it off? What contributed most to success? Answers to these questions are posted on chart paper. The team uses Pareto voting, described in Chapter 4, to narrow down the list to the most important elements. Next the team determines if any gaps exist between now and the desired future state. They then determine what opportunities they have to fill the gaps. The output of this team-building exercise is a set of team principles that are going to be used to fill the gaps and help get to the future state. Quality of team output and commitment are then discussed.

Throughout the paradigm workshop and the team-building exercise, it is necessary to stress and reaffirm that the competition exists outside of the company and not inside of the company or the team. Many times it is useful and fun to have the team select a team name. We have even given prizes or mementos such as team T-shirts or sweatshirts embossed with the team name.

The following responsibilities and operating guidelines are reviewed.

Process (VM) Consultant Responsibilities and Guidelines

1. Provides process model only.
2. Keeps the team on track.
3. Is not a content expert.
4. Will not "carry the team's monkey".
5. Will not provide all the energy to do the project!

Team Responsibilities

1. Will produce a deliverable (recommendations).
2. Will communicate with each other.
3. Agree to disagree.
4. Make decisions as a group.
5. Communicate as a group.
6. Respect each other's views.
7. Actively listen to each other.
8. Keep an open mind.
9. Focus on "What is right," not "who is right."
10. Won't be afraid of conflict and will use it to the team's/project's advantage.
11. Will have fun!

It is really best, if time permits, to have the team develop their own list of operating guidelines or vote on the most important from a prepared list such as above. Discussing these issues brings focus among all individuals involved. An overview of the VM process is then given to prepare the team for what they will be doing and the process they will be using. After the coffee break the following "Getting Started" questions are answered in detail:

1. Why are we doing this project?
2. When is it to be completed?
3. Who is the decision maker?
4. What is the scope? Include? Not include?
5. What country, market, segment, user?
6. What type of product? Evolutionary? Revolutionary?
7. Do we still have the right members on the team?

These topics have been discussed in more detail earlier in this chapter. Answering the questions starts to bring the team together to focus on the task. Answers to these questions may take longer than the process facilitator originally allows. However, the questions must be answered! So, flexibility is one important trait a process facilitator must have.

The team is now ready for lunch. After lunch or after the previous questions have been answered, the team begins the VM process by focusing on customer needs. This will last the rest of the day. Fifteen minutes are allowed at the end for team/project "administrivia." This is where future team meetings are blocked on the participants' calendars (which they were told to bring to the meeting).

All future team meetings will continue to follow the VM process. Members quickly become involved in the VM process to the point that the process becomes the vehicle for dialogue to carry the project to completion. Teams can work effectively or can waste time and make poor decisions (or none at all). The structure and momentum of the team is maintained by the VM process, methodology, and techniques. It provides the vehicle for overcoming obstacles. It helps to encourage rational conflict and minimize emotional conflict. It helps the team work at the right level of detail. To check the health of team meetings, commitment checks, or references to how well the team is following the team guidelines are very helpful to the facilitator and team to maintain human dynamics.

Team-building workshops conducted prior to doing the actual VM project are very effective and well worth the time. The purposes of the workshops are (1) to help the members develop communications skills and learn to work together as a group and (2) to use the group skills effectively to analyze problems, make decisions, and work as a team. Unfortunately the human relations/personnel relations practices in most companies do not support teamwork and social skills, nor do they hire people with those skills. There are some excellent sources on team building and process consultation. See the categorized list of references at the end of this chapter.

DEVELOPING A VM PROGRAM AND ACTIVITIES

The cardiac pace of change that is engulfing all organizations today is making planning, both reactive and proactive, a crucial process for survival. Value management, using our broader definition and generic application, can be used to develop and improve the input needed to make a more informed decision. To do this requires that VM be established for long-term success. In this regard: How does one get a VM project started? How is it introduced into the organization? Where does it belong in the organization? How does one put the VM plumbing in place? In considering these questions, the essential ingredients for integrating VM into an organization are top-down management support and participation, line and staff support across organization functions, development of a VM structure/organization, and training at all levels of the organization.

Training is started at the director/manager level with a 1-day seminar aimed at the basic concepts of VM and the efficient utilization of VM in the company. The 40-hr VM workshop is the most used method for training first-line supervision as well as, in many cases, the line work force. This workshop is aimed at formal training and first-hand application.

Once a critical mass of people have been trained, they have to be managed. A time-tested structure advocated by J.K. Fowlkes, chairman and founder of Value Analysis Inc., is to have a full-time value manager, a VM council and VM study teams.[13]

Mr. Fowlkes then recommends that the value manager report to the general manager if it is a company-wide program or a departmental or line of business manager if it is a segmented program. The value manager runs the VM business and also reports to the VM council, which is comprised of the general manager and his or her direct reports. The council approves projects and budgets and assigns responsibility for the implementation of recommendations. The value manager is the co-ordinator for all current and future VM projects, is a VM process resource to the study team, keeps current on the state of the art of VM technology, is the VM team's link to the VM council and develops, for VM council approval, the budgets/forecasts to support the VM teams and projects. The interdiscipline VM teams are commissioned by the VM council, with input from the value manager, to do the projects and are monitored and supported by the value manager. Their final recommendations are presented directly to the VM council.

We have also used Mr. Fowlkes' VM organization format with success. Upon reading numerous Japanese publications,[14-17] we find that the above VM organization structure and training format lies at the roots of numerous Japanese company's successful VM programs as well. Arthur Mudge, a seasoned veteran of VM, has some excellent publications regarding successful management of VM and elaborates in detail on this subject.[18,19]

We cannot reiterate enough that there is one prerequisite that has unanimous agreement across all VM practitioners, managers, and consultants as still being the most important: The VM program *must* have complete top-down management support and involvement. This is particularly critical whether it involves the start of a new VM program or the revamping of an existing one. Without senior management's full commitment to the program and in-depth involvement at all stages, VM has little chance for success.[20] Where VM fits and how projects are introduced, organized, and administered varies across each organization. L. B. Wilson[21] does note, based on a survey by the Society of American Value Engineers, that "the lower the value program is in the organization, the lower will be the return on the company's investment in its value activities. The worst place for the value program is down in the bowels of the organization, buried under 10 levels of management".

We realize that each company has its own personality and culture and that VM programs have to be tailored to each. However, the aforementioned format seems to be generic enough so that it can at least be used as a seedbed for ideas.

OTHER FACTORS AFFECTING APPLICATION OF VM

Group/Team Dynamics

The most successful VM teams we have worked with were those that

1. Had clear goals and knew where they were headed.
2. Had the freedom to work their own way without considerable interference (there was no punishment for failure).
3. Had free access to all needed data.

4. Had full responsibility for their deliverables.
5. Had a good balance between rational and intuitive thinking.

Finally, those teams that were able to respond most quickly and implement changes and recommendations the fastest were those teams whose members were empowered to make decisions without bureaucratic interference. This is extremely important because those companies who will be successful in the future are those who can respond the fastest to customer needs. Time, not quality, will be the battlefield. Team response and effectiveness provide the foundation for the company's success. The importance of the time factor will be discussed further in Chapter 15.

Benefits

Let us start first on the positive side with the benefits of using VM! If we can document these, they, in turn, can be used as factors to develop a positive mind-set for applying VM. First of all, why do value management at all? So,

1. We don't design something that is not needed or unnecessary.
2. We design what the customer wants.
3. We communicate better, internally/externally and vertically/horizontally.
4. We can minimize redesign and engineering changes.
5. We can build it right the first time.
6. We can improve design and reduce costs.
7. We can challenge and set realistic specifications.
8. We can manage costs on purpose and remain competitive.

This list is a beginning. You can certainly add more to it. The point is that they are factors affecting application of VM. Too often, though, they are overwhelmed by negative paradigms that are strongly entrenched in the company.

What are the benefits to the company?

1. Increased communications, vertically and horizontally.
2. Higher quality at lower cost.
3. Better value to customers; reduced value mismatch.
4. Simpler manufacturing methods.
5. Faster assembly.
6. Better material selection.
7. Just-In-Time type inventories.
8. Shorter development and ship-to-stock time.
9. More realistic specifications.
10. Reduced postintroduction product problem solving.

This list could be expanded to fill many pages. The type of benefits accrued depend on where and how the VM process is applied. In this respect, this list can also give you an idea where to apply the VM process. Actually, the application of VM should be "limited only to our imagination and time."[18] However, a common thread benefit running through all types of applications is increased company communications and a more customer-oriented value-added product at lower cost.

What are the benefits to management?

1. A means to have their people use a structured, focused approach to reduce costs and value mismatch.
2. A means to set an example for their people:
   (a) In being proactive in cost management.
   (b) In facilitating communications.
   (c) In encouraging a more customer-oriented approach to achieving value-added designs.
3. A connecting channel to be more interactive with subordinates.
4. A structured means to share strategic intent and company direction.
5. A way to tap the potential of subordinates.

What are the benefits to the VM practitioner or team member?

1. Personal development opportunity.
2. Ability to be seen by others as a creative individual.
3. Potential for high-level visibility and major impact.
4. Recognition from others.
5. Legitimate sounding board for ideas.
6. An opportunity to improve communications with colleagues, superiors, and clients.
7. An opportunity for another viewpoint toward design.
8. An opportunity to work interactively and in parallel with other departments rather than in a series.
9. An opportunity to view problems and opportunities from a total corporate perspective rather than shop floor tunnel vision.

Barriers

Even with all of the benefits of the application of the VM process and a VM program, it still can be difficult to get VM started within a company. Some of the most frequent barriers to VM are:

1. Lack of Management Support

Management often states they support VM, but too often it turns out to be lip service. Many teams get formed with a designated amount of time allocated for the project. The allocated time usually competes with other activities having higher emphasis or priority from management. This results in teams having variable attendance, and lack of continuity on the project. A person will spend their time on the most visible project that commands the most attention, or the one emphasized by management or supervision.

2. The Reward System

Performance appraisals and pay raises are based on history. They are based on how well you satisfy things like schedules or sales quotas, or how many problems you solve. Problem solving and fire fighting provide the grist for making heroes. Problems are usually well known internally and make problem solvers highly visible and well rewarded. This tends to encourage problem solving instead of problem prevention. People who prevent problems are often unknown and unrewarded. The VM process is both problem solving and future oriented. When asked to participate on a VM project, people are sometimes reluctant to do so, for fear of taking time away from working on their own problems and the real work they are rated against and paid to do. Because VM is also many times future oriented, it can also be perceived as threatening to historical elements of performance appraisals and discourage people from using it.

3. "We Don't Have Time"

Team members are pressured to meet schedules on current projects and see VM as a delay to these time goals. Benefits are not always visible at the beginning of a VM project. It is hard for teams and individuals to risk their schedule attainment without supporting proof that they should apply the process. People also have a tendency toward quick solutions to problems, and when learning about a 40-hr VM project, they develop considerable anxiety about participating. The time frame to final solution in a VM project is not congruent with their personal shorter time frame for solving problems.

4. Change

Change never comes easy. Most people resist change, especially product designers, manufacturing managers, production engineers, assembly people, and plant management. The VM process by the very nature of the function approach creates change, sometimes drastic change. This kind of change represents a threat to the above type of people, because VM seeks to change the product/process, which in turn, creates a cascade of changes down the line. The plant personnel are accustomed to managing day-to-day activities and immediate problems. This is the nature of their work and what they are rated on. By using VM, managers and supervisors must take a longer term view beyond the present. They aren't used to doing this and by doing so, they get overwhelmed with perceptions of change. They feel uncomfortable. The first tendency is to resist VM.

5. "We're Already Doing It"

Our immediate response to this barrier is, "Show us, we would like to learn from your experience!" Most companies have some form of design review (DR) process, which too many people misconstrue as the same as VM. There are elements of VM that are similar to DR. Both processes deal with design. Value management involves concept ideation, whereas DR involves concept review. Value management is not a review, as such, it is a process and philosophy to improve value through the study of function. Design review does not create, and it does not involve functions. Design review begins with a design, whereas VM usually starts with a clean slate of functions. One might argue whether or not VM involves review of design. The information phase of the traditional VE job plan does require critique of current data which is used later in function analysis and cost comparison in the preparation of recommendations.

CONCLUSION

Given the benefits of VM and despite the barriers, a VM program should be a must for a company to remain competitive in the future. Value management along with QFD, discussed in Chapter 8, are currently quoted by many companies as competitive weapons of the future. Those companies who successfully implement a VM program will have a formidable advantage of customer-oriented, high quality, value-added, low cost, product designs.

With the surge of total quality management activity, the designing and operating of teams, especially design teams, will be increasingly more important. The parallel team approach to product design and manufacture will become the standard way to function as opposed to the traditional throw-it-over-the-wall-to-the-next-guy serial approach.

Currently there are several terms that describe this parallel approach: *simultaneous engineering, concurrent engineering, multifunctional team design,* and others.[22] Using the parallel team approach will require many changes in organizational structure, roles, responsibilities, performance appraisals, and reward structures. Organizational goals and the team's purpose must be congruent. Empowering and staffing the team will become more critical to successful team operations. Changing the rewards to fit with concurrent engineering team design will present a real challenge. A few companies are now starting to tackle these concerns.

People and teams can complicate the VM process but are a necessary part of it. Behavioral science aspects must be treated as carefully as any other part of the VM process. Look ahead, predict the problems, and carefully plan to deal with them. Remember, the first team meeting should be considered as the first day of implementation.

HELPFUL READING

The following references are grouped by subject matter. The complete details of each reference can be found in the reference section at the end of this chapter.

Author	Reference Number	Bibliography Number
Managing and Organizing VM		
Barlow		2
Berne	5	
Fowlkes, Groothius and Hayes	13	
Japan Management Association	15	
Kawamura	14	
Marcon		8
Mudge	18	
Mudge	19	
Riegle	4	
Sehr		15
Wilson	21	
Initializing the VM Process		
Bachli		1
Boothe		2
Dillard	3	
Fiorelli	6	
Laschutka and Laschutka	20	
Parker	7	
Riegle	4	
Rivet		11
Schein		12
Schein		13
VM Project Selection		
Drozdal	9	
Fowler	11	
Mudge		9
Parks	10	
U.S. Department of Defense	8	
Team Member Selection, Team Building, Process Consultation		
Bachli		1
Boothe		3
Fiorelli	6	
Fordyce and Weil		4
French and Bell		5
Gomolak		6
Kiefer and Stroh		7
Parker	7	
Port, Schiller and King		10
Schein		12
Scholtes		14
Welter		22

REFERENCES

1. Warfield, J. N., "An Assault on Complexity," in *Batelle Monograph No. 3,* Chapter 1, pp. 1.1–1.8, Battelle Memorial Institute, Columbus, Ohio, April 1973.

2. Cook, T. F., "Determine Value Mis-Match by Measuring User/Customer Attitudes," in *Proceedings, Society of American Value Engineers,* Vol. 21, pp. 145–156, May 1986.

3. Dillard, C. W., "Value Engineering Organization and Team Selection," *Proceedings, Society of American Value Engineers,* **10,** pp. 11–12, May 1975.

4. Reigle, J., "Value Engineering: A Management Overview," *Value World,* 3(3), 4–8, 1979.

5. Berne, E., *Structure and Dynamics of Organizations and Groups,* Grove Press, New York, 1966.

6. Fiorelli, J. A., "Some Considerations in the Selection of Individuals for Team Problem Solving Efforts," *Proceedings, Society of American Value Engineers,* Vol. 17, pp. 132–136, 1982.

7. Parker, G. E., "Applying Understanding of Individual Behavior and Team Dynamics with the Value Engineering Process," *Proceedings, Society of American Value Engineers,* Vol. 24, pp. 224–229, June 1989.

8. U.S. Department of Defense, *Principles and Application of Value Engineering,* Vol. 1, U.S. Govt. Printing Office, Washington, D.C., 1968.

9. Drozdal, S. C., "Criteria for the Selection of Value Engineering Projects," *Proceedings, Society of American Value Engineers,* **13,** 102–106, May 1978.

10. Parks, R. J., "Project Selection—Key to Success," *Proceedings, Society of American Value Engineers,* **21,** 71–74, 1986.

11. Fowler, T. C., *Value Analysis In Design,* Van Nostrand Reinhold, New York, 1990.

12. Barker, J. A., *Discovering the Future: The Business of Paradigms,* video, Charthouse Learning Corp., 221 River Ridge Circle, Burnsville, MN 55337.

13. Fowlkes, J. K., Groothius, J. D., and Hayes, R. T., "How to Develop and Implement Value Management Activities," *Proceedings, Society of American Value Engineers,* Vol. 23, pp. 263–265, May 1988.

14. Kawamura, S., "Promoting Value Engineering at Fuji Film," *Fuji Film Research and Development,* No. 34, 158–163, 1989. [In Japanese]

15. Japan Management Association, *Canon Production System,* translated by Campbell, A. T., Editor: Dyer, C. E., Productivity Press, Stamford, Conn., 1987.

16. Society of Japanese Value Engineers,*SJVE, History, Organization and Activities,* Society of Japanese Value Engineers, Sanno Institute of Business Administration, Setagaya, Tokyo, Japan, April 1980.

17. Society of Japanese Value Engineers, "1982 SJVE Value Program Study Tour," SJVE, Sanno Institute of Business Administration, Setagaya, Tokyo, 1982.

18. Mudge, A. E., "The Ingredients of Successful Value Management," *Proceedings, Society of American Value Engineers,* **10,** 28–33, May 1975.

19. Mudge, A. E., *Successful Program Management,* J. Pohl Associates, Pittsburgh, Pa., 1989.

20. Lashutka, S., and Lashutka, S. C., "The Management of Team Development in Value Analysis/Engineering," *Value World,* 3(4), pp. 5–10, 1980.

21. Wilson, L. B., "Guidelines for Making Value Programs Work," *Proceedings, Society of American Value Engineers,* **15,** 211–293, 1980.

22. Welter, T. R., "How to Build and Operate a Product Design Team," *Industry Week,* pp. 35–50, April 1990.

BIBLIOGRAPHY

1. Bachli, L., "Cooperative Value Engineering Team Building," *Proceedings, Society of American Value Engineers,* **19,** 106–109, May 1984.
2. Barlow, C. M., "Organizing for Acceptance—Value Management as a Corporate Resource," *Proceedings, Society of American Value Engineers,* **13,** 283–289, May 1979.
3. Boothe, W., *Developing Teamwork,* Golle and Holmes, Minneapolis, July 1974.
4. Fordyce, J. K., and Weil, R., *Managing with People,* Addison-Wesley, Reading, Mass., 1971.
5. French, W. L., and Bell, C. H., *Organization Development,* Prentice-Hall, Englewood Cliffs, N.J., 1990.
6. Gomolak, G. J., "Team Building—Who Said It Would Be Easy?" *Proceedings, Miles Value Foundation,* **1,** 125–132, 1988.
7. Kiefer, C. F., and Stroh, P., "A New Paradigm for Developing Organizations," in *Transforming Work* (J. D. Adams, Ed.), Chap. 11, Miles River Press, Alexandria, Va., 1984.
8. Marcon, L. J., "The Optimum Application of the Value Disciplines—Integrated Value Management," *Proceedings, Society of American Value Engineers,* **6,** 161–173, May 1971.
9. Mudge, A. E., *Value Engineering: A Systematic Approach,* McGraw-Hill, New York, 1971.
10. Port, O., Schiller, Z., and King. R. W., "A Smarter Way to Manufacture," *Business Week,* pp. 110–117, April 30, 1990.
11. Rivet, G. J. C., "The Value Engineering Instructor," *Proceedings, Society of American Value Engineers,* **11,** 213–219, April 1966.
12. Schein, E. H., *Process Consultation: Its Role in Organization Development,* Addison-Wesley, Reading, Mass., 1969.
13. Schein, E. H., *Process Consultation,* Vol. II, *Lessons for Managers and Consultants,* Addison-Wesley, Reading, Mass., 1987.
14. Scholtes, P. R., *The Team Handbook,* Joiner Associates, Madison, Wisc., 1988.
15. Sehr, E. K., "How to Improve the Efficiency of Value Engineering (VE)," *Proceedings, Society of American Value Engineers,* **17,** 93–97, May 1982.

14

VALUE PLANNING

David J. De Marle

VALUE PLANNING

Value planning is an organized process for planning and creating new products and services.[1] It assesses the impact of the future on products and services, develops functional performance and cost targets for future product designs, and uses VE to conceptualize new product offerings. Value planning combines elements of technology forecasting,[2] VE and VM. In VP, value teams measure the value of current products and services and forecast the performance and price characteristics of future goods. Supply and demand curves are created to show the gap between existing and future product offerings. Based upon these studies, VE teams are assigned the task of designing new products and services. Value planning is based on the premise that only products and services of high value will survive future market competition.

RATIONALE FOR VALUE PLANNING

Chapter 2 described how functionally superior items replace their predecessors. Incandescent lights replaced gas lanterns, and corporations replaced feudal estates. The fittest products replace predecessors, which die. This process of technological and social evolution is heavily influenced by the environment. Frequently, substitutions are brought about by shortages in key raw materials. This need-driven value-based substitution process lies at the heart of VP. Because substitutions take time to unfold, a manager can use substitution phenomena as one basis for logical predictions of the future. By examining historical information, we can derive parameters and relationships that portend the future. Management can consider a set of futures that are "likelier than not" to occur. These futures form a basis for planning and decision making. Value planning, based as it is on a study of value-based substitution phenomenon, allows management to anticipate and initiate changes to improve value.

FORECASTING MODELS

Value planning is based on a study of the future. Today rational forecasts of the future based on studies of consumer needs (market pull) and functional substitutions (technology push) are replacing intuitive forecasts. Modern forecasting techniques, such as trend analysis, simulation modeling, and substitution analysis, are used to help forecast the future of current product offerings and to anticipate new products with greater value. Value planning often uses technology forecasting to forecast the future.

TECHNOLOGY FORECASTING—DEFINITION AND BACKGROUND

Technology forecasting has been defined as a process in which data are gathered and analyzed in order to predict the future characteristics of useful machines, procedures, or techniques.[3] It grew out of the need of the military to predict the characteristics of future weapons systems. Technology forecasting has since had widespread use in defense planning, and this use led to the spread of technology forecasting in industry.

In 1961, Professor James Bright developed a course on technology forecasting and planning at Harvard that was offered to research and development managers.[4] Five years later in Europe, Erich Jantsch published the results of a study of some 100 technological forecasting and planning activities conducted in 13 different European nations.[5] These events served to stimulate interest in this field and attracted new practitioners. Since the late 1960s, the field has seen the formation of consulting firms specializing in technology forecasting; the publication of numerous articles, books, and journals on the subject; and the development of university courses devoted to the subject.

As the technology forecasting discipline grew, a large array of modern forecasting methods were developed and came into play, allowing the forecaster to broaden the scope of the forecasts. Many of these used the calculating and word processing capabilities of the computer. These new tools have been used in market research and analysis, as well as in technical, political, and environmental forecasting. These forecasting methods are used in VP studies to predict the performance and price characteristics of future products and processes.

THE VALUE PLANNING PROCESS

Value planning, like VA and VE, follows an orderly and structured process. The process involves six phases.

1. Origination phase
   (a) Select a subject for study.
   (b) Bound the study.
       i. Determine the time frame of the study.
       ii. Define scope of the study.
       iii. Define the forecasting and planning mission.
   (c) Form a value planning team.
2. Data gathering phase
   (a) Identify data needs (subjects).
   (b) Identify data sources.
   (c) Plan data collection.
   (d) Collect data.
   (e) Establish data file.
3. Data organization and analysis phase
   (a) Measure product value.
   (b) Categorize data.
       i. Performance functions.
       ii. Function costs.
   (c) Establish product value chronology.
   (d) Determine value interrelationships.
4. Projection phase
   (a) Establish optimistic, pessimistic, and realistic reference frames.
   (b) Project trends.
       i. Supply and demand.
       ii. Performance functions.
       iii. Product cost.
   (c) Determine value gaps.
   (d) Model future value.
5. Planning phase
   (a) Develop function and cost criteria for future products.
   (b) Develop a technology road map for future products and services.
   (c) Select high value product concepts for VE studies and implementation.
6. Value management phase
   (a) Use VA teams to improve the value of low value products and activities.
   (b) Use VE teams to design future products and services.
   (c) Establish a monitoring program.
       i. Determine what to monitor.
       ii. Establish monitoring process.
       iii. Upgrade forecasts and plans on a periodic basis.

A detailed description of the steps in this process follows.

1. Origination Phase

The origination phase establishes the nature and depth of the effort. In the origination phase, it is essential to define in detail the scope of the effort. By asking questions about the intended study, a clearer understanding of its purpose and mission will emerge. A short, well-written mission statement will allow management to communicate the study's goals and objectives. Why is the study being done? What subjects will the study include and exclude? Who is the client for the study? When should the study be completed? Will a team be used to plan? Who will be on this team? Answers to these questions serve to define the scope of the study and clarify the mission of the planning effort.

Time Frame Considerable attention should be devoted to the time frame of the planning effort. Is a 5- or 20-year plan desired? Answers to these questions are very important, as they dictate the approach to the study and the type of data needed. Because the future starts at this moment and extends forever, the first thing that management needs to consider is the planning horizon—the period of time from now into the future that the plan will address. Earl Joseph, a staff scientist at Sperry Univac and author of the book, *Future Trends,* recommends that managers concentrate on a period from 5 to 20 years in the future. Joseph categorizes the future into five time horizons.[6]

1. From now to next year.
2. Next 1 to 5 years.
3. From 5 to 20 years.
4. From 20 to 50 years.
5. Fifty years and beyond.

The first time frame is usually frozen by actions that have already been implemented, and present decisions and actions have little or no effect upon it. In the second time frame, evolutionary advances that have been planned for can be implemented, but revolutionary changes can only be made after massive efforts. The 5- to 20-year period is almost completely controllable and decisions made today can be implemented by evolutionary or revolutionary means. Time frames beyond 20 years are largely unpredictable and uncontrollable, but events and decisions made today will sow the seeds for changes that may be beneficial in the distant future.

Often organizations set 5-year goals and objectives and develop plans to accomplish these. These goals usually focus on cost rather than performance. In many performance areas, this time frame may be too short to accommodate innovative approaches. Longer time frames allow an organization to plan for major innovations, yet introduce an element of risk into planning that stems from the uncertain nature of the longer time horizon.

Once sufficient attention has been given to the time frame and scope of the plan, people should be assigned to a VP team. Planning teams can be formed to study various areas of interest to an organization, such as products, organizations, manufacturing processes, and marketing operations. A team should be made up of about five people who have time to access and study information that is required. Team members need to be objective and capable of gathering data and sorting it in an unbiased manner. They should be imaginative and able to envision various futures. They will need to work together and will need to focus their individual efforts on the team's overall mission. The team should have a basic understanding of value measurement and forecasting techniques. Because forecasting often involves mathematical modeling, the team should have knowledge of this field, although they may rely on others for mathematical model building.

At the inception of a study, it is helpful to educate the team in the VP process. Knowledge of the VP process will improve the efficiency of the team and help it to be effective. Here, a consultant can assume the temporary role of teacher. Training may include elements of team building and goal setting. If proper attention is paid to these topics, the subsequent planning effort will be more productive.

2. Data Gathering Phase

Identification of the team's mission should lead directly to an identification of information that will be needed in the study. This information may come from different types of data.

Hart described three different types of information gathered in a forecasting study.[7] He included recorded information contained in published records or files, observational data such as data determined in laboratory or operational studies, and subjective data obtained by interviewing or questioning people. Searching computer files for information is an easy way to start the data gathering process.[8] Abstracts recorded in libraries or computer files are easy to access and search and are unbiased (different people recorded the abstracts at different times). They often lead to articles that offer rich detail. This recorded information will need to be augmented with other data. Properly conducted interviews will uncover additional information and help enlist the aid of other knowledgeable participants and users in the study. It may also be necessary to gather observational data as the study progresses. As data are gathered, they will influence the data-gathering process of the study and should be assembled in a central file for subsequent manipulation and study. It is important to search for any relevant information and not to bias data gathering by excluding data that do not fit preconceived concepts of what the future plan "will" or "should" be.

Forecasting Techniques of Value in Data Gathering Several techniques used in technology forecasting are very useful in the data gathering phase of a forecast. By interviewing, constructing Delphi questionnaires, and searching through electronic data bases, the value planner can rapidly collect a large amount of relevant data.

Interviewing Interviewing is used as a data-gathering mechanism. It generally involves qualitative, future-oriented data that are subjective and "soft" in nature. Proper planning for and conducting of interviews are essential. Questions to address in preinterview planning are: Who should be the interviewer and the interviewees? Are all of the necessary organizations/areas represented? Are the right disciplines represented? How many interviews? Usually interviews are interdisciplinary, representing many areas and backgrounds. The questions are generally open ended and unstructured to produce stream-of-consciousness-type answers. Generic examples are: What do you see lying ahead in the next—years that could affect the future of—? What do you see as significant opportunities (or threats) for—? Time horizons can be specific (e.g., next 10 years) or open to encourage creativity and imagination. Individual interviews work well. Group interviews should be used with caution to minimize peer pressure and personality conflicts. One interviewer is often used, although experience indicates that two interviewers are more effective. See References 9 to 12 for further information on interviewing.

Delphi Delphi is an opinion-taking, consensus-forecasting technique employing experts or knowledgeable people to forecast future events. It was originally designed to minimize the personality and psychological shortcomings of committees. The process is performed by conducting several carefully designed interrogations. Participants are asked to predict when events will occur. Each questioning round is followed by a feedback of information from the previous round, including a graph of predicted times. According to Coates,[13] the process is "applicable to any situation to which quantitative values may be assigned, whether these are dates, weightings or scaling, etc." The entire process is usually performed through a director or mediator.

Electronic Interactive Data Retrieval Services Several commercial interactive data retrieval services are available that are excellent sources of information. Such electronic data banks are generally used to search for certain pieces of information in order to answer specific questions. They should not be the sole source of data but should be considered as supplementing the other data-gathering methods. They may also be used to look at aggregate information and the rate of change of information availability.[14,15] This, in turn, provides an indicator of trends and activity in a given subject. Searching through abstracts may also reveal new ideas and subjects to be searched and refined. The procedure is often iterative and may lead into areas not dreamed of when the search began. Several commercially available information services are referenced.[16–18]

3. Data Organization and Analysis Phase

Data that have been organized and analyzed take on a new character—the character of information. Hence, in this phase of a VP study, data are categorized, and a data chronology is established. The data are examined for interrelationships as part of the categorization process, and trends are established by the data chronology. In this phase of the study, comparisons of the value of current products are made to "benchmark" future projections.

Measure Present Product Value The value measurement techniques described in Chapter 4 should be used to measure the present value of items and functions in the area under study. This will provide an index from which future plans can be developed. An excellent approach involves plotting the value of current products and processes on value control charts. The following example for a color television (TV) manufacturer illustrates this. A hypothetical value control chart for the primary functions in a television set is shown in Figure 14.1. The value index, discussed in Chapter 3, of functions in the TV set is plotted on a semilogarithmetic scale. Inspection shows that the present signal recording, frame, and enclosure assemblies have low value and should be targeted by management for VA team studies. The manufacturer could expect these studies to yield cost reductions near 20%. These studies should also improve the product's performance, quality, and reliability.

Further profits and competitive advantages would occur if the company analyzed and improved its present manufacturing operations. Value control charts of the manufacturing operation will reveal areas where VA teams could be formed to improve the value of the manufacturing operation. Figure 14.2 is a hypothetical value control chart of the manufacturing operation. The wiring, soldering, and maintenance operations should be studied by VA teams. As these VA studies proceed, they will produce immediate savings together with recommendations that could profit the corporation in the future.

Categorizing Performance and Cost Data Although the subject under study often dictates the categorization of data, it is useful to categorize data by functions and

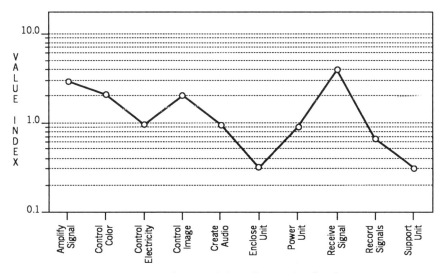

FIGURE 14.1 Value control chart of TV receiver functions.

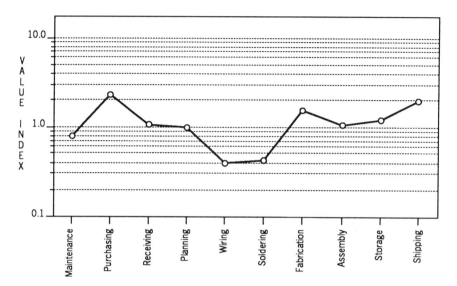

FIGURE 14.2 Value control chart of TV manufacturing operations.

function costs. Numerous data organizing methods are available, and several are listed here.

FAST Diagrams FAST diagrams are very useful in categorizing forecast information. Diagrams of this type can be constructed to depict both supply and demand functions at work. They are not limited to products, hardware, or manufacturing. An example of a FAST diagram used in VP is shown in Figure 14.3. This diagram describes broad functions involved in obtaining and using water in a water district. The demand side of the diagram, shown on the left-hand side answers why the water is obtained. Methods for meeting the water demand are shown on the right-hand side diagram and show how the water can be obtained. Common methods are shown at the top of the diagram and include collecting local water from dams and wells. As these supplies are consumed, the district needs to import water from adjacent water drainage basins. An arrow to the right of the diagram shows a function trend in which high-technology processes are substituting for older water supply methodologies. A FAST diagram like this can be used to help categorize the future. FAST diagrams are described in detail in Chapter 7.

Relevance Trees Landford[19] describes a relevance tree as "a way of setting forth interrelating variables in a coherent graphic form." Relevance trees may be quantitative, employing the use of relevance numbers described by Bright[20] and others.[21,22] They may be visualized as a vertical tree, branching out from a single unique function, objective, or parameter to form a hierarchical structure. Each level of the tree represents a certain level of detail that increases at successively lower levels. Relevance trees are a useful device for displaying large amounts of information in

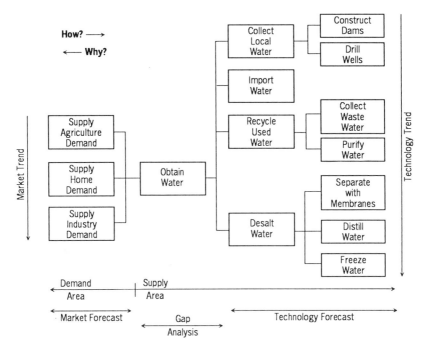

FIGURE 14.3 Value planning FAST diagram (water district).

concise structured form, searching for unforseen relationships, and discovering new ways to accomplish objectives and fulfill needs. The process of constructing the tree also serves as an excellent communications device. The tree itself, because it contains a wealth of information, may be considered what Landford[23] terms a *mini-information center.* Whatever scheme is used to categorize data, it should accommodate data on market pull and technology push. It should also integrate the data in a logical way and identify by its subject headings the main parameters at work in the subject area.

Other Techniques Many other categorization techniques can be used in this phase of the planning effort. Cause–effect diagrams,[24] morphological analysis,[25] numerical taxonomy,[26] conceptual maps,[27] and mission flow analysis[28] are all useful techniques that can be helpful at this point in the study. The affinity diagram discussed in Chapter 8 is one of the newest methods for aggregating data.

Establishing a Product Chronology Data take on new meaning when they are arrayed chronologically; written descriptions of events occurring in a subject area frequently reveal patterns that tell the planner much about the level of development of a technology. Life cycles are uncovered that reveal the level of development of a product idea. One idea may be in a speculative phase, whereas another may have passed through much of the research and development process and may soon be introduced as a new product. A mature product may be facing economic, social, or environmental pressures that will limit its growth. Figures 2.3, 2.6, and 2.8 are graphs that show the life cycle position of products.

An important step in establishing data chronology involves segregating "event" data from "trend" data. News items are events in the development of a product or technology, whereas numerical data arranged chronologically depict product or technology trends. Mathematically, future events are usually treated in terms of probability theory, whereas trends are often treated as material flows in a dynamic system or as vectors that interact with one another to produce resultant forces.

Determine Interrelationships When the data collected in the data-gathering step have been aggregated into distinct subject categories, it is necessary to determine interactions among the categories. To determine these interactions, matrices are often constructed in which subject parameters are arrayed against themselves (cross-impact analysis) or against external factors (impact analysis).

Function Matrices Function matrices are described and illustrated in Chapter 7. They show the interrelation of components and functions in a system and are useful when examining interrelationships.

Technology Road Maps Chapter 9 describes technology road maps that show the various ways a product or service can provide its basic functions. They are very useful in VP, often forming a template for future development.

Cross-Impact Analysis Cross-impact analysis as developed by T. J. Gordon and Olaf Helmer, is a computer simulation technique that uses a cross-impact matrix for input. Raters assign to each cell of the matrix: (1) the mode (direction of impact, positive, negative, or no effect), (2) the strength of impact between pairs of items, and (3) the time lag, how long it takes until the impact is realized. A quadratic equation relating the above three variables is programmed into a computer simulation program that computes the resulting probabilities of each pair-wise interaction of items, based on the matrix input. The output is a listing of the items, the initial probabilities, the probability shifts which occurred when the items were correlated, and the final probabilities. Items can then be ranked (by computer) by initial or final probability or by probability shift, each of which produces a different scenario. The process permits an elementary analysis of the potential interactions between items.[29,30]

Impact Analysis Shillito recommends the use of impact analysis to help determine data interrelations. He defines impact analysis as "a subjective quantification technique for evaluating the impact of external factors (influences that impinge on a set of items). It is a matrix scoring method and is similar to a force-field type of analysis wherein raters use (+)'s and (−)'s and zeros to quantify their perceptions of interactions of items. It is an effective technique for screening influences, identifying attributes, and locating sensitive variables in a system. It can be used in any situation where it is desirable to measure both the positive and negative forces that impinge on a system or set of alternatives."[31]

Other Techniques The scenario technique is also useful in ascertaining the inter-relationships and impacts of a forecast. This method, which will be described under the projection phase of VP, is an excellent method for showing cause–effect relationships. Similarly, PERT diagrams, so often used in planning and project management, are also useful tools for ferreting out detailed interactions. System dynamic models described in Chapter 5 model interactions present in complex systems.

4. Projection Phase of Value Planning

At this point, all of the necessary steps have been taken to allow the planner to project historical data into the future. If a mathematical model has been constructed, it is now run to simulate what may happen in the future.

Projecting the Future Planners use a variety of different techniques to project the future. In using these methods, it is desirable to project three different types of forecasts: optimistic, pessimistic, and realistic. In trend extrapolation, this might represent plus or minus deviations from the normal projection; in scenario writing, separate scenarios are written; in cross-impact or impact analysis, the model can be changed to simulate these different futures. The realistic future is, by definition, the most probable, yet all three projections should be made to help visualize a set of alternative futures. Each future will have a different impact on current activities, and these impacts need to be assessed in VP.

Trend Analysis Trend analysis entails plotting historical data and statistically extrapolating it into the future. Figure 14.4 is a learning curve which shows the rapid reduction in the unit cost of color TVs manufactured in Japan.[32] Its regression equa-

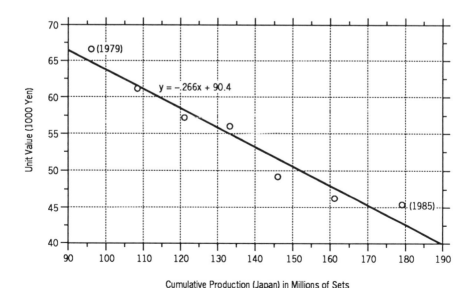

FIGURE 14.4. Price decline trend for color TV sets.

tion can be used to predict the price of these sets in the future. Such projections allow VP to estimate the future cost of color TVs. When cumulative production hits 250 million units, the cost of a color TV could be about 30,000 yen, based on the value of the yen in 1985. This simple trend extrapolation forecast would be augmented with other forecasting techniques to predict price declines in various segments of the color TV market.

Trend analysis involves time-series analysis and curve-fitting methods, as well as the traditional econometric forecasting methods. More elaborate trend extrapolation may involve envelope curve fitting and extrapolation, which involves analysis and extrapolation of a succession of several different technologies over time. An envelope curve of lighting efficiency is shown in Figure 2.3. The envelope curve serves as an indicator of capability improvement and technological progress. Many models can be used to fit historical data and provide valuable projections of the future. One point should be made, however. Trend extrapolation assumes that the forces acting to drive the trend in the past will still exist in the future. In using this technique, it is important to consider any limits that may cause the trend to deviate from its historical path. Many times, these limits can be determined and used to modify the trend line. Martino[33] elaborates in-depth on trend analysis methods.

In cases involving highly correlated trends, one product or technology may be a precursor to another. In these cases the precursor can be monitored as a leading indicator of the follower. Figure 14.5 is an example of precursor analysis. The graph shows the percentage of the total Japanese consumer electronics market devoted to the manufacture of different types of television products. The graph highlights changes that occurred in Japan's TV industry as black and white TV, color TV, and VCRs were developed.[34] Similar changes can be expected as video cameras and digital high-definition television (HDTV) develop.

Drawings Artists' drawings can furnish simple, nonmathematical views of the future. The diagram in Figure 14.6 illustrates this. It compares present TV with future

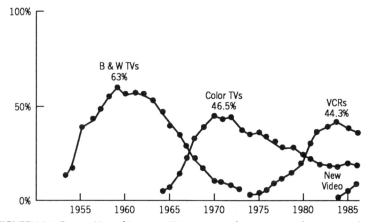

FIGURE 14.5 Composition of Japanese TV consumer electronic manufacturing market.
[Reprinted with permission of Dempa Publications, Inc.]

HI-VISION PRINCIPLE
Optimum Viewing and Listening Conditions

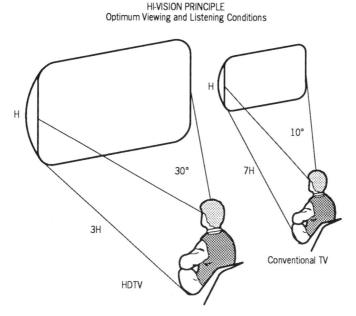

FIGURE 14.6 High definition and conventional TV viewing conditions.
[Reprinted with permission of Dempa Publications, Inc.]

HDTV. The diagram shows how the viewing angle of HDTV gives a person watching it a sense of being "in the picture." Japan has done a great deal of work developing an HDTV system.[35] This system, called *Hi-Vision*,[36] was introduced in Japan in 1987 and is far superior to current color TVs. The color TV screen is much larger, and image definition has been improved markedly. In VP, the rate of development of HDTV would be forecast and the performance features of future TV sets would be ascertained. Drawings are often used throughout product development to show how engineering concepts can be reduced to practice. An example is the engineering drawing in Figure 14.7. This schematic of a freeze desalination process shows how a future multistage freeze desalination plant would operate.[37] In this plant, a refrigerant (dark piping) continuously produces freshwater from seawater. Investigation of engineering drawings in patent sketches provides information about possible new product or process offerings well before the actual products enter the market.

Substitution Analysis The time-related techniques of trend analyses mainly involve quantifying the progress of past events. Substitution analysis,[38] on the other hand, entails quantifying the rate at which a new technology will expand into the market and replace an existing one. The technique is used mainly to predict the rate and the amount of substitution of one technology for another in order to estimate the market growth of new products, as well as the phaseout of existing ones. This method is finding increased use in planning and produces useful predictions of the rate at which a new technology will replace an older one. Substitution analysis provides a way of predicting functionally based replacements. Figure 14.8 is a plot of the rate of

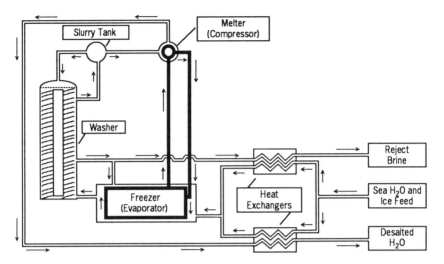

FIGURE 14.7 Freeze desalination process.

substitution of microwave ovens for all other ovens sold in the United States from 1973 to 1984. Note that the graph uses a semilogarithmic scale and shows the rate of substitution of microwave ovens in this market. The Fisher Pry numbers plotted on the y axis are equal to the ratio of new microwave oven sales to conventional oven sales. In 1981, microwave oven sales represented 55% of all new units sold, whereas in 1972, they represented less than 5% of all ovens sold.

Scenarios Another method for projecting the future involves scenario writing. Scenario writing is a qualitative forecasting technique used to project various futures

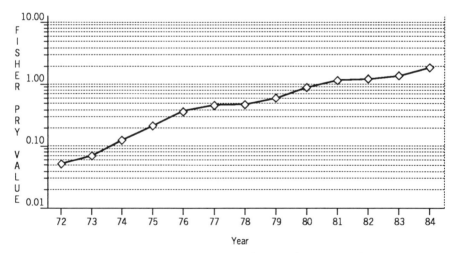

FIGURE 14.8 Microwave ovens—Fisher Pry substitution trend.

from present conditions based on stated assumptions. It is often used in conjunction with the more quantitative trend analysis models. Often these latter models cannot be constructed because numerical data simply do not exist. In this case, scenarios are very useful. Coates calls it "a major technique for exploring future implications."[39]

Scenarios may be written in many different forms, lengths, formats, and combinations.[40] Narrative forms are more popular than outlines, which lack detail and are easy to criticize. Generally, a set of several alternative scenarios, such as three or five, is constructed as opposed to a single most-likely scenario. These scenarios must be believable and relevant, and are usually based on data collected during the VP study. According to Hart,[41] scenarios can be a powerful means of forcing people to look at alternative futures, because scenarios presume that events, especially unpleasant ones, have occurred. This, in turn, helps reduce bias. Vanston *et al.* [42] and Mac-Nulty[43] relate the various steps in constructing alternative scenarios.

Gap Analysis A value gap results from an imbalance in supply and demand. Projections of consumer demand may point out an inability to meet this demand in the future. That is the case in Figure 14.9, which plots the demand for water in South Africa against the anticipated supply of water from natural sources. Two demand curves, one at 5% and another at 7%, are plotted over time. In the past, the amount of water impounded in dams and pumped from wells was increased to meet the needs of South Africa's growing population. However, this situation cannot continue, as the nation has already used up most of its available freshwater supply. The gap between the supply and demand curves represents a shortfall that must be filled to meet the future need for water. This gap represents an opportunity for innovation and will probably be filled by water reuse and new desalination technology (i.e., Figures 14.3 and 14.7).

Modeling Future Value Mathematical models can be used to simulate the operation of new products and processes. System dynamic models allow planners to test the effect of various policies on the future. They are discussed extensively in Chapter 5.

The operating characteristics of the freeze desalination process shown in Figure 14.7 were simulated by constructing a detailed system dynamic model of the process. Figure 14.10 is a computer-generated curve from this simulation model that shows the probable electrical cost of producing water for a proposed freeze desalination plant near Capetown, South Africa.[44] Energy costs are low during start-up when the plant operates primarily on a feedstock of natural ice. Energy costs rise as the feedstock of ice runs out and the plant has to operate increasingly on manufactured ice. Daily costs rise and fall as the plant switches from regular to off-peak electricity.

5. Planning Phase

Once a series of projections has been made, the task of appraising the impact of these projections on current activities and future plans begins. Function diagrams of a business can be used to select areas on which the future projections will have an impact. Refer to Figure 14.3, which arrays different needs for water against various

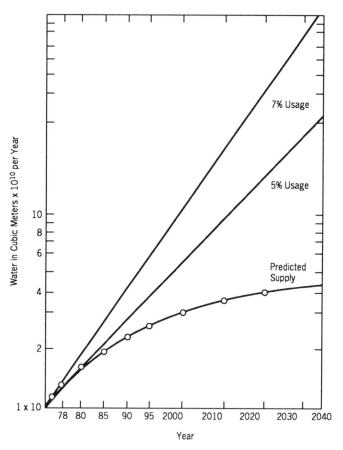

FIGURE 14.9 Gap analysis—South African water supply and use.

water supply technologies. Each of the functions shown on this diagram have specific future costs and performances associated with them. In planning to meet the future water needs of the country, the expected cost and quality of water can be examined by posting data to this diagram. The function cost approach can be applied to regions within the country where different situations exist. In doing this, plans can be generated that cover the need for further research and development, capital spending, manufacturing, sales, and/or labor force needs. This analysis will bring to light the need to invest in and develop new technologies to meet tomorrow's needs. In many areas like public works, major construction projects require long lead times. Value planning allows government agencies and industrial organizations sufficient time to plan for their client's needs.

Developing Function and Cost Criteria for Future Products As a consumer product matures, the percentage of money spent on basic user functions usually

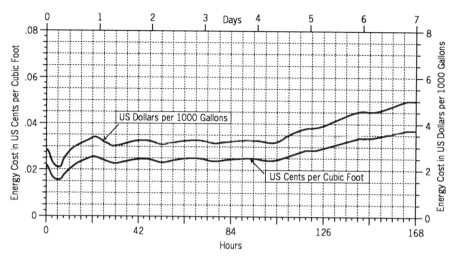

FIGURE 14.10 Simulated iceberg desalination electrical costs.

declines, whereas the amount of money devoted to new user functions increases. Money spent on basic functions represents a "price of admission" that a consumer must spend to obtain a device. Money spent on new and improved user functions covers sell features that were not present in the original product.

The sell features keep the product competitive and must be included in future product designs. Value analysis studies are useful in reducing the cost of the basic user function,[45] whereas VE studies optimize the value of new product designs.[46,47] In VP, an organized approach is used to ensure that new products contain a proper mix of functions and that the cost distribution of these functions matches the needs of future competition.

Figure 14.11 illustrates a planned distribution of costs to these functions for a 10-year time frame.[48] The cost of basic functions present in a product declines while the cost of new customer-oriented functions increases. Applied to future video technology, this distribution of function costs ensures that the value of new video products will meet the needs of ever more demanding customers.

Depending on when they are introduced to the market, future video products will need to meet specific function cost criteria. These criteria relate to the predicted price and performance features of specific classes of video products, such as color TV sets, video recorders, and video cameras. Once these performance and cost characteristics have been predicted, VP translates these data into specific function and cost goals for future designs. To do this, the annual projected cost of a color TV is multiplied by the function cost distribution factor for each future year.

By extrapolating the trend data shown in Figure 14.4, we find that the total cost of a TV set is anticipated to be 30,000 yen when the cumulative production reaches 250 million sets. Assuming this occurs in 1990 (shown as year 5 in Figure 14.11), then about 47% of the cost will be devoted to the basic functions of the set, whereas

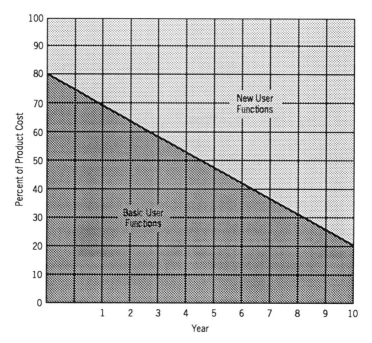

FIGURE 14.11 Value planning product functions.

53% will be spent on new user sell functions. In order to meet these function cost goals, it will be necessary to drastically reduce the cost of basic functions provided in the TV set. These functions were plotted in Figure 14.1 and are listed in Table 14.1 by value.

With these data, target costs can be calculated for each of these functions for a new color TV to be manufactured in 1990. To do this, we will need to spend 47% of 30,000 yen, or 14,100 yen, on these functions. To calculate specific function costs for a 1990 design, we convert the current cost percentages for each function into 1990 percentages, then multiply these costs by the value index to obtain 1990 target costs. These calculations are shown in Table 14.2. Note that the functions, control image, control color, amplify signal, and receive signal are not allocated more money. Instead this money is used to provide new user functions.

An additional 15,900 yen will be budgeted to design new features into the 1990 color TV sets (30,000 yen times the 53% allocated to new or enhanced functions). The TV manufacturer will distribute this money to a number of new features on the basis of their perceived customer importance. Some possible allocations might be those shown in Table 14.3. Note that money has been budgeted to all of the functions of the new TV set based on the anticipated performance and price characteristics of future television sets. This allocation allows designers to translate performance features into function descriptions such as "increase pixel density" or "record stereo sound." Similar targets are established in the manufacturing area to support the company's product design strategy.

TABLE 14.1 Hypothetical Value of Functions in a TV Set

Function	Value Index	% C	% I
Enclose unit	0.30	15.2	4.6
Support unit	0.30	17.0	5.1
Record signal	0.63	13.1	8.2
Power unit	0.87	14.7	12.8
Control current	0.94	8.1	7.6
Create audio	0.94	12.4	11.6
Control image	2.03	7.6	15.4
Control color	2.12	5.1	10.8
Amplify signal	3.00	3.8	11.4
Receive signal	3.91	3.2	12.5

Planning for New Product Substitutions The previous example illustrates the use of function cost design targets for conventional color TV sets. However, the future will see these sets replaced by HDTV sets (Figure 14.6). A manufacturer can plan for the introduction of HDTV by using function cost targeting for HDTV. The price of a HDTV set will be high enough to cover research and development costs and manufacturing costs and these costs will be allocated primarily to basic functions. But the process of planning for subsequent HDTV sets is identical to that described for a standard TV set. A manufacturer will need to develop function cost targets for conventional *and* HDTV sets. Substitution forecasts, using Fisher Pry graphs of the substitution of HDTV for standard and HDTV sets, will help companies plan the introduction of HDTV and the phaseout of standard color TV sets. The rate of substitution of color TV for black and white TV sets could be used as a precursor for

TABLE 14.2 TV Function Cost Targets

Function	Value Index	% C	1990 "Normal" Cost	1990 Theoretical Target Cost	1900 Accepted Target Cost
Enclose unit	0.30	15.2	2143	643	643
Support unit	0.30	17.0	2387	716	716
Record signal	0.63	13.1	1847	1164	1164
Power unit	0.87	14.7	2073	1804	1804
Control current	0.94	8.1	1142	1073	1073
Create audio	0.94	12.4	1748	1643	1643
Control image	2.03	7.6	1071	2174	1071
Control color	2.12	5.1	719	1524	719
Amplify signal	3.00	3.8	536	1608	536
Receive signal	3.91	3.2	451	1763	451

TABLE 14.3 TV New Function Cost Targets

Feature	% C	Yen
Enlarged picture screen	25	3975
Stereo sound system	15	2385
Improved picture clarity	10	1590
Picture in picture	22	3498
Improved unit decor	13	2067
Cable TV selector	5	795
Other functions	10	1590

the HDTV/conventional color TV substitution. A company would set function cost targets for a product mix of conventional and HDTV based on these substitutions.

Value Planning Future Water Supplies Planning for future water supplies also involves planning for functional substitutions. Figure 14.3 displayed several different water technologies that will likely be developed to meet future water needs. Figure 14.7 describes one of these technologies (freeze desalination) and Figure 14.9 shows the gap that will exist between future supply and demand in South Africa. How can a water district use VP to fund and design new water supply technologies?

A water district has to develop specific design plans for the major cities in its water district. Each of the major cities in a district have different populations and use different amounts of water. Projecting population growth into the future allows water planners to estimate future water needs for each city. Any city's water needs can theoretically be supplied by new water supply technologies. However, these technologies will not be equally applicable to each city. Coastal cities might use ocean desalination technologies, whereas inland cities cannot. A function matrix can be used to analyze the probable application of different technologies to each city.

Figure 14.12 is a function matrix that arrays five cities in South Africa against seven water supply technologies. The cities shown are A, Johannesburg; B, Capetown; C, Durban; D, Pretoria; and E, Port Elizabeth. The anticipated water needs of each city are shown for the year 2010. The hypothetical ability of each technology to meet these needs is indicated by rating numbers placed in the individual cells of the matrix. These ratings were obtained using the criteria analysis method described in Chapters 3 and 4. The sum of each column is the city water need, whereas the sum of each row is the hypothetical ability of a specific technology to meet the total water needs of these cities. The city with the greatest need is Johannesburg (287), which is closely followed by Capetown (274). The technology with the greatest potential for meeting these needs is water reuse (305), followed by desalination systems (297) and interbasin water transfer (154).

Note that the approach used in planning for new water supply technologies is similar to the TV manufacturing approach. In both plans, a functional approach was

	Cities				
	A	B	C	D	E
	Future Water Needs				
Water Technologies	Σ	287	274	253	112	92
Interbasin Transfer	154	60	0	56	21	17
Cloud Seeding	129	37	16	48	13	15
Water Reuse	305	103	36	91	50	25
Desalination by Distillation	64	23	25	8	8	0
Desalination by Membranes	139	43	47	20	19	10
Desalination by Freezing	94	20	63	3	0	8
Iceberg Utilization	133	0	88	28	0	17

FIGURE 14.12 Future water supply technologies.

used to estimate and plan for future technologies. Future needs and prices were projected and used to ascertain the contribution of each functional approach to the future.

6. Value Management Phase

The last phase of VP involves establishing a management process for implementing change. The forecasts and resulting plans produced up to this point need to be implemented. Now management needs to form teams to improve the value of the present business and to design new products and services to reduce the gap between present and future customer needs. Value management provides a practical way to accomplish this. Some companies have established a VM steering committee to help upper management coordinate value improvement. Composed of vice presidents responsible for managing key corporate functions, the VM committee establishes improvement goals that the company needs to undertake to remain competitive. Reuter reports, "Of the top twenty Japanese industrial firms, seventeen have VA executives at the vice-presidential level."[49] Figure 14.13 is an example graphing cost reductions that management desires in three areas of their business—products, manufacturing, and inventory. This hypothetical example has a 10-year time frame. Curves like these are often based on learning curves for individual segments of the company's business.

To achieve these goals, a VM council is created. This council consists of appropriate middle management people who report to the corporate VM steering committee. This council is responsible for managing VA/VE activities. It is responsible for managing the appointment of special VA/VE teams and reviewing their recommendations. The council is also responsible for instituting value training and for authorizing VA/VE studies. The council can use value control charts such as those shown in Figure 14.1 or 14.13 to monitor value within the business. Figure 14.14 is an organizational chart for VM.

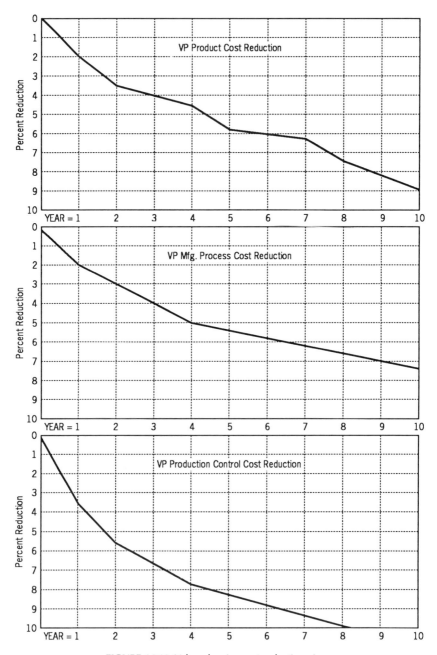

FIGURE 14.13 Value planning cost reduction aims.

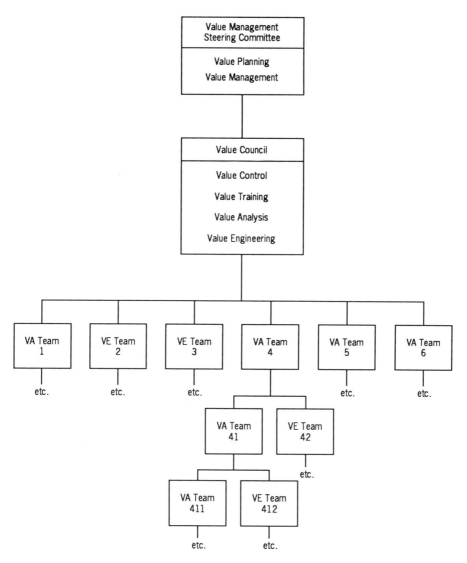

FIGURE 14.14 Value management organization chart.

Monitoring It is important to monitor developments both inside and outside the organization continuously. Life cycle analysis provides insights into the stage of development of many emerging technologies and is a valuable tool to use when following emerging technologies. A monitoring system should be established to follow future development as it unfolds. The basic need here is to establish an information storage and retrieval system and to assign responsibility for maintenance of this file to a clearly identified person. Using monitoring, external developments can be used to assure that VP stays on track.

Is demand holding constant? Are breakthroughs occurring in any of the new technologies? Monitoring is important because it involves what Bright calls "assessing events in-being."[50] It entails keeping current tabs on significant events in order to assess change. In addition to employing literature searches and environment scanning, monitoring also includes choosing the right parameters for observations, evaluating new information for significance, and presenting the information in an effective and timely manner. Data are collected from general literature, from reports and patents, and by personal observations or contacts with people in key strategic positions. Data once collected should be stored for subsequent rapid access.

When important new and unplanned events occur, it may be necessary to study some subjects in more detail, especially if they appear to be in a rapid state of flux. The results of such studies should then be incorporated into an updated value plan.

SUMMARY

The VP process is new and is evolving. It has proved of value to a number of organizations. In Japan, VP is often referred to as 0.5 Look VE. The VP techniques enable organizations to examine the value of their products and services and compare these with market trends. VP establishes customer needs in terms of product functions and allows designers to produce new cost effective products to meet future needs. It allows an organization to predict changes that may occur outside the organization that will affect the value of a company's products and services. Value management and VP insure that modern organizations will be able to compete in tomorrow's world.

ENDNOTES AND REFERENCES

1. In Japan value planning is called 0.5 Look VE, value engineering is called 1st Look VE and value analysis is termed 2nd Look VE.
2. Martino, J. R., *Technological Forecasting for Decision Making.* Elsevier, New York, 1972.
3. Martino, *Technological Forecasting for Decision Making.*
4. Bright, J. R., *A Brief Introduction to Technology Forecasting.* Permaquid Press, Austin, Tex., 1972.
5. Jantsch, E., *Technological Forecasting in Perspective.* OECD, Paris, 1967.
6. Joseph, E. C., "What Is Future Time?" *The Futurist,* August 1974.
7. Hart, J. L., "Technological Forecasting, from Board Room to Drawing Board," *Machine Design* **48**(3), 90–93, 1976.
8. De Marle, D. J., and Hart, J. L., "Searching for Signals Through Interactive Data Banks," paper presented at James R. Bright's Technology Forecasting Workshop, Castine, Maine, Industrial Management Center, Austin, Tex., June 1977.
9. Hart, J. L., "The Futures Study Applied to Manufacturing," paper presented at James R. Bright's Technology Forecasting Workshop, Castine, Maine, Industrial Management Center, Austin, Tex., June 1977.

10. Fordyce, J. K., and Weil, R., *Managing with People.* pp. 137–156, Addison-Wesley, Reading, Mass., 1971.

11. Johanson, R., and Ferguson, J. A., "Mapping Views of the Future in a Small Group," *Futures* **8**(2), 163–169, 1976.

12. Jones, J. E., "The Sensing Interview," in *The 1973 Handbook for Group Facilitators,* University Associates, San Diego, 1973, pp. 213–214.

13. Coates, J. F., "Some Methods and Techniques for Comprehensive Impact Assessment," *Technical Forecasting and Social Change,* Vol. 6, 1974, p. 349.

14. De Marle and Hart, "Searching for Signals through Interactive Data Banks."

15. Butler, J., Ball, D. F., and Pearson, A. W., "The Analysis of Technological Activity Using Abstracting Services," *R and D Management* **7**(1), 33–40, 1976.

16. Lockheed Information Retrieval Service, "Dialog-Chronologue," Lockheed Information Systems (D52-08, B/201), Lockheed Palo Alto Research Laboratory, Palo Alto, Calif.

17. SDC Search Service, Systems Development Corporation, 2500 Colorado Avenue, Santa Monica, CA 90406.

18. The Information Bank, One World Trade Center, New York, NY 10048 (a subsidiary of the New York Times Co).

19. Landford, H. W., *Technological Forecasting Methodologies,* American Management Association, 1969. New York, NY.

20. Bright, J. R., *Practical Technological Forecasting, Concepts and Exercises,* The Industrial Management Center, Austin, Tex. 1978.

21. Martino, J. R., *Technological Forecasting for Decisionmaking,* Elsevier, New York, 1972.

22. Bick, J. H., "Planning and Forecasting Using a Combined Relevance Analysis and Cross-Impact Matrix Method," Report RM-1757, General Research Corp., Westgate Research Park, McLean, Va., June 1974.

23. Landford, *Technological Forecasting Methodologies.*

24. Inoue, M. S., and Riggs, J. L., "Describe Your System with Cause and Effect Diagrams," *Industrial Engineering,* pp. 26–31, April 1971.

25. Bridgewater, A. V., "Morphological Methods—Principles and Practice," R. V. Arnfield, (Ed.), pp. 211–230, Edinburgh University Press, Edinburgh, U. K., 1969.

26. Sokol, R. R., and Sneath, J., *Numerical Taxonomy.* Freeman, San Francisco, 1963.

27. Jones, H., and Twiss, B. C., *Forecasting Technology for Planning Decisions.* pp. 103–111, PBI - Petrocelli Books, New York, 1978.

28. Linstone, H. A., "Mission Taxonomy," in *Technological Forecasting for Industry and Government* J. R. Bright, (Ed.), pp. 231–288, Prentice-Hall, Englewood Cliffs, N. J., 1968.

29. Turoff, M., "Cross Impact Analysis," *Technical Forecasting and Social Change,* **3**, 75–88, 1971.

30. Kane, J., "A Primer for a New Cross-Impact Language-KSIM," *Technical Forecasting and Social Change* **4**, 129–142, 1972.

31. Shillito, M. L., "Impact Analysis," paper presented at James R. Bright's Technology Forecasting Workshop, p. 36, Castine, Maine, Industrial Management Center, Austin, Tex., June 1977.

32. *Japan Electronics Almanac, 1987,* p. 167, Dempa Publications Inc., 400 Madison Ave. New York, N.Y.

33. Martino, *Technological Forecasting for Decisionmaking.*

34. *Japan Electronics Almanac, 1987,* p 167, Dempa Publications Inc., 400 Madison Ave. New York, N.Y.

35. *Japan Electronics Almanac, 1987,* p 196, Dempa Publications Inc., 400 Madison Ave. New York, N.Y.

36. The system is called "Hi-Vision," because it is broadcast from satellites in geo-synchronous orbit above Japan.

37. DeMarle, D. J. "Ice Enhanced Freeze Desalination," in *Proceedings of the ASHRAE-FRIGAIR 90, Conference on Refrigeration and Air Conditioning,* Pretoria, RSA, April 23–27, 1990.

38. Fischer, J. C., and Pry, R. H., "A Simple Substitution Model of Technological Change," *Technical Forecasting and Social Change* **3,** 75–88, 1971.

39. Coates, J. F., "Some Methods and Techniques for Comprehensive Impact Assessment."

40. Zentner, R. D., "Scenarios in Forecasting," *Chemical and Engineering News,* pp. 22–34, October 6, 1975.

41. Hart, "Technological Forecasting, from Board Room to Drawing Board."

42. Vanston, J. H., Frisbee, W. P., Lopreato, S. C., and Paston, D. L., "Alternate to Scenarios Planning," *Technical Forecasting and Social Change* **10**(2), 159–180, 1977.

43. MacNulty, C. A. R., "Scenario Development for Corporate Planning," *Futures* **9**(2), 138–137, April 1977.

44. DeMarle, "Ice Enhanced Freeze Desalination."

45. Shuichi Hoshino, Development of VE activities Utilizing "Worst Index Figure," *Proceedings, Society of American Value Engineers* Vol. 12, 1977, pp. 41–50.

46. Mitsugi Kanaya, "Value Creation as a Management Objective," *Proceedings, Society of American Value Engineers,* Vol. 14, 1979, pp. 13–17.

47. Masao Ogawa , "A Study of Strategic VE Target Costs Based on Cost Reduction Curves for Middle and Long Range Management Plans, *Proceedings, Society of American Value Engineers,* Vol. 20, 1985, pp. 231–239.

48. Takayuki Takubo, "Technology for Product Planning and Development," *Proceedings of the Society of American Value Engineers,* Vol. 20, 1985, pp. 240–250.

49. Reuter, V., Industrial Management, Nov.–Dec. 1983.

50. Bright, J. R., *Practical Technological Forecasting, Concepts and Exercises,* The Industrial Management Center, Austin, Tex., 1978.

15

VALUISM, VALUE MANAGEMENT AND THE FUTURE

David J. De Marle
M. Larry Shillito

EVOLUTION OF VA/VE AND VALUISM

Value analysis/value engineering is over 40 years old, and in many areas in the United States it is still being practiced much as it originally was. The reasons for this are many and include the success of the original methodology developed by Larry Miles and others at General Electric, the lack of research in the area of value (particularly in the United States), the special interests of VA/VE consultants trained in the original VA/VE methodology and a paucity of new publications on the subject.

We have attempted to put VA/VE in perspective. We examined VA/VE and its growth and evolution in the context of a variety of other value-centered methodologies. We did this to enlarge the VA/VE horizon beyond the entrenched paradigm of a Milesian 40-hr workshop. The growth of VA/VE technology has been encumbered by traditionalists who have not used new technology and who mistake the edge of the rut for the horizon.

The technology of VA/VE has had growth recently, and new and pioneering applications are beginning to occur in industry and government. The growth and contribution to the state of the art is coming primarily from the Pacific Rim, in particular Japan, Korea, and China. We are witnessing a paradigm shift in VA/VE technology, process, and application. New technologies such as QFD, technological forecasting, design for assembly and the elements of total quality management, and a multitude of others that have been discussed, are being integrated with the original VA/VE concept to produce hybrid VA/VE models and integrated applications. We are also witnessing an opposite effect, some VA/VE organizations and VA/VE practitioners are being subsumed by the TQM movement and some other technologies. These changes are part of an evolutionary process.

Every since Larry Miles developed VA at General Electric in 1947, the process has been evolving. Value engineering developed from VA (Chapter 12) and VM and VP (Chapter 14) developed from VA/VE. Value indices and graphs (Chapter 3) were developed. Value analysis has spawned value measurement, and value measurement techniques (Chapter 4) were used to quantify value (Chapter 1). As the methodological base of VA/VE expanded, new applications for the methodology were found. Today value studies are being used effectively in many different areas around the world. The focus of VA/VE on value improvement and its evolution from a hardware-oriented technique to a process that can improve the value of services and organizations places it at the center of the valuism movement, which we described in Chapter 11.

Valuism, born in the Industrial Revolution in England, is a worldwide movement with a broad methodological base that contains VA/VE, QC, industrial engineering and many other powerful technologies (Table 11.3). We expect that many of the methodologies of valuism will be combined and integrated into powerful new methodologies. This process is occurring today and will accelerate in the future. Chapters 6 to 10 illustrate some of the integrations that have already occurred: system dynamic value modeling, quality function deployment, technology roadmaps, and customer-oriented product concepting.

Continuing Growth of Valuism

Valuism will continue to grow as new management technologies are developed and added to its menu. Industry will aggregate many of the existing methods together and use them to dramatically increase productivity and lower costs. Many consulting businesses and professional societies will resist the integration of their methodologies with others, but they will be unable to prevent the development of integrated systems in industry. Value management will emerge as a dominant management practice in industry as it focuses on ways to improve the value of new products while eliminating low-value items by downsizing. Value measurement methods will be widely used to quantify value and to select items for value improvement and/or downsizing. Management will use special value teams to improve the value of products, services, and organizations.

Increasing deficits and taxes will lead to the downsizing and privatization of government. Valuism will take on political significance when used to restrain growth of government and improve the value of government services. Industrial engineering, quality control, and value management methodologies will be used to improve the performance and lower the costs of all levels of government. Improvements in the efficiency and effectiveness of government will revitalize and transform it. Value measurement and value planning will allow local, state, and federal government to target operations for downsizing. Privatization of government services will stimulate business growth, increase government revenues, and reduce taxes. A wide variety of value-centered methodologies will be used to ensure that public monies are only spent on services that offer the best value to all citizens.

EVOLUTION OF VA/VE INTO VALUE MANAGEMENT?

For some time SAVE has been having an identity crisis. Some members wanted to change the focus of the society from an engineering society to a management society. Others wanted to alter the cost-reduction focus of SAVE, whereas still others wanted to modify membership requirements. In 1974, W. W. Parks, then vice president of Vapor Corporation, recognized the need for SAVE to be flexible and adaptive. He stated:

> The obvious challenge we face is to reexamine our charters, our objectives, our scope, and our procedures and indeed, if need be, even our name and to determine if our value concepts and techniques have adequate relevance to the new problems of today and to dedicate ourselves to changing our directions to help in the solutions of these un-believable challenges.[1]

In the ensuing years SAVE has resisted all efforts to modify its name or its practice, although a name change will be considered at the 1992 SAVE conference. It instituted a certification plan based on traditional VA/VE practice. In 1989 it adopted a strategic plan that emphasized promoting the structured approach to VA/VE originally developed by Larry Miles in 1947. Understandably, consultants skilled in this methodology do not want SAVE to deviate from it. In our view this has reinforced the image that VA/VE is simply a cost-reduction technique, and this has prevented VA/VE from growing. VA/VE is not just an effective cost-reduction technique, but is a value improvement technique, of which cost reduction is merely one of the many benefits to be derived. Value management is a more generic name that describes a management process that utilizes VA/VE and does not carry this cost reduction connotation. It connotes value improvement and opens the door to other methodologies, many of which can combine traditional VA/VE with the other management processes. As a consequence, membership in SAVE has grown slowly over the years. This is not the case in Japan, where VA/VE has grown at a rapid pace. Today, some of the most significant advances in the state of the art as well as in application are coming from not only Japan, but Korea and China as well. In these countries VA/VE has been blended into the quality movement and formed the basis for new methodologies like QFD. In Japan a number of corporations have integrated QC, VE and industrial engineering practices, while in the United States we see value engineers being repositioned and reassigned to quality associated assignments within TQM programs. Here a number of VE departments have been subsumed by QC organizations where VA/VE is used as a design-to-quality method.

Thus as we look to the future of VA/VE, our crystal ball is cloudy. A great deal depends upon the willingness of SAVE to change. If it doesn't change, we expect VA/VE will continue to be viewed as a cost-reduction technique generally useful in lowering the cost of new or existing hardware items. As such it will have limited applications and will grow slowly. On the other hand if SAVE enlarges its scope to include value measurement and VM, we expect the resulting value management society would grow rapidly.

If SAVE does not focus on value measurement and VM methodologies, other societies will. As we noted in Chapter 11, the Institute of Industrial Engineers has already gone through a number of transformations, from the Taylor Society to the Institute of Industrial Engineers. In the process, it has assimilated a large number of specialized disciplines and since 1982 has devoted a chapter of its handbook to VE.[2] Because we believe that value measurement and management are essential elements in valuism, we will describe how some of the most important aspects of valuism will evolve.

A FORECAST OF CHANGES IN THE ELEMENTS OF VALUISM

Earlier we described the main components of valuism as

Consulting agencies
    Methodologies (proprietary and others)
    Consultants
    Management and consulting practices
Corporations
    Methodologies (Proprietary and others)
    Value methodologists
    Management and corporate practices
Professional societies
    Methodologies
    Members
    Management and charters
    Foreign affiliates
Colleges and universities
    Methodologies
    Faculty
    Administration and charters

How will valuism evolve in each of these organizations?

Consulting Agencies

Consulting agencies are an extremely important component in valuism. Individual consultants and large consulting firms offer a wide variety of management services to corporations around the world. They are the entrepreneurs of valuism, often creating new methodologies to assist their clients. The consulting business is very competitive and secretive and often proprietary. Consultants frequently copyright their procedures and software and then licence this material to their clients. Management consulting can be very lucrative and has been growing rapidly. From 1984 to 1989, the industry was growing at a double-digit rate but recently slowed as the economy cooled. *Consultant News,* a trade publication, reported that revenues dropped significantly in the first half of 1990. Large consulting agencies like Booze Allen, McKinsey & Company, Arthur Anderson & Company, and so on offer a wide variety of management services. Smaller agencies work in market niches. VA/VE consultants are an example of niche consulting.

Although consulting agencies compete with one another, they are often linked together through professional societies. A recent survey indicates that 60% of the management consultants in the United States depend on personal relationships and "old-boy" networks to obtain work, 33% rely on management seminars to attract customers, and 13% contact prospective clients by phone or through the mail.[3]

We expect management consultants to continue to act in a proprietary manner and to strongly resist efforts to integrate their methodologies. They are after all small businesses, which like any business jealously guard their proprietary processes. Unfortunately this restricts practice and delays widespread application and integration of new technology. It also leads to a great deal of duplication in practice and can perpetuate outdated methodologies. This protectiveness extends to professional societies that are controlled by consultants who want to restrict other consultants from practice in their field and who use the societies to attract new clients to an old boy network.

Corporations

Corporations are the clients of management consulting agencies. They range in size from multinational giants to small businesses with fewer than a 100 employees. These corporations are often linked together with small companies supplying components to larger organizations. The large companies often have large internal management consulting staffs, which provide specialized services throughout the corporation. Sometimes these services are extended to the companies clients and/or suppliers. Multinational corporations often extend these services to their foreign subsidiaries.

Like consulting agencies, corporations usually are secretive about their business practices and control the distribution of proprietary information. Information of a nonproprietary nature is usually passed on to clients, suppliers, and professional societies, which may benefit from it. Management practices vary from company to company and are based upon the company's individual needs and past experiences.

Companies keep abreast of developments in management science in a variety of ways, including old-boy networks, noncompetitive industrial exchanges, university contacts, publications of professional societies, and employment of management consultants. In large companies, this information is assimilated in internal management consulting departments and disseminated throughout the company. Top management often will use outside consultants to help in highly confidential operations like strategic planning or downsizing. Information in these areas is not shared within most companies.

Recently companies have begun to share their management expertise with non-competitive outside agencies, such as health agencies, schools, and government agencies. We expect this trend to accelerate as companies attempt to reduce the cost of their health care, educational, and retirement programs. This sharing is of growing importance as many government agencies have little experience in industrial business administration and/or downsizing.

Professional Societies

Professional societies differ from most consulting companies and colleges in the specialized nature of their professions. They often serve as centers for research into their profession. Different professional societies vary widely in size and scope, with membership ranging from hundreds to hundreds of thousands. Membership and practice are restricted to areas in the societies's charters. Although most societies are fairly specific as to their professional boundaries, it is not uncommon for a methodology that spans professions to be included in different societies. This is the case with VA/VE which is included in the professional practice of SAVE, IIE, SME, and several other societies.

Membership requirements differ from society to society and often include membership grades that acknowledge experience and/or training. Most professional societies have charters that restrict professional practice to members or specialists certified by the society as having sufficient knowledge and experience to practice within their profession. Border disputes among societies are not uncommon and act to constrain practice. Professional societies often maintain close liaisons with universities and colleges. These contacts are often designed to restrict entrance into the field and control business consulting practice. One consequence of this is that many colleges and universities confine their teaching to the basics and most colleges ignore teaching the proprietary management methods offered by private consultants. Only when a field has matured are diversified courses offered in a profession like industrial engineering or quality control.

Most professional societies have foreign affiliates. These offshore societies often have different characteristics. The SJVE is an example. In Japan, VA/VE centers on industrial practice with much less emphasis on private consultants. This emphasis has caused SJVE to grow rapidly, and SJVE has more members than SAVE. VA/VE practice is also different in Japan, because the society has been more open to change. SAVE has retained the Miles approach to VA/VE, whereas SJVE has blended this approach with QC and industrial engineering approaches in use in Japanese industry. The Value Engineering and Management Society of South Africa (VEMSSA) has placed special emphasis on the team-building aspects of VA/VE and has used VA/VE effectively to build interracial harmony.[4] Practice in other professional societies also varies from country to country. Quality control in Japan makes extensive use of QC teams and often differs from practice in the United States which is again based upon the use of QC consultants and specialists.

Colleges and Universities

Colleges and universities play a key role in teaching management practices. Both undergraduate and graduate programs are involved. At the undergraduate level, industrial engineering curricula at most large engineering schools furnish students with basic education in a large number of management science methodologies. Schools also offer majors in SQC and manufacturing engineering, which often emphasizes industrial management practices. Graduate programs in business administration offered by the more prestigious colleges often serve as a feedstock for consulting agencies and/or industrial corporations hiring. However, even at the graduate level these colleges refrain from teaching proprietary consulting techniques which are often only available to people working in industry.

Lately a number of colleges and universities have developed extensive continuing education programs. These agencies offer training in areas traditionally reserved by private consultants and have drawn the ire of many management consultants. Consulting agencies are seeking ways to limit tax-free agencies such as colleges or other nonprofit groups from managerial consulting.[5] In the future, we expect most colleges to play a major role in entry-level education and a minor role in managerial consulting.

INTEGRATION OF DIFFERENT MANAGEMENT CONSULTING PRACTICES

The parochial nature of each of the components in valuism is one of the main reasons why this process has not been described before. It is why it is one of the best kept secrets of the free enterprise system. Consider the difficulty of introducing valuism into Russia or other countries that lacked private management consulting companies, private corporations, and professional societies that focus on the efficient production of valuable goods and services. Little wonder that free enterprise is foreign to many areas of the world, such as Africa.

Yet the proprietary nature of management consulting has created a need for an integration of many of the consulting practices that it contains. We are convinced that industry and government need help in sorting through the many different methods that have been developed. They need ways to categorize these methods according to their specific needs. We believe that all of the different parties will benefit from this integration—consultants, societies, corporations, and universities. We are also convinced that VM techniques allow this to be done. Integration will eliminate duplication of efforts, retraining, and improper applications of technology.

As an effort toward this integration we describe three common industrial tasks below and show how different methodologies can be integrated into systems to solve these tasks efficiently. The first task involves new product planning and development, the second task involves new service/organization planning, and the last task involves downsizing.

Integrating Value Technologies in New Product Development

It is more important than ever for companies to plan for and shape their future. As Barker says, "You can and should shape your future. If you don't someone else surely will, and you better hope they will be good to you."[6] Figure 15.1 is our view of how many of the current business management methodologies fit into an integrated process for improving the value of a company's new product planning.

A critical input to company planning is what Hamel and Prahaiad[7] call *strategic intent.* Strategic intent captures "the essence of winning." Value management will help planning teams focus strategic intent via the VP process described in Chapter 14. Value managers should require planners to consider the *core competencies* of their businesses. Core competencies are the "collective learning of the organization" that provide "the roots of competitive advantage."[8] A variety of VA techniques such as FAST diagrams, function matrices, dysfunction analysis, and affinity and tree diagrams are very useful tools for organizing data to surface key issues. Future planning leads to business and market planning (see Figure 15.1). Here strategy maps as discussed in Chapter 9 (Figure 9.6) are a very useful tool to use to collect and

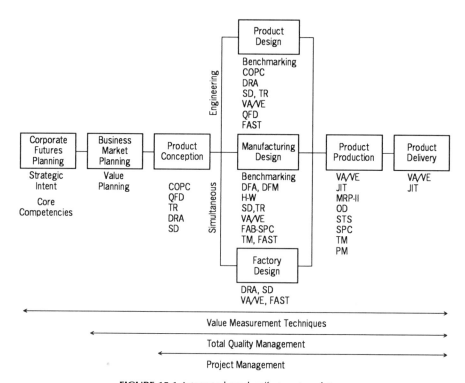

FIGURE 15.1 Integrated product/factory template.

interrelate the elements necessary to set strategy for families of products. Product concepting meshes with business and market planning. The COPC process described in Chapter 10 was developed specifically for product concepting. The elements of QFD and technology road mapping also work well in this kind of activity. Decision and risk analysis and system dynamic modeling (Chapter 5) provide a viewpoint toward a business case and the dynamics of value.

Once a product concept is formulated, it triggers a flurry of complex activity. Three functions are started simultaneously. They are product design, manufacturing design, and factory design. Although some may argue the point, we believe the management of these three activities is the essence of simultaneous engineering, also known as concurrent engineering. Every tool and process we have described to you directly apply to the simultaneous engineering of these three functions. Design for assembly[9] is an important tool to interface product design with assembly and manufacturing systems. Dick Bradyhouse has proposed a way to integrate VA/VE with design for assembly.[10,11] His work is a good example of the integration of two powerful techniques. It is this type of integration we have in mind when we consider the future survival of VM. Let us now consider other interfaces of VM that are just as important.

Competitive and strategic benchmarking is necessary for product and manufacturing design. COPC provides the grist to design a benchmarking program. The Hayes–Wheelwright approach[12] for aligning process and product structure through a relationship matrix details the guidelines for consideration in the analysis of existing and the development of new manufacturing systems. Value management methodology can be used with teams to assist in this activity. System dynamic simulations (Chapter 5) are invaluable for optimizing the value of current and future manufacturing systems and factory designs. Value planning teams can use value measurement techniques to create data for input into these simulation models. Traditional VA/VE can be used during simultaneous engineering to isolate and reduce the cost of the product and its manufacturing operations and even the cost of constructing a building or factory to produce it. VA/VE can also be combined with other techniques such as just-in-time[13] and manufacturing resource planning[14,15] during product and production design. Product delivery and distribution can benefit from VA/VE for developing distribution systems and sales forces as well as designing customer service functions. Production also benefits from VA/VE being used for process control.

Another excellent example of integration is FA-based statistical process control developed by Bopp and Grant.[16] It is an excellent technique for surfacing parameters to be control charted. The goal of FA-based statistical process control is reduced process variability through increased process (function) understanding. The process uses a FAST diagram where the how? direction identifies independent parameters and the why? direction indicates dependent parameters. It teaches you how to think in terms of process and function and not necessarily measurable characteristics. It is a powerful method! Project management will continue to play an indispensable role coordinating all of these activities and getting them to interface together so that the final product will be delivered on time and within budget. Total quality management is still evolving. It will involve a very large scope of activities to conceive, design, develop, deliver, and service world-class products and manufacturing systems. Because of its infancy we show it as a time arrow at the bottom of Figure 15.1. Value measurement techniques, on the other hand, are applicable throughout the entire template.

Integrating New Service and Organization Planning

According to Barker,[6] the three cornerstones of the 21st century will be anticipation, innovation, and excellence. He defines anticipation as the timing of the right thing, at the right place, at the right time. Anticipation informs us which innovation we should bring out. It starts to give us the competitive edge and protects us from surprise. Innovation gives us the competitive edge. Innovators lead the way and protect us from obsolete excellence. Excellence, although important, will not differentiate one company from another in the future and it will no longer guarantee a company's future. Excellence helps a company "get in the door." Excellence will be a hygiene factor in the 21st century, taken for granted by the customer. All products will have to have it just to stay in the game. If you don't have it, you will be hurt.

The critical strategic issues facing company organizations in the future will be timing, quality, and the flexibility to respond to changing customer needs.[17] There will be ever-increasing pressure to compress the product development cycle. Companies really will be competing against time. The battle will not be won with quality, but with time. Unfortunately time is one thing we cannot buy back. Stalk and Hout discuss this well in their article "Competing Against Time."[18] There are many management techniques that deal with inventory management, manufacturing times, and organizational structures. Figure 15.2 indicates how these techniques can be integrated into a system for organizational improvement.

Integrating and Downsizing Services

As with industry, government and service organizations will have to shape and plan their future. Again Hamel and Prahalad's strategic intent and core competencies will play a strategic role in planning (see Figure 15.2). Benchmarking and QFD will provide structure for input to policy planning. COPC, QFD, technology road mapping, and benchmarking will help focus new product and service concepts. Reducing the size of organizations in industry by downsizing and in government by privatization will be key activities. In these activities, the value index and FAST diagramming have an excellent fit. Simultaneous engineering (Figure 15.2) will encompass many of the same methodologies as previously seen in Figure 15.1. The only difference is that the three activities will be product service design, delivery design, and organization design. These three activities will also involve organization development (OD) and socio-technical system design. The delivery of the service can be improved through QFD and VA/VE. Value measurement techniques will be used throughout the template.

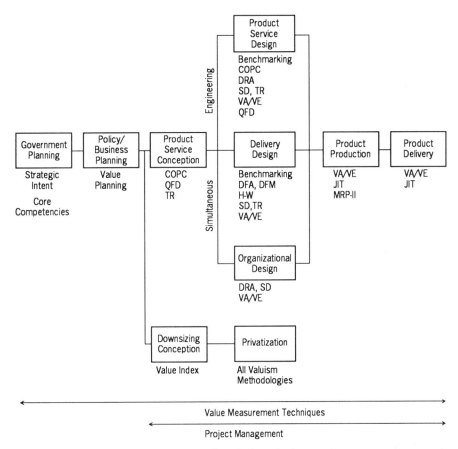

FIGURE 15.2 Integrated service/organization template.

Downsizing Activities

Downsizing is a very prevalent activity in industry and most likely will continue in the future. Privatization is an analogous practice in government. Figure 15.3 illustrates a template for such a restructuring. Again, the blend of most of the same techniques will be needed to be successful. Value measurement can apply again to the entire process. Downsizing will naturally require organizational restructuring. VA/VE can be used for organization restructuring and transformation. Fraser[19] describes this very nicely in her paper "Using Value Management Techniques for Organizational Change." Organization development and its behavioral science approach to group and organization design have been used since the 1960s. Because of its perceived "touchy-feely" nature, its acceptance by many organizations was/is limited due to management's reluctance to accept it. However, Fraser's model of VM "would increase the scope of OD, extending to the technical and strategic systems as well as the social and administrative."[19, p. 96] She describes the integration of VM with

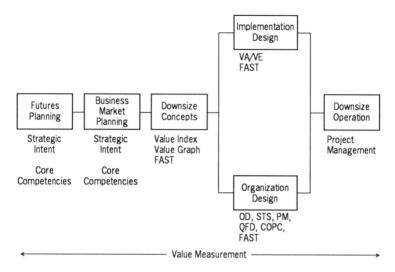

FIGURE 15.3 Down-sizing/privatization.

OD, outlines an approach, and provides some major principles for VM practitioners to follow. Davis[20] also discusses the connection between VM and organizational transformation. He states that "with minor modifications and changes to value tools and concepts, and with major modification to VA/VE practitioner thinking, VA/VE can emerge as the leader to transform organizations to higher levels of performance."[20, p. 86]

VALUISTS

As the various management consulting practices are integrated, we expect that a new profession will arise. Just as physicists are experts in physics, and chemists are experts in chemistry, valuists will be tomorrow's value experts. Valuists will be professional people trained in value measurement and management. They will have broad knowledge of the many methodologies that can be used to improve value. Graduates of college curricula that focus upon value measurement and management or on industrial or government career paths that train people in these methods, will use a wide variety of value management techniques to improve value. Skilled change agents with an in-depth knowledge of value improvement techniques, valuists will assist management in determining the applicability of these techniques to an organization's needs, and will help value teams apply these techniques to these needs.

Valuists will expand the study of value, integrating value methodologies into a professional discipline that will span a variety of today's parochial disciplines and expand the application of value methodologies in industry and government. The value measurement and management methodologies that we have described will be the nucleus for the development and evolution of valuists and valuism. Today,

VA/VE lies at the methodological center of valuism, but it needs to open itself to a rapid integration of external methodology, if it is to stay in touch with valuism and its evolution.

Several years ago SAVE began certifying "value specialists." Today a value specialist is one skilled in the original Milesian value process. A value engineer needs to pass a test based on his or her knowledge and practice of traditional VA/VE. Although we respect the dedication and knowledge of many of our colleagues who are value specialists, we believe they need to become conversant with much of the course matter taught to students in industrial and/or manufacturing engineering. SAVE requires no formal educational requirements in economics or engineering for certification as value specialists. Today many value specialists are not conversant with much of the subject matter in this book. They are not conversant with VM techniques, value modeling, or with new processes and techniques, such as QFD, that are evolving and gaining quick acceptance in industry. Nonetheless, value specialists trained in the value methodologies give a group who has never been together a way to look at a problem in a new way. They improve the effectiveness of teams and help clarify goals, objectives, and alternatives. Value specialists and VM technology can provide structure and direction to a wide variety of ad-hoc value teams. If the knowledge and experience base of these value specialists were upgraded they could be the first valuists. Alternately, some MBA programs could be altered to produce the first valuists.

INTEGRATING VALUE MANAGEMENT WITH OTHER TECHNOLOGIES

Figures 15.1, 15.2 and 15.3 are three templates that show how different methodologies can be combined with VA/VE. Using the full process (job plan) of VA/VE and the 40-hr workshop will limit integration. Integration will be easier if portions of VA/VE, such as FA and value graphs, are detached from the process and integrated with appropriate parts of the other technologies.

A host of methodologies such as total QC, statistical process control, Taguchi methods, just-in-time for manufacturing, design for assembly, design for manufacturability, QFD, integrated learning, simultaneous engineering, performance management, Hoshin Kanri planning, manufacturing resource planning, decision and risk analysis, sociotechnical systems, organizational development, function analysis-based statistical process control (FAB-SPC) offer opportunities for integration with VA/VE. The future will see hybrid processes developed that are a blend of several techniques.

The most significant upgrades of value methodology will be through integration of value measurement and VM with other technologies. This should involve integration of methodologies as well as applications. We have already seen some excellent articles regarding value integration by Verzelletti,[21] Togawa,[22] and Nagoya.[23] Kaufman[24] also discusses the VM approach to total quality management and describes how it can be used to convert quality goals to actions. Frank[25] described the connection between total quality management and VA/VE back in 1988. Park points to the need for this integration in his article "Value Engineering in the Modern World of Taguchi, QFD et al."[26] In a different vein, Bacheli[27] described how VA can be used to analyze and select the most appropriate management methods from the many processes and systems now available. The methodologies described in this book need to be integrated into VA/VE. This will not be easy. When these methods are shown to many long-term VM practitioners, there is almost mild resentment and resistance. The first comment they make is, "That's not VA/VE! Where's the FAST diagram?" Anything other than pure Milesian VA/VE is generally rejected and considered a travesty on VA/VE. Although we believe that traditional Milesian VM will be with us in the future, the fate of traditional VA/VE is of little consequence. If it progresses in its traditional locked-step adherence to the Miles job plan, it will quickly become gaslight technology in a new age of valuism. Indeed many managers who are not familiar with modern VM have already relegated VA/VE to the dustheap of yesterday's management fads.

How Might We Accomplish Integration?

We can think of several different ways companies can hasten integration. They include the use of value teams, job rotation, and the development of internal management consulting staffs.

1. Value Teams

Value teams provide a medium for integrating various value-centered disciplines. Depending on the task at hand, people with expertise in different methodologies could work together on a value team. Each individual would be asked to contribute to the team's task and in so doing would expose others on the team to new methodologies. Thus, a manufacturing team might contain an expert on simultaneous engineering, a value engineer, a quality control expert, an industrial engineer, and so on. The team would work together to solve a specific problem and then be disassembled. Subsequently the individual team members would be reassigned to a new project requiring new methodologies. In time people skilled in a variety of different disciplines would become expert valuists who could lead new value teams. Value teams will last anywhere from a few days to a few years.

Today, both government and industry deploy autonomous work teams to design and develop new products and services. Many industrial organizations have turned to task forces and teams to cope with change. We believe that value-centered teams will be the essence of tomorrow's organizational structure. These teams, already in widespread use in VA/VE, will fit the "Age of Ad-hocracy" described by Toffler 20 years ago. In his book *Future Shock*, Toffler devoted a chapter to Organization: The Coming of Ad-Hocracy. He predicted,

> man will encounter plenty of difficulty in adopting to this new style organization. Instead of being trapped in some unchanging, personality smashing niche, man will find himself liberated, a stranger in a new free form world of kinetic organizations. In this

alien landscape, his position will be constantly changing, fluid and varied. His organizational ties will turn over at a frenetic and ever-accelerating pace.[28, p. 125]

For over 40 years VA/VE teams have been used to design and develop new products and services. Business and government have since developed dissolvable throw-away-type organizations to create new products and services or to downsize existing organizations that no longer provide sufficient value. We expect these temporary organizations to act as training centers for managers skilled in the methodologies of valuism.

2. Job Rotation

In Japan, industry has developed a system of job rotation where new hires and existing personnel are rotated to different disciplines at regular intervals (usually every 2 to 4 years). Professional employees follow a career path in which people occupy a series of different desks. That is, a person occupies the "VE desk" for x years and then moves to the "QFD desk" for x years and so on. Desks take the place of titles and integrate the skills of the professional work force. The constant rotation to new desk disciplines is an interesting way to foster integration. Employees are given an opportunity to grow and the companies develop a work force that has a wide spread understanding of basic methodologies. This increases productivity by improving communications.

3. Internal Staffs

Many large corporations employ a large number of different professionals in centralized internal consulting groups. Typical examples include industrial engineering groups, QC groups, and computer systems groups. Individuals in these groups consult in their area of expertise throughout the corporation. Aggregation of professionals in these staffs increases competency in the profession and allows individuals to progress through a series of jobs in their chosen career. In professions like industrial engineering or computer science, individuals can progress from basic entry level jobs like time study or computer programming into more demanding jobs such as inventory analysis and control or systems design and architecture. In this case integration occurs within a profession and seldom extends outside of this professional area.

In addition to these business-centered processes, professional societies and colleges or universities also can help integration.

4. Professional Society Expansion

As mentioned previously, many professional societies grow by expanding their venue. Societies often sponsor research at colleges and universities that enlarges their knowledge base. They also invite their members to publish papers on new methods they have developed or new applications they have found. Joint meetings with other societies can lead to research on new methodologies that enlarges the societies' professional base. SAVE works in concert with the Miles Value Foundation to enlarge its foundations. We have published a large number of papers in conferences sponsored by this society. Any society can enlarge its scope to include new methodologies. Control over such expansion rests in the rules and regulations of the society and its management. A recent example of such outreach and integration was the October 1990 SAVE-sponsored seminar, "Quality and Value: A World Class Partnership." The SAVE constitution also includes bylaws to achieve cooperation with other societies and SAVE's strategic plan encourages closer association with other disciplines.[29]

5. Academia

Colleges constantly introduce new methodologies into the curriculum. This is especially true in science and engineering fields, which are very dynamic. However, management science is an area where most U. S. colleges teach few new methodologies. In this area business schools often use case studies to teach "modern" business administration methods. Much of the teaching in this area is based on economic methods like cash flow and break-even analysis and is more financial than value oriented. This is not the case in Japan, where academia is active in VA/VE and has integrated VA/VE with QC. In the U. S. private consultants who licensed new methodologies make it difficult for colleges to teach students new methods. Such licensing stagnates programs and is a giant step backwards in putting America in the competitive forefront. However, colleges can do much more than they now do in this field. Few colleges offer courses in VA/VE or in value measurement or value management.

EXPANDING VALUISM AND VALUE MANAGEMENT TO GOVERNMENT

Government needs to adopt many of the methods industry uses to control its growth and improve its efficiency and effectiveness. The federal deficit is a case in point. On October 1, 1990, when the Gramm-Rudman sequester was scheduled to go into effect because Congress and the administration were unable to agree upon cuts in the federal budget, an automatic sequester would have cut federal spending by $85.4 billion. Yet this $85.4 billion represents only 6.5% of the federal budget of $1.3 trillion, an amount *less than* the increase in federal spending Congress had already scheduled. In other words even with a sequester of $85.4 billion, federal spending would still significantly exceed federal spending in 1989![30] To avoid this sequester, Congress, after failing to approve a bipartisan budget proposal and unable to override a presidential veto of a continuance resolution, finally approved a budget cut of less than half the $85.4 billion sequester. To pay for increases in the new budget, which remained higher than the 1989 budget, Congress cut Medicare entitlements and enacted the largest new tax package in its history.

Many Americans wonder how their government can be so out of control. The reasons are numerous, and it is clear that something has to be done to bring the situation under control. In our view, the government needs to adopt measures used by industry to control spending and to optimize the value of government services. The situation in many state governments is similar to that at the federal level. New York State has a deficit of over $40 billion and many other state and local governments are also in the red! The time is ripe for valuism to act in the public as well as the private sector.

Some worthwhile efforts have already been taken. For a number of years VA/VE has been used to reduce the cost of equipment sold to the U. S. Department of Defense. The Department of Defense VE policy is contained in the Armed Services Procurement Regulation ASPR, 1-1701 which reads in part that,

> In order to realize fully the cost reduction potential of value engineering, provisions which encourage or require value engineering shall be incorporated in all contracts, including letter contracts, of sufficient size and duration to offer reasonable likelihood for cost reductions. All such provisions shall offer the contractor a share in cost reductions ensuing from change proposals he submits under contracts. While most proposals will result from the contractor's value engineering efforts, any proposal submitted by the contractor which meets the documentation and other requirements of the value engineering clauses may be rewardable. In addition, the value engineering provisions are intended to induce major prime contractors to utilize value engineering techniques. Finally, to realize the cost reduction potential of value engineering, it is imperative that value engineering change proposals be processed by all parties as expeditiously as possible.[31,32]

In 1987 SAVE successfully lobbied Congress to expand the use of VA/VE to all branches of the government. Senate hearings conducted by Senator Carl M. Levin, deputy chairman of the Office of Federal Procurement Policy, led to the Office of Management Budget circular A-131 that requires all federal departments and agencies to have a VE program. The results of this program are shown in Table 15.1.[33]

Although it is gratifying to know that VA/VE has been expanded to all branches of the federal government and is saving U. S. taxpayers billions of dollars per year, it is disconcerting to see that savings in Department of Defense dropped significantly from 1988 to 1989. The cause of this drop is under investigation by SAVE and the Department of Defense at this time. Preliminary results point to the fact that some of the Department of Defense's VE professionals have been subsumed by the Department of Defense QC programs. It has also been suggested that the required use of certified value specialists in government contracts has reduced the number of people who can provide VA/VE assistance to the government.

Clearly these results show that VA/VE can have a significant impact in reducing the cost of government. Other valuism disciplines undoubtedly can also help lower costs and improve efficiencies. Methods study used by industry is one example. We believe that industrial engineering studies of paperwork flow could dramatically reduce costs and improve performance in most government agencies. Some of this is being done but much more needs to be done. At the present time most of the value centered methodologies reside in industry and a technology transfer to government is sorely needed.

TABLE 15.1 1988 and 1989 savings attributed to VA/VE use in the U. S. government[34]

Department	FY 1988 Savings	FY 1989 Savings
Agriculture	0	5,114,590
Arms Control and Disarmament	*a	—
Commerce	2,400	0
Commodity Futures Trading	*a	—
Commission on Civil Rights	*a	—
Energy	23,323,729	59,400,739
Federal Emergency Management Agency	*a	0
Air Force	658,417,188	488,643,661
Army	830,260,609	412,817, 661
Army (Corps of Engineers, Civil Works)	*a	(66,985,000)
Defense Logistic Agency	108,550,000	123,320,000
Navy	845,928,000	423,100,000
General Services Agency	684,189	59,259,622
Health and Human Services	11,226	20,650
Housing and Urban Development	—	
Interior	12,183,100	2,293,232
Justice	0	314,255
Labor	0	0
National Archives	—	0
National Science Foundation	*a	0
Railroad Retirement Board	0	0
NASA	0	826,000
Occupational Safety and Health Commission	*a	0
Panama Canal Commission	*a	0
State Department	4,100,000	4,087,542
Transportation[b]	5,457,378	11,992,320
Treasury	0	0
Consumer Product Safety Commission	0	0
Veterans Administration	9,549,952	8,625,988
Total	$2,498,467,831	$1,599,806,336

[a]No report submitted in 1988.
[b]Transportation reported additional VE savings of $42,473,848 achieved by state and local governments through grants and federally assisted programs.
Reprinted with permission of the Society of American Value Engineers.

THE NEED FOR LEADERSHIP

To integrate and expand valuism in business and to increase its application to government will require leadership. Leaders are needed who can bridge an environment of self-interest and parochialism. Leaders in industry, government, and the professions need vision, courage, and determination to transform their organizations. These leaders also need procedures to help them transform these organizations.

Baguley and Machen describe this well: "In accepting change the organizational culture must include the pressure for change, a clearly shared vision of the result, a capacity to accept change and an achievable plan to prevent haphazard efforts and false starts." [35,p.12] Together, VP and VM can provide a new shared vision of the future and a team-oriented process for implementing organizational change.

SUMMARY

This chapter contains forecasts of the future of valuism and VM. These forecasts are based on a continuation of the trends we described in Chapters 2 and 11. We expect valuism, with its emphasis on the efficient production of value, to become the dominant management movement in industry in the 21st century. As valuism expands its operations from industry to government, it may become a political movement. We believe that governments will begin to adopt some of industry's methodologies to privatize and downsize government operations. We believe that value measurement and management are essential elements of valuism, and we expect that traditional VA/VE will continue to grow at a slow rate unless augmented with value measurement and value management methodologies. William D. Hitt[36] in his book *The Leader-Manager* defines the four common functions and core characteristics of leadership across culture and organization: (1) having a clear vision of what the organization (or department or group) might become; (2) the ability to communicate the vision to others; (3) the ability to motivate others to work toward the vision; and (4) the ability to work the system to get things done.

We have given our vision of the future of VA/VE and valuism. It is a prescription for VA/VE and its integration into the 21st century. The future will provide an opportunity to place VM at the center of valuism. The opportunity exists to expand VA/VE and integrate it with other methodologies. This requires a change from the Miles paradigm. The choice of leadership is yours.

REFERENCES

1. Parks, W. W., "Value Management in Changing Times," luncheon address, Society of American Value Engineers International Convention, April 1974.
2. De Marle, D. J., and Shillito, M. L., "Value Engineering," in G. Salvendy, Editor, *Handbook of Industrial Engineering,* Chapter 7.3, pp. 7.3.1–20, Wiley, New York, 1982.
3. Bennet, A., "Consulting Concerns, Competing Hard, Learn the Business of Selling Themselves," *Wall Street Journal,* p. B1, September 27, 1990.
4. Behrens, R. A., "Value Management in South Africa: A Land of Many Cultures," in *Proceedings, Miles Value Foundation Value Engineering Conference,* Vol. 1, pp. 255–261, 1988.
5. Bowers, B., "Small Businesses Fight Untaxed Rivals," *Wall Street Journal,* p. B1, September 27, 1990.

6. Barker, J. A., *Discovering the Future: The Business of Paradigms,* ILI Press, St. Paul, Minn., 1985.

7. Hamel, G., and Prahalad, C. K., "Strategic Intent," *Harvard Business Review,* pp. 63–76, May–June 1989.

8. Prahalad, C. K., and Hamel, G., "The Core Competence of the Corporation," *Harvard Business Review,* pp. 79–91, May–June 1990.

9. Boothroyd, G., and Dewhurst, P., "Design for Assembly," University of Massachusetts, Amherst, Mass., 1983.

10. Bradeyhouse, R. G., "Design for Assembly and Value Engineering: Helping You Design Your Product For Easy Assembly," *Proceedings, Society of American Value Engineers* **19,** 14–23, 1984.

11. Bradeyhouse, R. G., "Blending VA and Design for Assembly to Leverage the Benefits of Two Powerful Techniques," *Proceedings, Society of American Value Engineers,* **25,** 247–255, 1990.

12. Hayes, R.H., and Wheelwright, S. C., *Restoring Our Competitive Edge: Competing Through Manufacturing,* Wiley, New York, 1984.

13. Schonberger, R. J., *World Class Manufacturing, the Lessons of Simplicity Applied,* Free Press, New York, 1986.

14. Wright, O. W., *The Executive's Guide to Successful MRP II,* Prentice-Hall, Englewood Cliffs, N. J., 1985.

15. Wallace, T. F., *MRP II: Making It Happen,* Oliver Wright Publications, Essex Junction, Vermont, 1985.

15. Bopp, A. L., and Grant, R. P., "Statistical Process Control Based on Function Analysis," *TAPPI Journal* pp. 77–79, April 1989.

17. Port, O., Schiller, Z., and King, R. W., "A Smarter Way to Manufacture," *Business Week,* pp. 110–117, April 30, 1990.

18. Stalk, G., and Hout, T. M., "Competing Against Time," *Research and Technology Management* pp. 19–24, March–April 1990.

19. Fraser, R. A., "Using Value Management Techniques for Organizational Change," *Proceedings, Society of American Value Engineers* **22,** 96–101, 1987.

20. Davis, S. O., "Transforming Organizations to Higher Performance Levels: An Opportunity for Value Engineers to Lead, Follow, or Get Out of the Way," *Proceedings, Society of American Value Engineers,* **22,** 85–95, 1987.

21. Verzelletti, G., "Value Measurement, a Comprehensive Method to Support the Conception and the Development of New Product and Processes," *Proceedings, Society of American Value Engineers* **25,** 213–226, 1990.

22. Togawa, S., "Value-Creating V.E.C. Based on the Establishment of a Practical Method to Satisfy Customer Needs," *Proceedings, Society of American Value Engineers* **25,** 208–212, 1990.

23. Nagoya, Y., "A Study of Marketing VE Based on Evaluation of Customer Needs," *Proceedings, Society of American Value Engineers* **25,** 197–207, 1990.

24. Kaufman, J. J., "Total Quality Management a New Value Management Initiative," *Proceedings, Society of American Value Engineers* **25,** 235–243, 1990.

25. Frank, G. A., "Total Quality Management," *Proceedings, Society of American Value Engineers* **25,** 49–56, 1988.

26. Park, R. J., "Value Engineering in the Modern World of Taguchi, QFD et el," *Proceedings, Society of American Value Engineers* **24,** 81–86, 1989.

27. Bachli, L. J., "MRP-II, CIM, JIT/TQC, and Other Fascinating VA Applications," *Proceedings, Society of American Value Engineers* **24,** 3–8, 1989.

28. Toffler, A., *Future Shock,* p. 125, Random House, New York, 1970.

29. Willingham, G., "Education Plan Blooms with Quality, Value Seminar," *Interactions* **15**(9), September 1990.

30. "Beltway Armageddon," editorial, *Wall Street Journal,* p. A16, September 26, 1990.

31. Department of Defense ASPR, 1-1701.

32. *Principles and Applications of Value Engineering,* pp. 8.4–8.5, Superintendent of Documents, U.S. Gov. Printing Office, Washington, D. C., 1968.

33. Tufty, H. G., "Good News in Government," *Interactions,* **15**(9), 1–2, 1990.

34. Tufty, "Good News in Government."

35. Baguley, D. J., and Machen, R. W., "Doing Things Better with Value Management," in *Proceedings, Seventh Conference on Electric Power Supply Industry,* Vol. 1, pp. 1–13, Queensland Electrical Commission, Brisbane, Australia, 1988.

36. Hitt, W. D., *The Leader-Manager,* Battelle Press, Columbus, Ohio, 1988.

BIBLIOGRAPHY

Barker, J. A., *Future Edge,* William Morrow, New York, 1992.

Siorek, R. W., "Value Analysis—Does It Have A Place In The Sun?" *Value World,* pp. 2–4, July/August/September, 1991.

Smith, P., and Reinertsen, D., *Developing Products In Half The Time,* Van Nostrand Reinhold, New York, 1991.

Wu, B., "Total Quality Management: A New Challenge To Value Engineers," *Proceedings, Society of American Value Engineers,* 26, 128–133, 1991.

INDEX